# ANNALS OF THE NEW YORK ACADEMY OF SCIENCES

Volume 773

EDITORIAL STAFF

*Executive Editor*
BILL M. BOLAND

*Managing Editor*
JUSTINE CULLINAN

*Associate Editor*
TRUMBULL ROGERS

The New York Academy of Sciences
2 East 63rd Street
New York, New York 10021

NEW YORK ACADEMY OF SCIENCES
(Founded in 1817)

BOARD OF GOVERNORS, September 1995–September 1996

HENRY M. GREENBERG, *Chairman of the Board*
MARTIN L. LEIBOWITZ, *Vice Chairman of the Board*
RODNEY W. NICHOLS, *President and CEO* [ex officio]

HENRY A. LICHSTEIN, *Treasurer*

*Governors*

| | | |
|---|---|---|
| ELEANOR BAUM | BARRY R. BLOOM | D. ALLAN BROMLEY |
| LAWRENCE B. BUTTENWIESER | EDWARD COHEN | |
| | | SUSANNA CUNNINGHAM-RUNDLES |
| BILL GREEN | SANDRA PANEM | RICHARD A. RIFKIND |
| DOMINICK SALVATORE | WILLIAM C. STEERE, Jr. | SHMUEL WINOGRAD |

JOSHUA LEDERBERG, *Past Chairman of the Board* WILLIAM T. GOLDEN, *Honorary Life Governor*
HELENE L. KAPLAN, *Counsel* [ex officio]   SVETLANA KOSTIC-STONE, *Secretary* [ex officio]

# WAVES IN ASTROPHYSICS

ANNALS OF THE NEW YORK ACADEMY OF SCIENCES
Volume 773

# WAVES IN ASTROPHYSICS

*Edited by J. H. Hunter, Jr. and R. E. Wilson*

The New York Academy of Sciences
New York, New York
1995

*Copyright © 1995 by the New York Academy of Sciences. All rights reserved. Under the provisions of the United States Copyright Act of 1976, individual readers of the* Annals *are permitted to make fair use of the material in them for teaching and research. Permission is granted to quote from the* Annals *provided that the customary acknowledgment is made of the source. Material in the* Annals *may be republished only by permission of the Academy. Address inquiries to the Executive Editor at the New York Academy of Sciences.*

Copying fees: *For each copy of an article made beyond the free copying permitted under Section 107 or 108 of the 1976 Copyright Act, a fee should be paid through the Copyright Clearance Center, Inc., 222 Rosewood Drive, Danvers, MA 01923. For articles of more than 3 pages, the copying fee is $1.75.*

∞ *The paper used in this publication meets the minimum requirements of American National Standard for Information Sciences—Permanence of Paper for Printed Library Materials, ANSI Z39.48-1984.*

Cover photograph (paperback only): *An optical photograph of NGC 1398 reproduced from the Hubble Atlas. North is up and east is to the left.* (See page 329.)

### Library of Congress Cataloging-in-Publication Data

Waves in astrophysics/edited by J. H. Hunter, Jr. and R. E. Wilson
   p.  cm.—(Annals of the New York Academy of Sciences, ISSN 0077-8923  ; v. 773)
   "The papers in this volume were presented at the Tenth Florida Workshop in Nonlinear Astronomy... which was held on September 22-24, 1994, in Gainesville, Florida"—P. [vi].
   Includes bibliographical references and index.
   ISBN 0-89766-957-6 (cloth  ;  alk. paper).—ISBN 0-89766-958-4 (pbk  ; alk. paper)
   1. Astrophysics—Congresses.  2. Magnetohydrodynamic waves--Congresses.  3. Stellar oscillations—Congresses.  I. Hunter, J. H. (James H.)  II. Wilson, R. E. (Robert E.), 1937–  .
III. Florida Workshop in Nonlinear Astronomy (10th  :  1994  : Gainesville, Fla.)  IV. Series.
   Q11.N5   vol. 773
   [QB460]
   500 s--dc20
   [523.01]
                                                                           95-48345
                                                                             CIP

S/PCP
*Printed in the United States of America*
**ISBN 0-89766-957-6 (cloth)**
**ISBN 0-89766-958-4 (paper)**
**ISSN 0077-8923**

ANNALS OF THE NEW YORK ACADEMY OF SCIENCES
Volume 773
December 28, 1995

# WAVES IN ASTROPHYSICS[a]

Editors
J. H. HUNTER, JR. and R. E. WILSON

Conference Organizers
G. CONTOPOULOS, J. H. HUNTER, JR., and R. E. WILSON

## CONTENTS

| | |
|---|---|
| Preface. *By* J. H. HUNTER and R. E. WILSON | vii |
| Chaos in Observational Data of Variable Stars—Irregularity from the Nonlinear Interaction of Standing Waves? *By* J. ROBERT BUCHLER, Z. KOLLÁTH, and THIERRY SERRE | 1 |
| Unstable Long Wavelength Magnetohydrodynamic Waves on Highly Collimated Outflows. *By* PHILIP E. HARDEE | 14 |
| Interface Instabilities in the Interstellar Medium. *By* J. H. HUNTER, JR., R. W. WHITAKER, R. V. E. LOVELACE, and C. SIOPIS | 32 |
| Alfvén Waves and Interstellar Turbulence. *By* S. SRIDHAR | 44 |
| Sinuous Modes and Steady Warps of Polytropic Discs. *By* N. J. BALMFORTH and E. A. SPIEGEL | 55 |
| Linear and Nonlinear Waves in Shock-Bounded Slabs. *By* ETHAN T. VISHNIAC | 70 |
| Normal Modes and Continuous Spectra. *By* N. J. BALMFORTH and P. J. MORRISON | 80 |
| Vorticity and Mixing in Disks. *By* PHILIP YECKO | 95 |
| Bending Waves in Flattened Stellar Systems. *By* C. HUNTER | 111 |
| On Global Wave Patterns in Galaxies: Their Generation and Maintenance. *By* C. C. LIN and G. BERTIN | 125 |
| Invariant Spectra of Dynamical Systems. *By* G. CONTOPOULOS, N. VOGLIS, C. EFTHYMIOPOULOS, and E. GROUSOUZAKOU | 145 |
| Chaos and Order in Time-Periodic Potentials and the Problem of Structural Stability. *By* HENRY E. KANDRUP, ROBERT A. ABERNATHY, M. ELAINE MAHON, and BRENDAN O. BRADLEY | 168 |

[a]The papers in this volume were presented at the Tenth Florida Workshop in Nonlinear Astronomy, entitled Waves in Astrophysics, which was held on September 22–24, 1994, in Gainesville, Florida.

Spectra and Lyapunov Numbers in Pulsating Systems. *By* HAYWOOD SMITH, JR. and GEORGE CONTOPOULOS ............................................... 189

Arnold Diffusion and Equipartition in an Oscillator Chain. *By* ALLAN J. LICHTENBERG ............................................................................. 205

Universal Properties of Escape. *By* CHRISTOS V. SIOPIS, HENRY E. KANDRUP, G. CONTOPOULOS, and RUDOLF DVORAK .......................................... 221

Counterrotating Bars. *By* CHAD L. DAVIES and JAMES H. HUNTER, JR. .......... 231

Shadowing and Noise in Nonhyperbolic Systems. *By* DAVID E. WILLMES ...... 242

Low-Frequency Modes of Pulsation of Relativistic Accretion Disks. *By* JAMES R. IPSER ................................................................................... 256

Density Waves and Warps Generated by Tidal Perturbation of a Gaseous Disk. *By* J. C. B. PAPALOIZOU, D. G. KORYCANSKY, and C. TERQUEM ....... 261

Nonlinear Waves in Magnetized Accretion Disks. *By* R. V. E. LOVELACE and M. M. ROMANOVA ........................................................................ 277

Structural Stability and $\lambda$-Transitions. *By* DIMITRIS M. CHRISTODOULOU, DEMOSTHENES KAZANAS, ISAAC SHLOSMAN, and JOEL E. TOHLINE ............ 285

Some New Understandings on Nonlinear Stellar Pulsation. *By* YANQIN WU ... 296

Tital Perturbations, Gravitational Amplification, and Galaxy Spiral Arms. *By* GENE BYRD ................................................................................. 302

Gaseous Vortices in Barred Spiral Galaxies. *By* MARTIN N. ENGLAND and JAMES H. HUNTER, JR. .................................................................. 320

The Pattern Speed of the Barred Spiral Galaxy NGC 1398. *By* E. M. MOORE and S. T. GOTTESMAN ................................................................... 329

Index of Contributors .......................................................................... 345

**Financial Assistance was received from:**
- COLLEGE OF LIBERAL ARTS & SCIENCES, UNIVERSITY OF FLORIDA
- DIVISION OF SPONSORED RESEARCH, UNIVERSITY OF FLORIDA

The New York Academy of Sciences believes it has a responsibility to provide an open forum for discussion of scientific questions. The positions taken by the participants in the reported conferences are their own and not necessarily those of the Academy. The Academy has no intent to influence legislation by providing such forums.

# Preface

J. H. HUNTER AND R. E. WILSON

*Department of Astronomy*
*University of Florida*
*Gainesville, Florida 32611*

One could scarcely think of a workshop more encompassing than Waves in Astrophysics. Thetopic was the idea of George Contopoulos, who was—true to form—right again. The subject could have been dismissed as too broad, but we are most pleased with the interactions that flowed from the contributors, both at the meeting and in the subsequent mutual refereeing. Areas included stellar pulsation, plasmas, galactic dynamics, accretion disks, gravitational waves, Alfvén waves, solutions, order–chaos, and the various natural companion topics of those areas. We may be prejudiced, but it seemed to us that the participants had stimulating and enjoyable experiences. We thank them all for attending and for creating this book.

The following list of *Annals* comprises the Second Florida Workshop through the Ninth Florida Workshop.

1. BUCHLER, J. R. & H. EICHHORN, Eds. 1987. Chaotic Phenomena in Astrophysics. Ann. N.Y. Acad. Sci. **497**.
2. BUCHLER, J. R., J. R. IPSER & C. A. WILLIAMS, Eds. 1988. Integrability in Dynamical Systems. Ann. N.Y. Acad. Sci. **536**.
3. BUCHLER, J. R., S. T. GOTTESMAN & J. H. HUNTER, JR., Eds. 1989. Galactic Models. Ann. N.Y. Acad. Sci. **596**.
4. BUCHLER, J. R., & S. T. GOTTESMAN, Eds. 1990. Nonlinear Astrophysical Fluid Dynamics. Ann. N.Y. Acad. Sci. **617**.
5. BUCHLER, J. R., S. L. DETWEILER & J. R. IPSER, Eds. 1991. Nonlinear Problems in Relativity and Cosmology. Ann. N.Y. Acad. Sci. **631**.
6. DERMOTT, S. F., J. H. HUNTER, JR. & R. E. WILSON, Eds. 1992. Astrophysical Disks. Ann. N.Y. Acad. Sci. **675**.
7. BUCHLER, J. R. & H. E. KANDRUP, Eds. 1993. Stochastic Processes in Astrophysics. Ann. N.Y. Acad. Sci. **706**.
8. KANDRUP, H. E., S. T. GOTTESMAN & J. R. IPSER, Eds. 1995. Three-Dimensional Systems. Ann. N.Y. Acad. Sci. **751**.

# Chaos in Observational Variable Star Data—Irregularity from the Nonlinear Interaction of Standing Waves?[a]

J. ROBERT BUCHLER, Z. KOLLÁTH, AND THIERRY SERRE

*Physics Department
University of Florida
Gainesville, Dlorida 32611*

## INTRODUCTION

It was Eddington who first explained stellar variability as the excitation of normal modes of oscillation in the star.[1] In fact, to phrase it simply, the stellar pulsations are self-excited standing waves in a spherical quarter-wavelength "tube." In the region of the Hertzsprung–Russell diagram, called the instability strip, where the radially pulsating stars are located, the physical conditions are suitable for the modes or waves to be energized by the interaction of the pulsations with the outward heat flow. In modern thermodynamical parlance the star acts as a natural heat engine[2] in which the temperature dependence of the opacity behavior provides the driving.[1]

Observations and theory have shown that in the classical variable stars, such as the Cepheids and RR Lyrae stars, the pulsations are often periodic to good accuracy, or when they are multiperiodic they involve just a couple of frequencies, and thus of standing waves. In contrast, the pulsations of many of their more luminous and metal poor, Population II (Pop. II) cousins are irregular for pulsation "periods" in excess of 15 days. This was already known observationally at least as far back as the beginning of the century.[3] For a long time, however, the nature of this irregularity remained a mystery, even though some early numerical hydrodynamical modeling managed to reproduce irregular pulsations.[4] It is only in the last decade that a systematical numerical hydrodynamical survey of W Virginis models[5,6] has shown that the pulsations of these Pop. II objects are in fact *chaotic* in the dynamical systems sense of the word.[7,8] The dominant clue as to the chaotic nature of these pulsations comes from the fact that, as a control parameter, namely the equilibrium effective temperature of the models is varied, the pulsations display a period-doubling cascade that is a clear signature of a chaotic dynamics (for a review, see reference 9).

---

[a]This research was supported in part by National Science Foundation Grant AST92-18068, in part by an RDA grant at the University of Florida, in part by the French Ministère pour la Recherche et l'Espace, and in part by an RCI grant from IBM through the University of Florida.

However, an analysis of observational data to corroborate the chaotic nature of these pulsations has been sorely missing. Fortunately such analyses have recently become possible. First, suitable observational data sets are now available, and, second, new methods of nonlinear time-series analysis have been developed that can be adapted to astronomical data. In this paper we review these recent developments and the results that have been obtained.

## THE NONLINEAR ANALYSIS

In astronomy we generally have only one observed quantity at our disposal, namely the magnitude (luminosity) of the star; other possible observed quantities may be the radial velocity, for example, but they usually have a smaller signal-to-noise ratio. The important question then arises *which, if any properties of the physical phase space of the underlying dynamics we can infer from this single observed quantity.*

It is well known from the study of fluid dynamics (e.g. reference 10) that in spite of the infinite number of degrees of freedom a fluid can undergo low-dimensional behavior when dissipative effects are present. When such is the case, only a few degrees of freedom are dominant, and the remaining degrees of freedom are enslaved to the former. The dynamics thus lies in a low dimensional phase space, called an *inertial manifold*,[11] and is described by a flow, that is, a set of ordinary differential equations,

$$d\mathbf{Y}/dt = \mathbf{G}(\mathbf{Y}), \tag{1}$$

where $\mathbf{Y}$ is the $d$-dimensional vector of the physical phase space (or dominant) variables.

It is the nature and properties of this reduced phase space that we are trying to uncover. Fortunately, the mathematical, numerical, and experimental work of the last couple of decades has shown that this is an achievable goal,[12-14] even when we can measure or observe only a single quantity. In astronomy, however, the problem is harder than in experimental disciplines where data can be gathered under controlled conditions. Observations, because of weather, location, telescope time scheduling, politics, or other reasons cannot be obtained *ad libitum*. Astronomical data generally therefore have gaps, and they are obtained at irregular time intervals. Furthermore, noise is almost always significant. Notwithstanding, in the study of real systems, interestingly it is in astronomy, and more specifically in the solar system[15] and in pulsating stars, that the notions of chaos lead to useful practical results.[16,17]

### Data Preparation

The first step in any analysis of real data is to extract a "signal" from the noise. To some extent this process has arbitrariness associated with it. It may be clear, in principle at least, what constitutes extraneous noise (e.g., in the electronics, in the sky transparency). Convection and turbulence are necessarily present in a star, especially one that is undergoing violent pulsations. It is well known that turbulence is a high dimensional phenomenon (many degrees of freedom)[8] whose effect is bound to appear in the observed data with properties similar to noise. Obviously, only when we are able to filter out such high dimensional structure can we hope to extract a low dimensional dynamics from the data. This data smoothing, and a concomitant interpolation to generate equally spaced data, is therefore a delicate and potentially dangerous procedure. Too much smoothing destroys the information about the low dimensional dynamics, while with too little filtering the dynamics remains buried in the noise. Some types of filterings can also change the dimension of the attractor.[14] In recent papers that analyze data sets of astronomical interest we discuss in some length the effects of the data preparation on the final analyses.[18-20]

### Global Phase-Space Reconstruction

The actual core of our analysis falls under the category of *global phase space reconstructions* and the details are documented elsewhere.[18-20] Here we just give the gist of it. The analysis is based on the Takens embedding theorem and its extensions.[12] If the pulsations are generated by a flow of the form of (1), where $Y$ is the $d$-dimensional vector of the physical phase space variables, then there exists a $d_e$-dimensional variable $X$ that satisfies a nonlinear equation (map) of the form

$$X^{n+1} = F[X^n], \qquad (2)$$

and an embedding dimension, $d_e$ of at most $2d + 1$ is required. Actually, it is sufficient that $d_e > 2 \times$ the *box-counting dimension* of the attractor.[12] In practice often the dimension required to reconstruct the attractor is less than this upper limit.

The set of $d_e$-dimensional vectors

$$X^n = \{g(t_n), g(t_n - \Delta), g(t_n - 2\Delta), \ldots, g(t_n - (d_e - 1)\Delta)\} \qquad (3)$$

can be constructed *from the observed scalar variable* $\{g(t_n)\}$, where the $\{t_n\}$ are equally spaced times. This variable is a smooth function of the physical phase space variable $Y$, viz., $\{g(t_n)\} \equiv \{g(Y(t_n))\}$.

This theorem, which establishes a one-to-one correspondence between the attractor in the physical phase space and the one reconstructed from the embedding variable, is extremely powerful because some properties are preserved in the embedding.[12,14] It thus allows one to *infer quantifiable properties of an unknown underlying dynamics from the observations of a single quantity*, namely the magnitude in our case.

The best choice for the global map $F$ is of course unknown. It is natural to try a polynomial expansion in our case because radial stellar pulsations are generally weakly nonlinear.[21] Thus we assume

$$F(X) = \sum_k C_k P_k(X), \qquad (4)$$

where the $P_k(X)$ represent polynomials of order up to $p$ that are orthogonal on the data set.[22] Rather than constructing these polynomials by a Gram–Schmidt procedure,[22] we have found it more stable numerically and about ten times faster to obtain the coefficients of the polynomial map with a singular value decomposition (SDV) method.[18]

## R SCUTI AND AC HERCULIS

Two stars of the RV Tauri (RV Tau) class, namely R Scuti (R Sct) and AC Herculis (AC Her) have been observed rather faithfully by a myriad of independent amateur astronomers around the globe, and a large data base has been accumulated and processed by the American Association of Variable Star Observers (AAVSO).[23]

Our data preparation (averaging, cubic spline smoothing and interpolation, and low-pass filtering)[17,19,20] produces a smoothed signal with equally spaced, daily values.

In FIGURE 1 we display typical short segments of the light curve data for R Sct and AC Her. The scatter in the observations (dots) illustrates the noise level. Superposed is the smoothed signal (solid line). Both light curves show the molar tooth structure that is typical of RV Tau stars. R Sct is not only much brighter than AC Her, but it also has a higher amplitude. It has therefore a much better signal-to-noise ratio. The cycle "periods" are approximately 71 days and 37 days, respectively, for the two stars. R Sct therefore also has more observed points per cycle than AC Her. On the other hand, R Sct has a rather complicated light curve in which there are sizable long-term modulations in the overall amplitude in addition to very strong alternations of shallow and deep minima.

It is intuitively clear that if we want to capture the dynamics we need data that are *typical* and furthermore we need a sufficient quantity of them. Unless the data

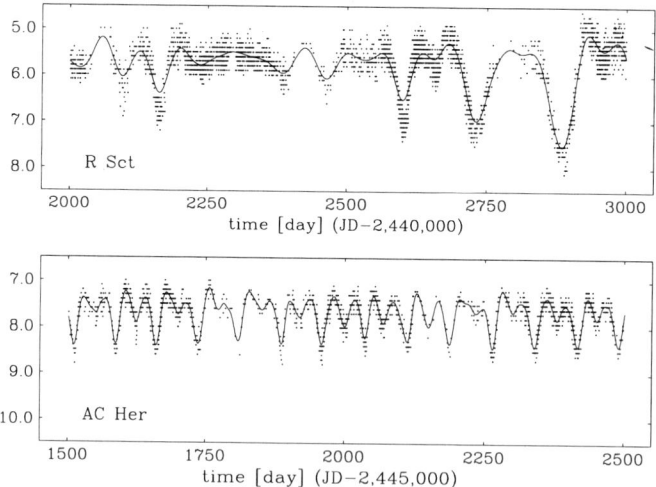

**FIGURE 1.** Typical observed light curve segments for R Scuti and for AC Herculis. **Dots:** the individual observations; **line:** the smoothed signal.

set samples all regions of the attractor, we cannot expect to capture its complete dynamics.

In the case of R Sct, data are available over a century and a half,[23,24] but the early data are of poor quality, and sparse, and have many gaps. We therefore have restricted ourselves to the last 15 years. This data set is of sufficiently good quality that a nonlinear map can capture the underlying dynamics.[17,19]

In the case of AC Her the largest source of data is again the AAVSO. Here we have 12 years of data at our disposal.

The smoothed light curves that are used in our analysis are displayed on top in FIGURES 2 and 3 for the two stars. We work with the magnitude rather than the "more physical" luminosity because the errors in these visual observations are found to be Gaussian, presumably because of the logarithmic response of the human eye.

We have been able to construct four-dimensional (4D) global polynomial maps of the form (2) that capture the overall appearance of the observed pulsations both for R Sct and AC Her.[17,19] It is important to stress that we have *not* been able to represent the data with a map in a dimension lower than 4, that is, in a three-dimensional (3D) reconstruction the synthetic signals bear no resemblance to the observed light curves. We add the caveat that this does not necessarily mean that the embedding dimension cannot be 3. Indeed, it is possible that maps more complicated than polynomial could do the job. For AC Her[20] it has been possible to find a map in three dimensions, but the map is considerably better in four dimensions.

Once a nonlinear map has been obtained, synthetic light curves can then be generated by iterating this map with some initial condition. Examples of such *synthetic signals* that have been generated with 4D maps are displayed at the bottom of FIGURES 2 and 3. Here we simply show the light curves, even though better repre-

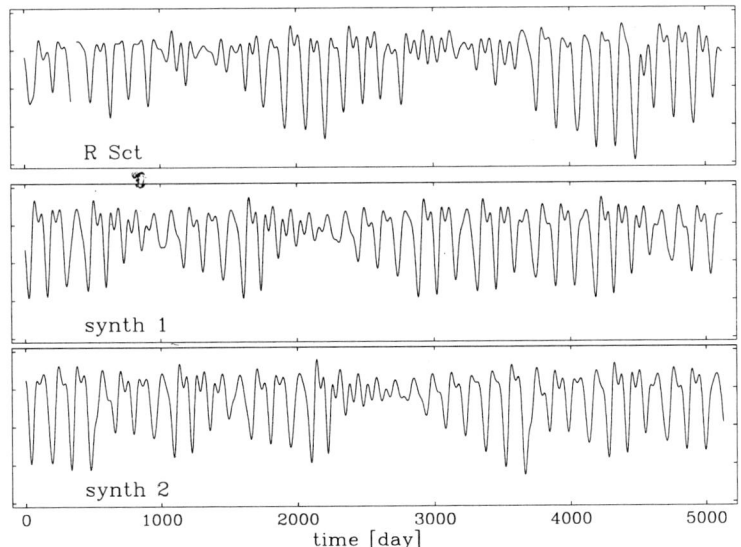

**FIGURE 2.** R Sct. **Top:** observed data (Julian Day 2,440,970 +); **below:** two sections of synthetic signal.

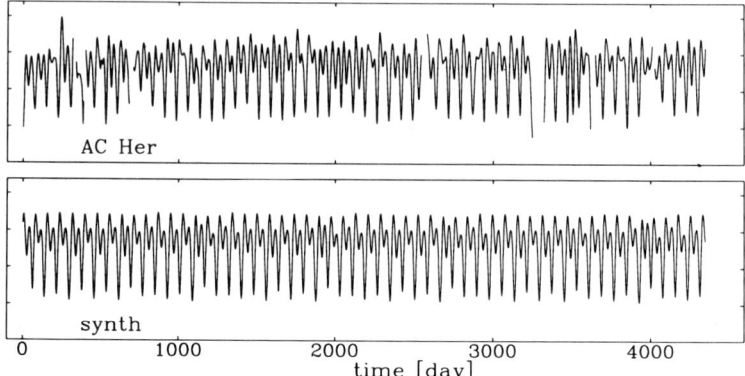

**FIGURE 3.** AC Her. **Top:** observed data (Julian Day 2,445,000 +); **below:** a section of synthetic signal.

sentations exist for such comparisons, as, for example, Broomhead–King coordinates.[8] We merely note here that the Fourier spectra of these synthetic signals also have a close resemblance to those of the observed data.[17,19,20]

For R Sct the agreement of the synthetic signals with the data is excellent. The map captures both the small alternations and the large amplitude modulations. For AC Her the agreement is good, but the synthetic data are not quite as irregular as the data seem to suggest. We think that the sample is not long enough nor typical enough to capture the larger alternations and their irregularity. As already noted the data are also noisier and less dense.

Because of the shortness of the observed data base we are unable to perform statistical comparison tests between the data and the synthetic signals. We cannot either compute Lyapunov exponents or fractal dimensions from the data, but such calculations are possible for the synthetic signals that can be made as large as desired.

For the R Sct synthetic signals we find Lyapunov exponents $\lambda \approx \{1.9 \times 10^{-3}, 0, -1.3 \times 10^{-3}, -5.0 \times 10^{-3}\}$, yielding a Lyapunov dimension[8] of $d_L \approx 3.1$. On the other hand, for AC Her one finds $\lambda \approx \{2.5 \times 10^{-3}, 0, -9.3 \times 10^{-3}, -1.8 \times 10^{-3}\}$ and $d_L \approx 2.3$. In both cases the largest, positive exponent is relatively robust with respect to changes in the data preparation, and the small positive exponent is always close to zero. The third, negative exponent is somewhat sensitive to the data preparation, especially for AC Her. AC Her is much less irregular than R Sct, and one might therefore have expected a smaller dimension (see also the first return maps below). We note that the presence of a positive Lyapunov exponent confirms the chaotic nature of the two dynamics.[5,6,9]

From the fact that the embedding dimension of R Sct is 4, while the fractal dimension is greater than 3, we infer that the dynamics evolves in a 4D space. For AC Her, as we have seen, the embedding dimension is 3, or perhaps 4, and the fractal dimension is slightly greater than 2. The dimension of the dynamics is therefore 3 or 4.

We do not want to mislead the reader into believing that it is easy to generate nonlinear maps that produce synthetic signals similar to the data. Sometimes, instead of a chaotic signal, one may obtain a periodic or multiperiodic one. Small changes in the smoothing, the filtering, the interpolation, the selection of the gaps, the polynomial order $p$, or the delay $\Delta$ can change the synthetic signal from chaotic to periodic, or to unstable. *Prima facie* this may appear very disturbing. However, one recalls that even for the one-dimensional (1D) logistic map periodic cycles are intimately close to chaotic solutions as the parameter of the map is varied.[8] One therefore also expects and indeed obtains a strong sensitivity to the coefficients of the polynomial 4D map. The important point is that *within a reasonable range of the previously mentioned parameters chaotic solutions exist* and that they are very robust, at least for R Sct. By this we mean that, on the one hand, they resemble the data (e.g., in Broomhead–King projections) and have similar Fourier spectra, for example, and, on the other, that the Lyapunov exponents $\{\lambda\}$ and Lyapunov dimension $d_L$ do not vary much in some parameter neighborhood. This then leads to a practical methodology for searching for a good map. We construct maps and generate synthetic signals for a large number of points in the parameter space of the smoothing parameter, the delay, and the SVD eigenvalue cutoff. These signals are then compared to the original signal, and the region of good solutions is identified. If such a region exists and is sufficiently broad, then we claim success.

Finally, we briefly address stability. In general, for good maps, with almost all seed points we can generate synthetic signals of hundreds of thousands of points before the iteration blows up. Sometimes it happens, though, that the map is so unstable that the iteration almost immediately blows up, no matter what the seed point is. There are always regions of high divergence in the attractor, and instability is caused by the unavoidable low density of data points in such regions. The nonlinear coefficients are not determined accurately enough to prevent the orbit to be kicked away from the attractor. (A similar divergence is well known even in a linear context from the application of autoregressive moving average (ARMA) methods[13] to multiperiodic signals—the eigenvalues instead of being all imaginary generally acquire small real parts which, when positive, cause the synthetic signal to diverge.)

We want to stress that it is not possible to produce synthetic signals that resemble the data with *linear* maps or representations. For example, an ARMA scheme fails miserably. On the other hand, it is possible to represent the data with reasonable accuracy with a Fourier fit with the 35 most significant frequencies. However, when this Fourier representation is extrapolated, that is, if one wants to create a synthetic signal, it also fails very badly. This could be anticipated from an inspection of FIGURE 1, which shows a strong asymmetry in the signal, which in turn implies very strict phase relationships over the time-interval described with a Fourier decomposition. This asymmetry cannot be preserved in an extrapolation unless different strict phase relationships are imposed, or a much larger number of frequencies is included. This clearly shows that the Fourier representation is merely an interpolation and that it does not capture the dynamics.

In the literature *first return maps*[7,8] have been plotted for RV Tau stars in the hope that they might yield a clue as to the possible chaotic nature of these objects.[25,26] These maps turn out to be more or less scatter diagrams, as can be seen in FIGURES 4 and 5 where, on the left, we reproduce the first return maps for the (smoothed) observed data of R Sct and for AC Her, respectively. We have disregarded the points for which one of the minima was not well defined observationally.

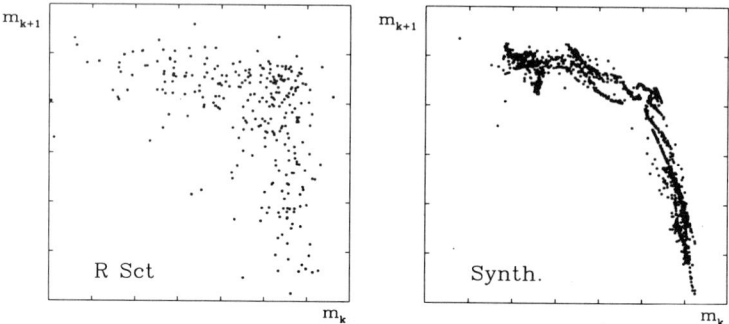

**FIGURE 4.** First return maps on the light curve minima for R Sct. **Left:** observed data; **right:** synthetic signal.

The first return maps for the maxima are even more scattered because of the relative smallness of the alternations compared to the noise.

The reason for this scatter becomes apparent from an inspection of the right-side diagrams where we have plotted the first return maps for our best synthetic signals for the two stars. It is now clear that the scatter in R Sct is in fact not merely due to observational noise, but is inherent in the signal. The mathematical reason is that "clean" first return maps can only be found when the attractor has a fractal dimension that exceeds 2 by very little. Otherwise the first return maps are necessarily very fuzzy. One notes that the synthetic AC Her signal with a $d_L$ of 2.3 has a much less fuzzy return map than its larger period sibling R Sct.

The numerical hydrodynamical models of W Vir stars[5,6] showed already a progression from clean first return maps to fuzzy ones as the luminosity of the models, or equivalently their period, was increased (at fixed mass). In the lower luminosity models the first return map was a thin curve (akin to that found for the Rössler band, e.g., reference 7), whereas a lot of structure developed for the more luminous

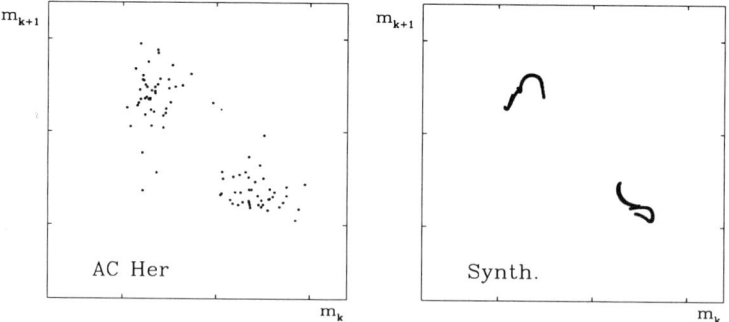

**FIGURE 5.** First return maps on the light curve minima for AC Her. **Left:** observed data; **right:** synthetic signal.

models. The analysis of the observational data on the even more luminous RV Tau stars shows that this trend continues.

Our analyses have been performed with *constant* maps F. We have thus made the implicit assumption that the system, that is, the star, does not evolve over the time interval spanned by the data set. It is difficult to see how one could fruitfully proceed without such an assumption and we will take this constancy as a necessary working hypothesis unless we are forced to abandon it when more observational data become available. R Sct data are available from the last 150 years.[24] The earlier data are very gapped, are poorly sampled, and have larger errors. They are not suitable for an analysis, but they are good enough to show trends. One finds sections that resemble our data set rather closely, but there are also relatively long time spans when the star has a much smaller amplitude. In fact, it appears that the star has again entered such a regime[19] (for $t >$ JD 2,445,130, i.e., the last 5 years). When we shift the time span of our analysis to include, for example, this small region or if we enlarge our data set to include this last region, the maps then produce synthetic signals that are still chaotic, but not as close to the data.[19] We can see two possible explanations for this deterioration of the map. The first is that in the observed data sample, there are not enough regions of low-amplitude behavior to capture the additional feature. A more disturbing possibility is that the star may be evolving slowly. Various estimates from stellar evolution have put the length of the RV Tau phase between several hundreds to several thousands of years.[27] If we recall the sensitive dependence of the synthetic signals to small changes in the map, we unfortunately cannot rule out this possibility with the available observations.

We also want to address the question of whether a flow can be constructed directly from the data, that is, whether one can find a function H whose evolution generates $\{\mathbf{X}^n = \mathbf{X}(t_n)\}$

$$d\mathbf{X}/dt = \mathbf{H}(\mathbf{X}). \tag{5}$$

A good approximation of the time derivative is obtained with an Adams predictor–corrector scheme.[28] The large number of points per cycle in our data guarantees that we obtain a good estimate. This then leads to an SVD minimization problem that is very similar to the one associated with the map (equation 2), especially if a polynomial expansion is also assumed for H.

It is comforting that we can find a flow H that gives rise to synthetic signals with properties similar to those of the discrete map. This is perhaps reassuring because one may question (J. Goodman at the workshop) whether there is necessarily a flow associated with the discrete maps that we have found. But then this is perhaps not astonishing in view of the short time step of the map (many points per cycle). It might also have been anticipated from the fact that the maps have one Lyapunov exponent that is always very close to zero (in the flow it must be exactly zero).

We now would like to mention another development that may turn out to become useful in the analysis and classification of attractors, namely a study of the topological organization of the unstable orbits that make up a chaotic attractor.[29] From a recent application[30] to the numerical hydrodynamical results of a chaotically pulsating W Virginis (W Vir) model[6] some interesting results have already emerged. First, it has been shown that the underlying attractor is akin to the Rössler band,[7] but has an additional global twist. Second, it confirmed what an astrophysicist may have taken for granted, namely that the luminosity is typical in the sense that it carries the same topological information as the temporal variations of the surface radius or the radial velocity. *The observed light curves are thus generic and suitable for use in a phase space reconstruction.*

The topological approach is based on knot and braid theory, which would seem to limit it *a priori* to three dimensions. However, it can be extended to higher dimensional phase space[29] provided that the fractal dimension is less than 3 and sufficiently close to 2. Physically, this means that there has to be enough dissipation in the dynamics to collapse the latter to thin bands in 3D, as is the case for example in the Rössler attractor[7] or in the W Vir hydromodels.[8]

The topological method has not yet been applied to observed variable star data. From the fact that we find a fractal Lyapunov dimension $d_L \gtrsim 3$ for R Sct and from the appearance of the first return map (FIGURE 4) we are somewhat skeptical that the method in its current form will be useful for the study and classification of the longer period RV Tau stars. However, since the hydrodynamical models of the lower period W Vir stars[5,6] and the short period RV Tau stars such as AC Her that we have analyzed here, give a dimension $< 3$ it is quite possible that the method may find a use for such objects provided good enough observational data sets can be obtained.

## EPILOGUE

The reader may say that it is all very nice that we have shown the data to be chaotic, and to have been generated by a chaotic dynamics of low dimension, for example, 4D for R Sct, but "so what?" What is the significance of this finding?

For the physicist it may appear astonishing in fact that a real fluid gives rise to such *simple* behavior, namely, in 4 dimensions and with only weak nonlinearity. After all this self-gravitating blob of gas is pulsating wildly with changes in energy output (luminosity) exceeding factors of 40. The radial velocity curves, determined from the spectra, indicate radial surface velocities up to 50 km/s and the presence of strong shock waves.[31] In addition, there must be violent internal motions of ionization fronts, as well as convection and turbulence.

It has been known from numerical modeling in the early 1960s that the pulsations of the classical variable stars (Cepheids, RR Lyrae) are only weakly nonlinear. This is evidenced, for example, by the small finite amplitude correction to the linear period, and by the similarity of the nonlinear radius, velocity, density, temperature, and luminosity profiles with the corresponding eigenvectors that are derived from a linear analysis. This weak nonlinearity of the pulsations conveniently allows then a description of the pulsations in terms of *modes* or *standing waves*. Furthermore, only very few modes are found to play a role in the pulsations. For example, many classical Cepheids either pulsate in the fundamental or in the first overtone mode. (Here when one identifies the *nonlinear pulsation* with a *mode*, this identification is based on the closeness of the frequency and the resemblance of the spatial structure of those of a *linear mode*, e.g., the fundamental or first overtone eigenmode.) Thus the Cepheids show a secondary bump or shoulder on their light curves and on their radial velocity curves that has earned them the name bump-Cepheids. The bump appears in the broad vicinity of a period of 10 days (Hertzsprung progression). This feature can be studied quantitatively with a Fourier analysis,[9,32] and it has been explained as due to the synchronized excitation of the second overtone, which is due to the $2\omega_0 \approx \omega_2$ resonance. These Cepheids, though periodic, in effect thus have two standing waves excited. There is also a class of beat-Cepheids[1] in which the fundamental and the first overtones are simultaneously excited with constant amplitudes and phases. This nonlinear two-wave interaction with two incommensurate frequencies results in steady beat oscillations. Essentially the same behavior is found in the low period Pop. II (low metallicity) Cepheids that carry the name BL Herculis stars.[33] The RR Lyrae stars similarly pulsate in either the fundamental or first overtone mode, or in a beat combination thereof. The so-called Blazhko effect, which is a long-term modulation of the pulsation amplitudes in RR Lyrae has so far not been explained, but it may involve a resonant interaction with an additional mode. The point is that *none of the pulsations of the classical variable stars seem to involve more than two or three modes.* These classical Pop. I variable stars are all weakly dissipative; in other words, all their excited modes have small growth or decay rates compared to their frequency of oscillation. We just mention that it is the weak dissipation and the weak nonlinearity that guarantee the existence of a *center manifold*.[9] This has led to the development of *amplitude equations* or *normal forms* with which the systematic behavior of the pulsations can be described very simply in terms of a few variables, namely the amplitudes of the dominant modes.[33]

We turn now to the more luminous Pop. II Cepheids. Previous numerical hydrodynamical modeling work,[5,6,34] albeit on lower luminosity (W Vir type) stars has shown that the irregular variability arises from the nonlinear interaction of a linearly unstable mode and a resonant, linearly stable overtone.[9,35] In these models the linear growth rates are large, comparable to the frequencies of oscillation. Comparable values of the period and of the time scale of amplitude modulation are, of course, a prerequisite for the occurrence of chaotic pulsations (but certainly not sufficient). Again, the pulsation is the result of the nonlinear interaction of two dominant modes.

On the basis of this theoretical work one may think that even for the luminous, RV Tau stars only a small number of modes is involved in shaping the pulsation. Our analysis of the observational data shows that the dynamics takes place in a 4D subspace for R Sct. This implies that four independent variables are necessary to describe this dynamics. The (complex) amplitudes of modes of oscillation are again the most natural choices of these variables. From a study of linear models of RV Tau stars we infer that the basic pulsation most likely occurs in the fundamental mode of oscillation. What are then the remaining 2 variables? In principle, they could either be one (complex) vibrational mode or two (real) thermal modes. The R Sct maps has been found to have a fixed point with two spiral roots, one unstable with imaginary part equal to $2\pi f_0$, the other stable with $\simeq 2.3 \times 2\pi f_0$ [reference 19], implying that the additional mode is vibrational. On the basis of our theoretical work on W Vir models we are led also to believe that the second mode is a vibrational mode that is coupled to the fundamental mode through a resonance. In the W Vir models this was the $5\omega_0 \approx 2\omega_2$ resonance with the second overtone. Could it be the same resonance that is operative in R Sct? This conjecture finds some support in the presence of a broad peak at $\approx 2.3f_0$ in the Fourier spectrum of R Sct.[17,19] This remains, however, a conjecture that cannot be confirmed without a thorough numerical hydrodynamical survey of RV Tau models.

The origins of irregular variability had remained a mystery for a long time. It is true that some *ad hoc* mechanisms had been suggested, such as erratic convective overshoots, beating between radial and nonradial modes, for example. As mentioned in the introduction, recent modeling shows that the irregularity has a well-defined cause, and is a manifestation of low-dimensional chaos. From an astrophysical point of view it is thus interesting to finally confirm a viable explanation for the irregular behavior of the classical variable stars.

From an astronomical point of view it is now worthwhile to gather new data on these types of irregular variable stars. It is exciting that the possibility exists of extracting constraints on the stellar parameters from otherwise random looking data. Since these stars are as bright as 10-day clasical Cepheids, it is conceivable that, once we understand them better, they may also became standard candles and help calibrate the cosmological distance scale.

## REFERENCES

1. Cox, J. P. 1980. Theory of Stellar Pulsation. Princeton Univ. Press. Princeton, N.J.
2. WHEATLY, C. W., G. W. SWIFT & A. MIGLIORI. 1986. Los Alamos Sci. **14**: 2.
3. LUDENDORFF, H. 1928. Handb. Astrophys. **6**: 49.
4. CHRISTY, R. F. 1972. Astrophys. J. **172**: 419.
5. BUCHLER, J. R. & G. KOVÁCS. 1987. Astrophys. J., Lett. **320**: L57–L62.
6. KOVÁCS, G. & J. R. BUCHLER. 1988. Astrophys. J. **334**: 971.
7. BERGÉ, P., Y. POMEAU & C. VIDAL. 1984. Order Within Chaos. Wiley. New York.
8. OTT, E. 1993. Chaos in Dynamical Systems. Cambridge Univ. Press. Cambridge, England.
9. BUCHLER, J. R. 1990. Ann. N.Y. Acad. Sci. **617**: 17.
10. DUONG-VAN, M. 1987. Chaos 87. North-Holland. Amsterdam.
11. CONSTANTIN, P., C. FOIAS, B. NICOLAENKO & R. TEMAM. 1989. Integral Manifolds and Inertial Manifolds for Dissipative Partial Differential Equations. Applied Mathematical Sciences, **70**. Springer-Verlag. New York.
12. SAUER, T., J. A. YORKE & M. CASDAGLI. 1991. J. Stat. Phys. **65**: 579.
13. WEIGEND, A. S. & N. A. GERSCHENFELD. 1994. Time Series Prediction. Addison-Wesley. Reading, Mass.
14. ABARBANEL, H. D. I., R. BROWN, J. J. SIDOROWICH & L. S. TSIMRING. 1993. Rev. Mod. Phys. **65**: 1331.
15. WISDOM, J. 1987. Proc. R. Soc. London. **A413**: 109.
16. RUELLE, D. 1994. Phys. Today. July 24.
17. BUCHLER, J. R., T. SERRE, Z. KOLLÁTH & J. MATTEI. 1995. Phys. Rev. Lett. **74**: 842.
18. SERRE, T., Z. KOLLÁTH & J. R. BUCHLER. 1995. submitted for publication in Astron. Astrophys.
19. BUCHLER, J. R., Z. KOLLÁTH, T. SERRE & J. MATTEI. 1995. submitted for publication in Astrophys. J.
20. KOLLÁTH, Z., J. R. BUCHLER, T. SERRE & J. METTEI. 1996. in preparation.
21. BUCHLER, J. R. 1990. NATO ASI Series C302: 1. Kluwer. Dordrecht, The Netherlands.
22. BROWN, R. 1993. UCSD–INLS. Preprint.
23. MATTEI, J., M. SALADYGA, E. O. WAAGEN & C. M. JONES. 1988. AAVSO Monograph 3.
24. KOLLÁTH, Z. 1990. Mon. Not. R. Astron. Soc. **247**: 377.
25. VELDHUIZEN, T. & J. R. PERCY. 1989. J. AAVSO **18**: 97.

26. SAITOU, M., M. TAKEUTI & Y. TANAKA. 1989. Astron. Soc. Japan **41:** 297.
27. GINGOLD, R. A. 1976. Astrophys. J. **204:** 116.
28. BROWN, R., N. F. RULKOV & E. R. TRACY. 1994. Phys. Rev. E. **49:** 3784.
29. MINDLIN, G. B. & R. GILMORE. 1992. *Physica* **D58:** 229.
30. LETELLIER, C., P. DUTERTRE & F. SOUFI. 1994. "Charactérisation topologique d'une cephéide de type II modélisée par un code hydrodynamique". private communication.
31. GILLET, D., A. DUQUENNOY, P. BOUCHET & C. GOUIFFES. 1989. Astron. Astrophys. **215:** 316.
32. BUCHLER, J. R. 1993. *In* Nonlinear Phenomena in Stellar Variability, M. Takeuti and J. R. Buchler, Eds. Kluwer. Dordrecht, The Netherlands. Reprinted from 1993. Astrophys. Space Sci. **210:** 1.
33. BUCHLER, J. R. & N. E. G. BUCHLER. 1994. Astron. Astrophys. **285:** 213.
34. AIKAWA, T. 1990. Astrophys. Space Sci. **164:** 295.
35. MOSKALIK, P. & J. R. BUCHLER. 1990. Astrophys. J. **355:** 590.

# Unstable Long Wavelength Magnetohydrodynamic Waves on Highly Collimated Outflows[a]

PHILIP E. HARDEE

*Department of Physics and Astronomy*
*The University of Alabama*
*Tuscaloosa, Alabama 35487*

## INTRODUCTION

Extragalactic radio sources are observed to be produced by highly collimated outflows from the centers of galaxies and quasars. More locally, highly collimated outflows are observed to be associated with protostellar systems (e.g., HH34).[1] Such highly collimated outflows are subject to the Kelvin–Helmholtz instability. Observation of extragalactic jets provide examples where highly collimated jets propagate tens of times their diameter before bending or flaring (e.g., NGC6251),[2] or develop sinusoidal bending after only a short distance (e.g., 3C449).[3] It is hoped that observed jet structures can be related to jet velocity, density, and magnetic fields via stability theory and modeling. This is of particular importance to extragalactic jets where the continuum nature of synchrotron emission provides no direct measure of these quantities. In this paper we study the magnetohydrodynamic (MHD) waves driven unstable by the highly collimated jet flow, study the instability properties of two- and three-dimensional flows, and compare the predicted results to several numerical simulations.

## LINEARIZED STABILITY ANALYSIS

Let us model a jet as a slab of half-thickness $R$, or cylinder of radius $R$, having a uniform density, $\rho_{jt}$, a uniform internal axial magnetic field, $B_{jt}$, and a uniform velocity, $u$. The external medium has a uniform density, $\rho_{ex}$, and contains no magnetic field. The jet is in static total pressure balance with the external medium where the total static uniform pressure is $p_{ex}^* = p_{ex} = p_{jt}^* \equiv p_{jt} + B_{jt}^2/8\pi$. The general approach to analyzing the stability properties of this system is to linearize the one-fluid MHD equations of continuity and momentum along with an equation of state within each medium where the flow velocity $u = 0$. The flow velocity is then reintroduced when solutions are matched at the jet boundary. In each medium we assume

---

[a]The author acknowledges support from the National Science Foundation through Grant AST-9318397 and EPSCoR Grant OSR-9108761 to the University of Alabama.

that the perturbations are adiabatic in nature. The linearized ideal magnetohydrodynamic equations that are relevant to our model become

$$\frac{\partial \rho_1}{\partial t} + \nabla \cdot (\rho_0 u_1) = 0,$$

$$\rho_0 \frac{\partial u_1}{\partial t} = -\nabla p_1 + \frac{[(\nabla \times B_0) \times B_1 + (\nabla \times B_1) \times B_0]}{4\pi}$$

$$\frac{\partial p_1}{\partial t} = -\Gamma p_0 (\nabla \cdot u_1),$$

where the density, velocity, pressure, and magnetic field are written as $\rho = \rho_0 + \rho_1$, $u = u_1$, $p = p_0 + p_1$, $B = B_0 + B_1$, and subscript 1 refers to a perturbation to the equilibrium quantity.

## *Slab Geometry*

A slab jet is spatially resolved along two Cartesian axes and is infinite in extent in the third dimension. In slab geometry we look for perturbations $\rho_1$, $u_1$, $p_1$, and $B_1$ of the form

$$f_1(x, z) = f_1(x) \exp[i(kz - \omega t)], \tag{1}$$

where $z$ is the flow direction, and $x$ is perpendicular to the flow direction with the flow bounded by $|x| = R$. In slab geometry $y$ is perpendicular to the flow in the plane of the jet. In (1), and in the following analysis we only consider perturbations with wavevector parallel to the flow direction. A more general analysis has been performed by Hardee et al.[4] With this form for the perturbations a differential equation for the dependence of the total pressure perturbaton $p_1^* \equiv p_1 + (B_1 \cdot B_0)/4\pi$ within each medium as a function of $x$ can be written in the form

$$\frac{\partial^2 p_1^*}{\partial x^2} + \beta^2 p_1^* = 0, \tag{2}$$

where in general

$$\beta \equiv \left[ -k^2 + \frac{(\omega - ku)^4}{(\omega - ku)^2(a^2 + V_A^2) - k^2 V_A^2 a^2} \right]^{1/2}. \tag{3}$$

In (3), $a = (\Gamma p/\rho)^{1/2}$ is the sound speed, and $V_A = (B^2/4\pi\rho)^{1/2}$ is the Alfvén speed in the appropriate medium. The solutions are best written in the form

$$p_{1jt}^*(x) = C_{as} \sin(\beta_{jt} x) + C_s \cos(\beta_{jt} x) \tag{4}$$

inside the jet, and

$$p_{1ex}^*(x) = C_{\pm} e^{\pm i\beta_{ex} x} \tag{5}$$

outside the jet where the $\pm$ sign is chosen so that $p_{1ex}^*(x) \to 0$ as $|x| \to \infty$. In (4) and (5)

$$\beta_{jt} = \left[ -k^2 + \frac{(\omega - ku)^4}{(\omega - ku)^2(a_{jt}^2 + V_A^2) - k^2 V_A^2 a_{jt}^2} \right]^{1/2}$$

and

$$\beta_{ex} = \left[-k^2 + \frac{\omega^2}{a_{ex}^2}\right]^{1/2}.$$

At $|x| = R$ we require that the total pressure be continuous across the boundary, that is, $p^*_{1jt}(R) = p^*_{1ex}(R)$, and that the fluid displacement inside and outside the jet in the $x$ direction must be equal at the jet boundary, that is, $\xi_x^{jt}(R) = \xi_x^{ex}(R)$, where the fluid displacement in the $x$ direction is given by

$$\xi_x^{jt} = \frac{1}{\chi_{jt}} \frac{\partial p^*_{1jt}}{\partial x} \quad \text{and} \quad \xi_x^{ex} = \frac{1}{\chi_{ex}} \frac{\partial p^*_{1ex}}{\partial x}. \tag{6}$$

In (6) $\chi_{jt} = \rho_{jt}[(\omega - ku)^2 - k^2 V_A^2]$, and $\chi_{ex} = \rho_{ex}\omega^2$. The requirement that the two jet boundaries be coupled, along with (4), (5), and (6), and the boundary conditions on the pressure and the displacement can be shown to result in two dispersion relations

$$\frac{\beta_{jt}}{\chi_{jt}} \tan(\beta_{jt} R) = -i \frac{\beta_{ex}}{\chi_{ex}}, \tag{7}$$

and

$$\frac{\beta_{jt}}{\chi_{jt}} \cot(\beta_{jt} R) = +i \frac{\beta_{ex}}{\chi_{ex}}. \tag{8}$$

Equation 7 describes the symmetric pinch, $C_{as} = 0$ in (4), and (8) describes the asymmetric sinusoidal, $C_s = 0$ in (4), normal modes of a slab jet.

*Modes in the Low-Frequency Limit*

The symmetric pinch mode dispersion relation admits a surface (fundamental) wave and multiple body (reflection) wave solutions. In the low-frequency limit we have found that the real part of the pinch surface wave solution can be shown to be

$$\frac{\omega}{k} \approx u \pm \left\{\frac{1}{2}\left(V_A^2 + \frac{V_A^2 a_{jt}^2}{a_{ms}^2}\right) \pm \frac{1}{2}\left[\left(V_A^2 + \frac{V_A^2 a_{jt}^2}{a_{ms}^2}\right)^2 - 4\frac{V_A^4 a_{jt}^2}{a_{ms}^2}\right]^{1/2}\right\}^{1/2}, \tag{9}$$

where $a_{ms} \equiv (a_{jt}^2 + V_A^2)^{1/2}$. These solutions are related to fast (+) and slow (−) magnetosonic waves propagating with and against the jet flow speed $u$, but strongly modified by the jet–external-medium interface. The unstable growing solution is associated with the backwards moving slow magnetosonic wave. The pinch body wave solutions in the low-frequency limit are found to be given by

$$kR = \frac{(m - 1/2)\pi}{[(M_{ms}^2/1 - (M_{ms}/M_{jt} M_A)^2) - 1]^{1/2}}, \tag{10}$$

where $m \geq 1$ is an integer, $M_A \equiv u/V_A$ is the Alfvén Mach number, $M_{jt} \equiv u/a_{jt}$ is the sonic Mach number, and $M_{ms} \equiv u/a_{ms}$ is the magnetosonic Mach number.

The asymmetric sinusoidal mode dispersion relation also admits a surface (fundamental) wave and multiple body (reflection) wave solutions. In the low-frequency limit the sinusoidal surface wave solution can be shown to be[4]

$$\frac{ku}{\omega} \approx \frac{1}{1 - V_A^2/u^2} \left\{ 1 \pm i \left[ \left( 1 - \frac{V_A^2}{u^2} \right) \frac{X^2}{\eta} - \frac{V_A^2}{u^2} \right]^{1/2} \right\}, \quad (11)$$

where

$$X^2 \approx \frac{i}{[(u/a_{ex})^2 - (ku/\omega)^2]^{1/2} \omega R/u}$$

and $\eta \equiv \rho_{jt}/\rho_{ex}$. For supermagnetosonic jets unstable growing waves correspond to the solution with Re $[(ku/\omega)^2] > (u/a_{ex})^2$. In the dense jet limit, that is, $\eta \to \infty$, (11) becomes $\omega/k \approx u \pm V_A$, and the sinusoidal surface wave is revealed to be related to Alfvén waves propagating with and against the jet flow speed $u$, but strongly modified by the jet–external-medium interface. The unstable growing solution is associated with the backwards moving Alfvén wave. The sinusoidal body wave solutions in the low-frequency limit are found to be given by[4]

$$kR = \frac{m\pi}{[(M_{ms}^2/1 - (M_{ms}/M_{jt} M_A)^2) - 1]^{1/2}}, \quad (12)$$

where $m \geq 1$ is an integer.

*Growth Rate*

With the exception of the pinch surface wave, which has a broad plateau in the growth rate, all body waves and the sinusoidal surface wave have a relatively sharp maximum in the growth rate when a jet is supermagnetosonic. In the supermagnetosonic limit a maximum in the growth rate is achieved when[4]

$$\omega_m^* R/a_{ex} \approx m\pi + (\pi/2)_{as},$$

$$\lambda_m^* \approx \frac{2\pi}{\omega_m^* R/a_{ex}} \frac{M_{ms}}{1 + M_{ms}/M_{ex}} R, \quad (13)$$

$$v_{ph}^* \approx \frac{M_{ms}/M_{ex}}{1 + M_{ms}/M_{ex}} u,$$

where $m = 0$ gives the sinusoidal surface wave, $m \geq 1$ gives the pinch and sinusoidal body waves $(\pi/2)_{as}$ is added only for the sinusoidal mode, and $M_{ex} \equiv u/a_{ex}$ is the sonic Mach number in the external medium. With the exception of the pinch surface wave, the maximum spatial growth rate of these waves is approximately given by[4]

$$k_I^* \approx (2M_{ms}R)^{-1} \ln \left[ 4 \frac{\omega_m^* R}{a_{ex}} \right]. \quad (14)$$

When the jet is transmagnetosonic and super-Alfvénic the growth rate of the sinusoidal surface wave has no maximum, but increases proportional to the frequency. When the jet is sub-Alfvénic the sinusoidal surface wave is stable, but the pinch surface wave can remain unstable. The low-frequency form for the body modes indicates that these modes exist provided the denominator of (10) and (12) is real. The behavior of the denominator expressed as

$$f(u) = \frac{M_{ms}^2}{1 - (M_{ms}/M_{jt}M_A)^2} - 1$$

is shown in FIGURE 1. The body waves exist and are unstable when $f(u) > 0$, and this occurs if the jet speed exceeds the fast magnetosonic speed, or if the jet speed is slightly below the slow magnetosonic speed.

*Mode Transverse Dependence*

The total pressure, radial velocity, and radial displacement perturbations inside the jet associated with the sinusoidal modes of the slab jet are given by

$$p_1^*(x) = p_{1s}^* \frac{\sin(\beta_{jt}x)}{\sin(\beta_{jt}R)} \exp[i(kz - \omega t)]$$

$$u_{x1}(x) = u_{x1s} \frac{\cos(\beta_{jt}x)}{\cos(\beta_{jt}R)} \exp[i(kz - \omega t)] \qquad (15)$$

$$\xi_x(x) = \xi_{xs} \frac{\cos(\beta_{jt}x)}{\cos(\beta_{jt}R)} \exp[i(kz - \omega t)],$$

where $p_{1s}^*$, $u_{x1s}$, and $\xi_{xs}$ are the surface amplitudes of the total pressure, the radial velocity, and the radial displacement perturbations, and $\beta_{jt}(\omega, k)$ where $\omega$ and $k$ are

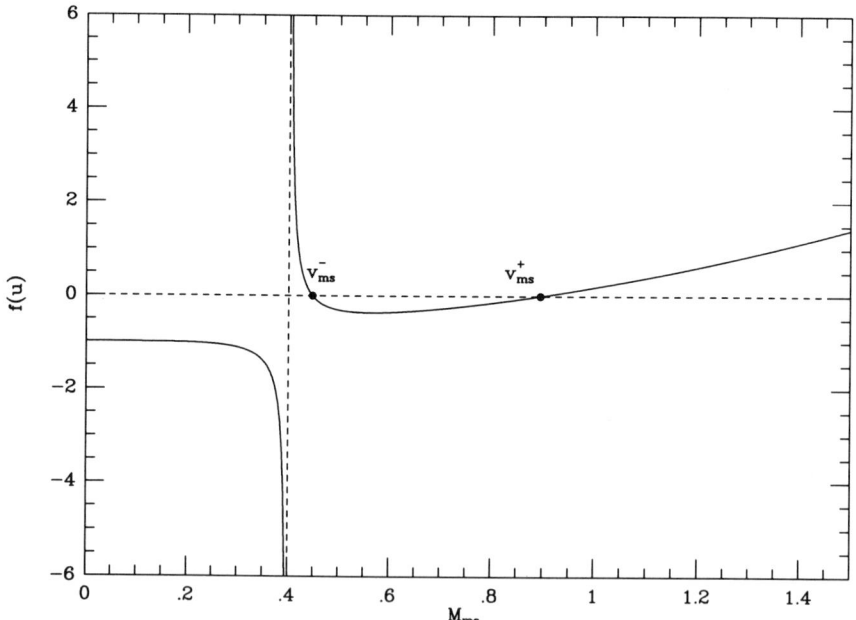

**FIGURE 1.** Typical behavior of the function $f(u)$. The zeros occur at the slow and fast magnetosonic speeds.

normal mode solutions of the asymmetric dispersion relation. Note that $\partial \xi_x/\partial t = u_{x1}$ and therefore $\xi_x(x) \propto u_{x1}(x)/\omega$. In (15) $[\sin(\beta_{jt}x)]/[\sin(\beta_{jt}R)]$ is replaced by $[\cos(\beta_{jt}x)]/[\cos(\beta_{jt}R)]$, and $[\cos(\beta_{jt}x)]/[\cos(\beta_{jt}R)]$ is replaced by $[\sin(\beta_{jt}x)]/[\sin(\beta_{jt}R)]$ for the symmetric mode. The $\beta_{jt}(\omega, k)$ are in general complex, and the transverse behavior of perturbations associated with the normal modes inside the jet can be written in the form $f_1(x) = A(x)e^{i\phi(x)}f_{1s} \exp[i(kz - \omega t)]$. There is both an amplitude and phase dependence of the perturbation internal to the jet surface relative to the perturbation at the jet surface. The amplitude dependence leads to null surfaces between the jet center and the surface at which no fluid displacement occurs, one null surface for the first body wave, two null surfaces for the second body wave, and so forth. The behavior of the fluid displacement associated with the first body sinusoidal wave is illustrated in FIGURE 2. The first sinusoidal body wave deforms an inner portion of the jet into a sinusoid with displacement approximately opposed to the surface displacement. Because motions in the transverse direction are not likely to exceed the jet magnetosonic speed, that is, $u_{x1}(x) \leq a_{ms}$, and $\omega_m^*$ increases for higher order body waves, the largest displacements at the maximum growth rate $\xi_x^*(x) \propto u_{x1}(x)/\omega_m^*$ should be associated with the surface wave and the lowest order body wave.

*Modes in the High-Frequency Limit*

In the high-frequency limit the real part of the solutions to the symmetric and asymmetric dispersion relations for all surface and body waves is given by[4]

$$\frac{\omega}{k} \approx u \pm \left[\frac{1}{2}(a_{jt}^2 + V_A^2)\left\{1 \pm \left[1 - \frac{4a_{jt}^2 V_A^2}{(a_{jt}^2 + V_A^2)^2}\right]^{1/2}\right\}\right]^{1/2}. \tag{16}$$

Equation 16 describes fast, $\omega/k \approx u \pm \max(V_A, a_{jt})$, and slow, $\omega/k \approx u \pm \min(V_A, a_{jt})$, magnetosonic waves. The unstable growing solution is associated with the backwards moving fast magnetosonic wave.

## Cylindrical Geometry

Axial magnetic fields in cylindrical geometry were initially investigated in several articles.[5,6] In cylindrical geometry we look for perturbations $\rho_1$, $u_1$, $p_1$, and $B_1$ of the form

$$f_1(r, \phi, z) = f_1(r) \exp[i(kz + n\phi - \omega t)], \tag{17}$$

where $z$ is the flow direction, and $r$ is in the radial direction with the flow bounded by $r = R$. In cylindrical geometry $n$ an integer is the azimuthal wavenumber, and for

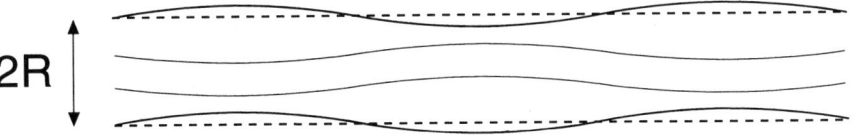

**FIGURE 2.** Illustration of the fluid displacement produced by the sinusoidal first body wave on a slab jet.

$n > 0$ waves propagate at an angle to the flow direction. With this form for the perturbations the differential equation for the dependence of the total pressure perturbation $p_1^* = p_1 + (B_1 \cdot B_0)/4\pi$ within each fluid as a function of $r$ can be written in the form

$$r^2 \frac{\partial^2}{\partial r^2} p_1^* + r \frac{\partial}{\partial r} p_1^* + [\beta^2 k^2 r^2 - n^2] p_1^* = 0, \tag{18}$$

where $\beta$ is given by (3). The solutions that are well behaved at jet center and at infinity are

$$p_{1jt}^*(r) = C_{jt} J_n(\beta_{jt} kr) \tag{19}$$

inside the jet, and

$$p_{1ex}^*(r) = C_{ex} H_n^{(1)} \beta_{ex} kr) \tag{20}$$

outside the jet, where $J_n$ and $H_n^{(1)}$ are Bessel and Hankel functions, respectively, and $\beta_{jt}$, $\beta_{ex}$ are the same as for the axially magnetized slab jet.

At $r = R$ we require that the total pressure be continuous across the boundary, that is, $p_{1jt}^*(R) = p_{1ex}^*(R)$, and that the radial fluid displacement inside and outside the jet be equal at the jet boundary, that is, $\xi_r^{jt}(R) = \xi_r^{ex}(R)$, where the fluid displacement in the radial direction, $\xi_r(r)$, is given by

$$\xi_r^{jt} = \frac{1}{\chi_{jt}} \frac{\partial p_{1jt}^*}{\partial r} \quad \text{and} \quad \xi_r^{ex} = \frac{1}{\chi_{ex}} \frac{\partial p_{1ex}^*}{\partial r}. \tag{21}$$

In the preceding expressions, $\chi_{jt}$ and $\chi_{ex}$ are the same as for the slab jet. Equations 19, 20, and 21, along with the boundary conditions on the pressure and radial displacement, result in a dispersion relation

$$\frac{\beta_{jt}}{\chi_{jt}} \frac{J_n'(\beta_{jt} R)}{J_n(\beta_{jt} R)} = \frac{\beta_{ex}}{\chi_{ex}} \frac{H_n^{(1)\prime}(\beta_{ex} R)}{H_n^{(1)}(\beta_{ex} R)} \tag{22}$$

describing the normal modes of a cylindrical jet. In (22) the primes denote derivatives of the Bessel and Hankel functions with respect to their arguments. The $n = 0, 1, 2, 3, \ldots$ modes involve pinching, helical, elliptical, triangular, and so on, distortions of the jet, respectively.

*Modes in the Low- and High-Frequency Limit*

The dispersion relation admits a surface (fundamental) wave and multiple body (reflection) wave solutions for each wave mode $n$. In the low-frequency limit we have found that the real part of the pinch surface wave solution is identical to that found for the symmetric pinch surface wave and is given by (9). In the low-frequency limit all surface waves of higher order modes ($n > 0$) are described by (11) with $X^2 = 1$. This difference is entirely the result of the different propagation angle of the wave in cylindrical geometry, and numerical solution of the slab jet and cylindrical jet dispersion relations reveals slightly different behavior of the solutions. The body wave solutions in the low-frequency limit are found to be given by

$$kR = \frac{(n + 2m - 1/2)\pi/2}{[(M_{ms}^2/1 - (M_{ms}/M_{jt}M_A)^2) - 1]^{1/2}},\quad (23)$$

where $n$ is the mode number and $m \geq 1$ is an integer. In the high-frequency limit the real part of all modes is given by (16), and the growing unstable solutions are associated with a backwards moving fast magnetosonic wave.

*Growth Rate*

With the exception of the pinch mode surface wave, which has a relatively broad plateau in the growth rate, all body waves independent of the mode, and all surface waves of higher order modes, have a relatively sharp maximum in the growth rate when a jet is supermagnetosonic. In the supermagnetosonic limit we have found that a maximum in the growth rate is achieved when

$$\omega_{nm}^* R/a_{ex} \approx (n + 2m + 1/2)\pi/4,$$

$$\lambda_{nm}^* \approx \frac{2\pi}{\omega_{nm}^* R/a_{ex}} \frac{M_{ms}}{1 + M_{ms}/M_{ex}} R, \quad (24)$$

$$v_{ph}^* \approx \frac{M_{ms}/M_{ex}}{1 + M_{ms}/M_{ex}} u,$$

where $n$ gives the wave mode, $m = 0$ gives the surface wave, and $m \geq 1$ gives the body waves. With the exception of the $n = 0$, $m = 0$, pinch mode surface wave, the maximum spatial growth rate is approximately given by

$$k_I^* \approx (2M_{ms}R)^{-1} \ln\left[4\frac{\omega_{nm}^* R}{a_{ex}}\right]. \quad (25)$$

When the jet is transmagnetosonic but super-Alfvénic the growth rate of the helical and higher order surface waves has no maximum but increases proportional to the frequency. When the jet is sub-Alfvénic, the helical and higher order surface waves are stable,[6] but we find that the pinch surface wave can remain unstable. The body waves exist and are unstable when the jet speed exceeds the fast magnetosonic speed or if the jet speed is slightly below the slow magnetosonic speed.[6]

*Mode Radial Dependence*

The pressure, radial velocity, and radial displacement perturbations associated with the normal modes of the cylindrical jet are given by[7]

$$p_1^*(r) = p_{1s}^* \frac{J_n(\beta_{jt} r)}{J_n(\beta_{jt} R)} \exp[i(kz \pm n\phi_s - \omega t)],$$

$$u_{r1}(r) = u_{r1s} \frac{J_n'(\beta_{jt} r)}{J_n'(\beta_{jt} R)} \exp[i(kz \pm n\phi_s - \omega t)], \quad (26)$$

$$\xi_r(r) = \xi_{rs} \frac{J_n'(\beta_{jt} r)}{J_n'(\beta_{jt} R)} \exp[i(kz \pm n\phi_s - \omega t)],$$

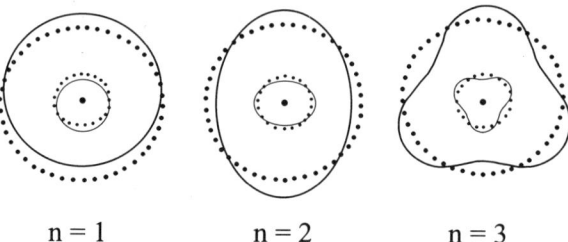

n = 1    n = 2    n = 3

FIGURE 3. Illustration of the fluid displacement produced by the first body wave associated with helical, elliptical, and triangular modes. The *inner contour* illustrates the inner displacement caused by the first body wave at the radius where the displacement is rotated in azimuth by 180°, i.e., $\phi(r) - \phi_s = 180°$, relative to the surface displacement.

where $p_{1s}^*$, $u_{r1s}$, and $\zeta_{rs}$ are the surface amplitudes of the total pressure, the radial velocity, and the radial displacement perturbations. The primes on the Bessel functions denote derivatives with respect to the arguments, and $\beta_{jt}(\omega, k)$ where $\omega$ and $k$ are normal mode solutions of the dispersion relation. In general $J_n'(\beta r)/J_n'(\beta R)$ and $J_n(\beta r)/J_n(\beta R)$ are complex, and the total pressure, radial velocity, and radial displacement perturbations associated with the normal modes can be written in the form $f_1(r) = A(r)e^{i\phi(r)}f_{1s} \exp[i(kz \pm n\phi_s - \omega t)]$. There is both an amplitude and an azimuthal phase angle dependence of the perturbation inside the jet relative to the perturbation at the jet surface. The dependence of the radial fluid displacement inside the jet of a surface mode is approximately given by $\xi_r(r) \propto (r/R)^{n-1}$, where $n$ is the mode number.[7] In fact, the surface waves show a somewhat faster falloff in amplitude relative to the surface amplitude than that predicted by the analytical approximation as a result of significant rotation in azimuth of, say, the maximum internal displacement relative to the maximum surface displacement.[8] At a constant azimuth the body waves show a reversal in fluid displacement at null surfaces between jet center and surface, one null surface for the first body wave, two null surfaces for the second body wave, and so forth. Only the helical body waves produce nonzero displacement at the jet center. The behavior of the fluid displacement associated with the first body helical, elliptical, and triangular modes is illustrated in FIGURE 3. The first helical body wave deforms an inner portion of the jet into a helical twist with displacement approximately opposed to the displacement of the jet surface associated with this body wave. Because motions in the radial direction are not likely to exceed the jet magnetosonic speed, that is, $u_{r1}(r) \leq a_{ms}$, and $\omega_{nm}^*$ increases with $n$ and also increases for higher order body waves, the largest displacements at the maximum growth rate $\xi_r^*(r) \propto u_{r1}(r)/\omega_{nm}^*$ should be associated with the lowest order surface modes, and the accompanying lowest order body modes.

## NUMERICAL ANALYSIS AND SIMULATIONS

Simulations were performed using the three-dimensional MHD code ZEUS-3D based on the ZEUS code described in a number of articles.[9-11] In the ZEUS code the second-order accurate monotonic van Leer[12] interpolation algorithm is used for the advective terms in the conservation equations. Shocks are handled by a von Neumann–Richtmyer artificial viscosity. The MHD algorithm is based on the constrained transport formalism of Evans and Hawley,[13] but uses some new innovations developed by David Clarke. A rigorous MHD test suite has been used to calibrate the code.[14] The following numerical simulations were carried out in collaboration with David Clarke.

### Two-Dimensional Simulations

Two-dimensional simulations are initialized by establishing a wedge jet with radius $R = r\psi$, where $\psi = 0.03$ radian is the half-opening angle, across a two-dimensional grid in cylindrical coordinates $r$ and $\theta$. In the $r$ direction 500 (650 in simulation A) scaled grid zones span a distance of $200R_0$ between an inner radial boundary $r_{inner} = 33.3R_0$, and an outer radial boundary $r_{outer} = 233.3R_0$. A total angular distance of 1.8 radian is spanned by 300 (400 in simulation A) grid zones with 60 (80) uniform angular grid zones centered on the jet spanning twice the jet diameter, and with an additional 120 (160) ratioed angular grid zones on each side of the jet. Outflow boundary conditions are used except where the jet enters the grid. The pressure in the external medium is matched to the total jet pressure of an assumed adiabatically expanding constant velocity jet. The jet and external medium have an adiabatic index $\Gamma = 5/3$. The initial jet Mach numbers are given by

$$M_{jt} = M_{jt0}(r\psi/R_0)^{1/3}$$
$$M_A = M_{A0}(r\psi/R_0)^{1/2} \qquad (27)$$
$$M_{ms} = \frac{M_{jt} M_A}{[(M_{jt})^2 + (M_A)^2]^{1/2}},$$

where $M_{jt} \equiv u/a_{jt}$, $M_A \equiv u/A_A$, and $M_{ms} \equiv u/(a_{jt}^2 + V_A^2)^{1/2}$. In the isothermal unmagnetized pressure matched external medium the jet Mach number is $M_{ex} \equiv u/a_{ex}$ and constant, and the density ratio between the jet and the external medium can be written in the form

$$\eta \equiv \rho_{jt}/\rho_{ex} = \rho_{jt0}/\rho_{ex0} \left[ \frac{1 + (P_{jt0}^b/P_{jt0}^{th})}{1 + (P_{jt0}^b/P_{jt0}^{th})(r\psi/R_0)^{-1/3}} \right] (r\psi/R_0)^{2/3}. \qquad (28)$$

The subscript 0 indicates values at the inner radial boundary where $r\psi/R_0 = 1$, and $P_{jt0}^b$ and $P_{jt0}^{th}$ refer to the jet magnetic and thermal pressures, respectively. The atmosphere is held steady by an appropriate artificial gravitational potential that is introduced for the purpose of maintaining a static atmosphere.

Three simulations were initialized with conditions at the inner radial boundary given in TABLE 1, which correspond to (A) an unmagnetized supersonic jet, (B) a magnetic pressure-dominated but supermagnetosonic jet, and (C) a magnetic pressure-dominated jet that makes a transition from sub-Alfvénic to super-Alfvénic transmagnetosonic flow at $r = 75R_0$. In each simulation the jet is perturbed at the origin by a sinusoidal oscillation of the jet velocity relative to the radial direction with frequency $\omega R_0/a_{ex} = 0.5$, and maximum amplitude $0.01u$.

TABLE 1. Conditions at the Inner Radial Boundary

| Run | $\rho_{jt}/\rho_{ex}$ | $u/a_{ex}$ | $a_{jt}/a_{ex}$ | $v_A/a_{ex}$ | $P^b_{jt}/P^{th}_{jt}$ | $M_{jt}$ | $M_A$ | $M_{ms}$ |
|---|---|---|---|---|---|---|---|---|
| A | 1.000 | 4.0 | 1.00 | 0 | 0 | 4.0 | $\infty$ | 4.00 |
| B | 0.348 | 4.0 | 1.00 | 1.50 | 1.87 | 4.0 | 2.67 | 2.22 |
| C | 0.029 | 4.0 | 2.00 | 6.00 | 7.69 | 2.0 | 0.67 | 0.63 |

Solutions to the slab jet dispersion relations appropriate to simulations (A) and (B) at $r = 75R_0$ are shown in FIGURE 4. The most important of the predicted instability properties in the supermagnetosonic regime are (1) that the growth rate of the pinch surface wave is considerably reduced by the axial magnetic field, and (2) that the maximum growth rate of the sinusoidal surface wave and all body waves associated with the symmetric and asymmetric modes increases approximately proportional to $M_{ms}^{-1}$. In particular, a detailed computation of the growth of sinusoidal oscillation at the wavelength corresponding to the driving frequency used in the simulations predicts growth more than a factor of 2 faster in simulation (B) than in simulation (A).

Solutions to the slab jet dispersion relations appropriate to simulation (C) at $r = 175R_0$ where the jet is transmagnetosonic are shown in FIGURE 5. The jet is predicted to be stable to sinusoidal oscillation interior to the Alfvén point at $r = 75R_0$. The most important of the predicted stability properties in the transmagnetosonic regime relative to the supermagnetosonic regime are (1) decrease in the growth rate of the pinch surface wave, (2) increase in the growth rate of the first pinch body wave, (3) decrease in the growth rates of the higher order pinch body waves, (4) decrease in the growth rates of all sinusoidal body waves, (5) disappearance of a maximum in the growth rate associated with the sinusoidal surface wave, and (6) an overall increase in the growth rate of the sinusoidal surface wave. Thus, stability theory predicts a very rapid amplitude growth of transverse oscillation beyond the Alfvén point, and possible growth of pinching associated with the first pinch body wave.

Density images from simulations (A), (B), and (C) that illustrate the effects of magnetic fields are shown in FIGURE 6. In simulation (A) large transverse amplitudes associated with sinusoidal oscillation at a wavelength $\lambda \approx 33R_0$ corresponding to sideways motions of the jet material at greater than the jet sound speed occur at a distance of about $r = 155R_0$ and the initial highly collimated flow broadens and slows significantly beyond this point. In simulation (B) large transverse amplitudes associated with sinusoidal oscillation at a wavelength $\lambda \approx 21R_0$ corresponding to sideways motions of the jet material at about the jet magnetosonic speed occur at a distance of about $r = 75R_0$, and the initial highly collimated flow broadens significantly and is slowed by about a factor of 2 by the outer edge of the computational grid. In both simulations the observed sinusoidal oscillation wavelength is within 10 percent of that predicted by theory to correspond to the driving frequency at the origin. Large sideways oscillation and jet broadening in simulation (B) occur about $40R_0$ from the inner boundary and about $120R_0$ from the inner boundary in simulation (A), and this confirms the theoretical prediction of faster growth as the magnetosonic Mach number is decreased. A detailed computation reveals that the

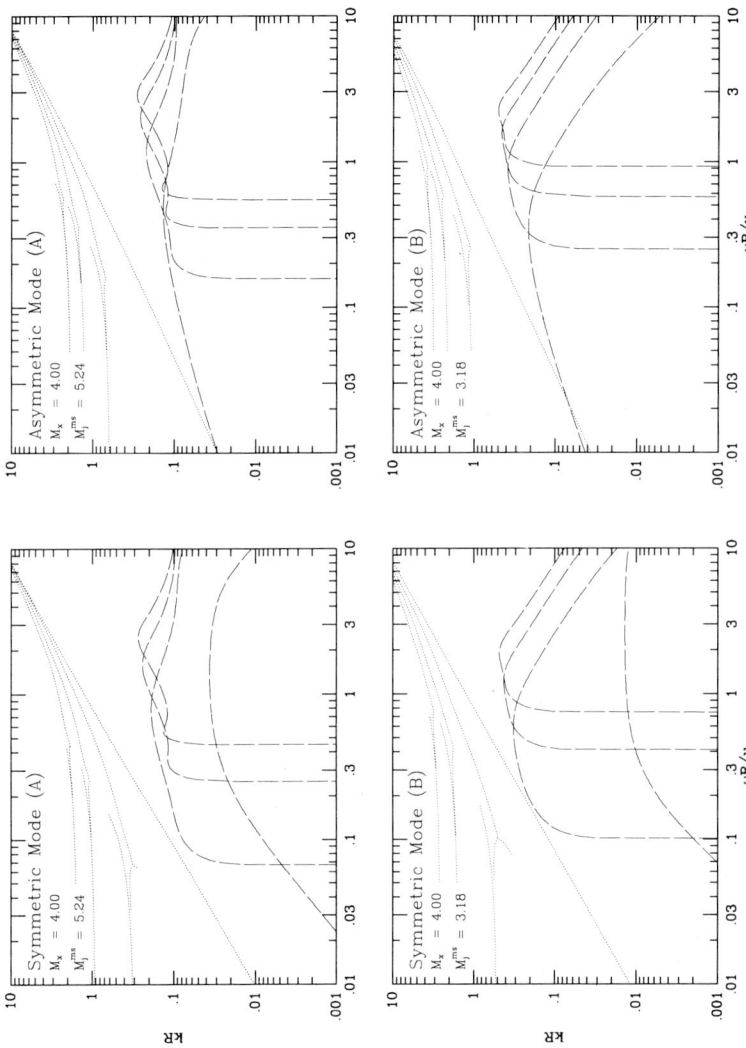

**FIGURE 4.** Surface and first three body wave solutions to the symmetric and asymmetric dispersion relations for simulations A and B at $r = 75R_0$. Dashed lines give the imaginary part of the wavenumber.

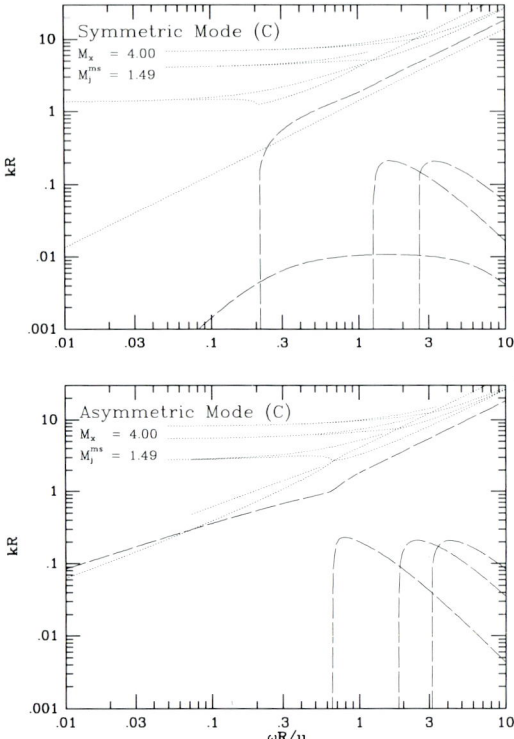

**FIGURE 5.** Surface and first three body wave solutions to the symmetric and asymmetric dispersion relations for simulation C at $r = 175R_0$.

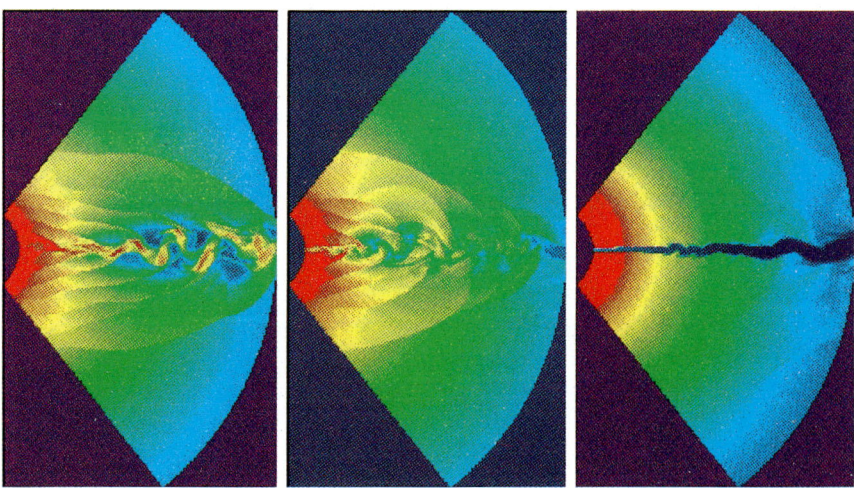

**FIGURE 6.** From left to right density images at late times in simulations A, B, and C. High densities are red and low densities are blue in these images.

observed point at which maximum transverse oscillation is achieved is at about the point where the theoretically predicted transverse amplitude of oscillation would be equal to the jet radius.

In simulation (C) the jet is observed to be stable as predicted out to $r \approx 75R_0$ at which point the jet destabilizes abruptly with a relatively short wavelength sinusoidal oscillation of wavelength $\lambda \approx 11R_0$. The observed oscillation wavelength is somewhat longer than the wavelength $\lambda \approx 8R_0$ predicted to correspond to the driving frequency, and this result indicates only a weak coupling between the perturbation at the origin and the destabilization point. Transverse velocities reach about 70 percent of the jet magnetosonic speed at $r \approx 100R_0$, or only $25R_0$ from the destabilization point, and the axial velocity drops by a factor of 2 at this point. The rapid growth of transverse oscillation, and observed destabilization is approximately in agreement with the theoretical prediction. At larger distances magnetic tension resulting from the strong axial magnetic field has prevented the flow from broadening significantly as occurred in simulations (A) and (B).

### *Three-Dimensional Simulation*

The simulation is initialized by establishing a cylindrical jet across a three-dimensional Cartesian grid of dimension $130 \times 130 \times 325$. Thirty uniform zones span the jet diameter along the transverse Cartesian axes. Outside the jet the zones are ratioed so that the 130 transverse zones span a total distance of $30R_0$ along the transverse Cartesian axes. In the axial direction 225 uniform zones span a distance of $30R_0$, and an additional 100 ratioed zones span an additional $30R_0$ for a total distance of $60R_0$ in the axial direction. In this simulation the external medium is unmagnetized, the jet is axially magnetized, and the jet magnetic and thermal pressures are comparable. The external medium is pressure-matched to the jet and has a density equal to that of the jet. The relevant Mach numbers are $M_{ms} = 4.92$ and $M_{ex} = 5.14$. The jet is perturbed at the origin by a periodic precession of frequency $\omega R/a_{ex} = 1.6$ at an angle of 0.005 radian relative to the axial direction. Outflow boundary conditions are used except where the jet enters the computational grid.

Solutions to the cylindrical jet dispersion relation for the pinch ($n = 0$), helical ($n = 1$), elliptical ($n = 2$), and triangular ($n = 3$) normal modes appropriate to this simulation are shown in FIGURE 7. The most important of the predicted stability properties in the supermagnetosonic regime are (1) the low maximum growth rate of the pinch mode surface wave, (2) the slightly higher maximum growth rate of the body modes compared to the maximum growth rate of the helical mode surface wave, and (3) the higher maximum growth rate of higher order surface waves compared to the maximum growth rate of the body modes. The predicted mode radial dependence and the computed growth rates lead us to predict that significant dynamical structures will be produced by the lowest order pinch and helical body waves, and by the helical, elliptical, and triangular surface waves.

Previous purely fluid three-dimensional simulations found velocity structures consistent with the helical, elliptical, and triangular surface waves, and the first helical body mode.[8,15] The inclusion of a magnetic field in the present simulation has allowed the generation of a line-of-sight integration through the data cube

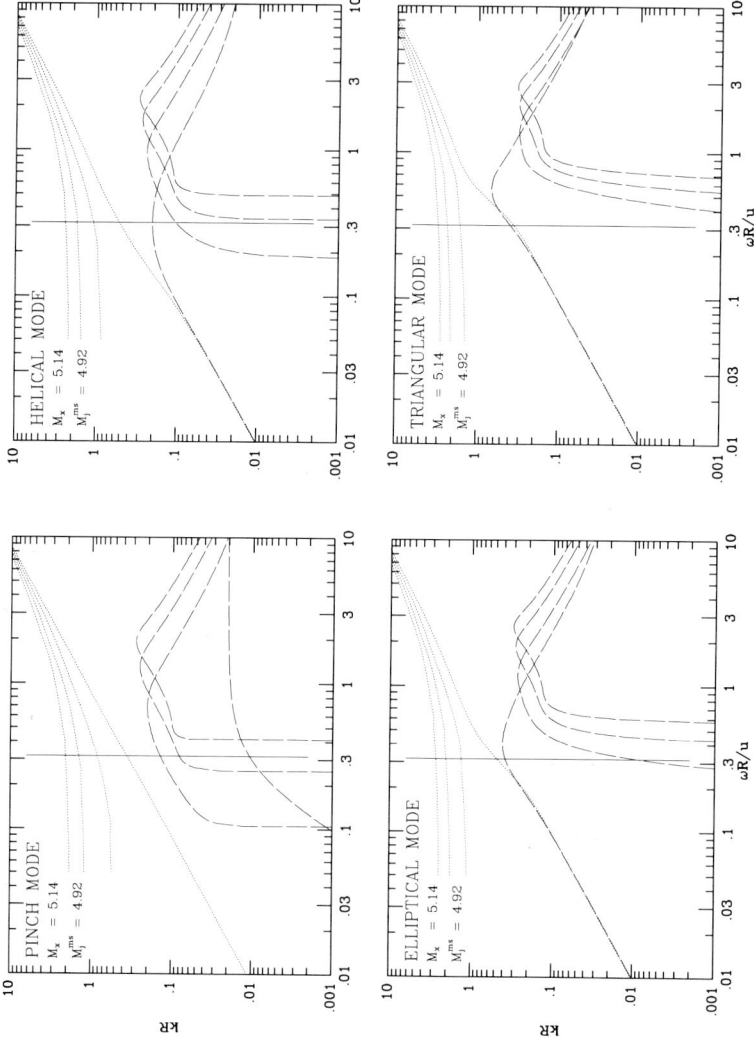

**FIGURE 7.** Surface and first three body wave solutions of the dispersion relation for four normal modes and parameters appropriate to the numerical simulation. The solid vertical line indicates the location of the precessional frequency used in the simulation.

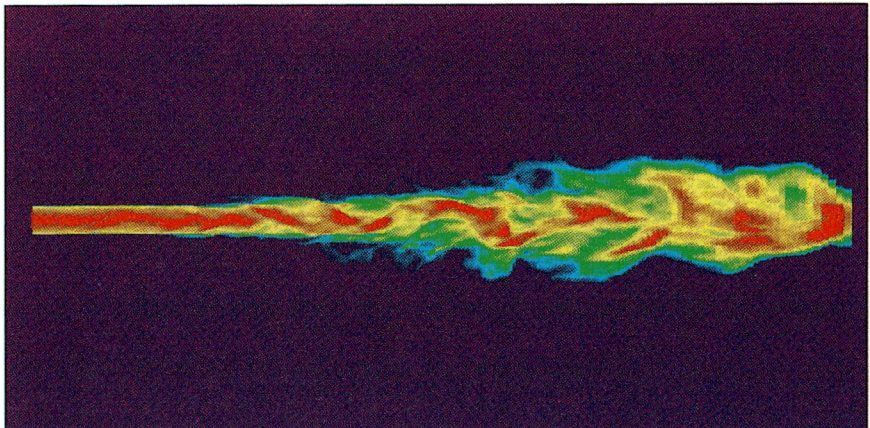

**FIGURE 8.** Integrated line-of-sight emission image from the three-dimensional simulation.

giving the emission image shown in FIGURE 8. This image suggests a short wavelength dual filamentation not visualized in the previous fluid simulations at distances less than $30R_0$. The initial collimated flow can be seen to begin to broaden significantly at a distance of about $30R_0$. More information is provided in FIGURE 9, which shows the density, sonic Mach number, and transverse velocity structure down the jet axis. Significant broadening at distances beyond $30R_0$ appears to be the result of transverse motion arising from the helical twist having grown to about 80 percent of the jet magnetosonic speed. This is in marked contrast to previous fluid simulations of jets less dense than the surrounding medium[8,15] in which significant broadening occurred as the result of jet filamentation and subsequent entrainment accompanied by shock heating.

TABLE 2 contains a summary of the wavelengths obtained from the emission image, and from the density and velocity structure, and various wavelengths calculated from the stability analysis. The observed helical twist is consistent with the wavelength predicted to correspond to the precessional frequency at the origin. The observed density and sonic Mach number oscillation wavelengths appear to result from the second (axial distance $< 15R_0$) and first ($30R_0 >$ axial distance $> 15R_0$) pinch body waves. As predicted larger fluctuation is associated with the longer wavelength and lower order body wave. A short wavelength transverse velocity oscillation on the jet axis evident in the velocity plot, and the twisted dual filamentation evident in the image between axial distances of $15R_0$ and $30R_0$ appear to result

TABLE 2. Observed and (Calculated) Wavelengths

| $\lambda_1^p/R$ | $\lambda_2^p/R$ | $\lambda_0^h/R$ | $\lambda_1^h/R$ | $\lambda_1^e/R$ |
|---|---|---|---|---|
| 3.8 (4.6$^a$) | 2.2 (2.4$^a$) | 11.5 (12.6$^b$) | 3.7 (3.3$^a$) | 3.9 (2.6$^a$) |

$^a$Value at maximum growth.
$^b$Value at $\omega R/a_{ex} = 1.6$.

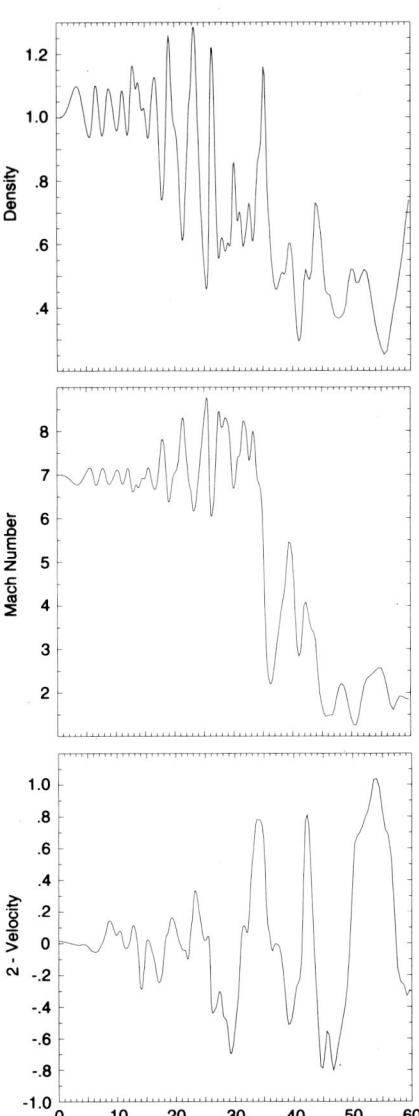

**FIGURE 9.** Quantities along the jet axis at the same time as the image in FIGURE 8. Axial distance is scaled relative to the jet radius, and the transverse velocity is scaled relative to the external sound speed.

from the first helical and first elliptical body waves, respectively, although the wavelengths are longer than that associated with the maximum growth rate of these waves. There is some suggestion that these wavelengths are shorter at axial distances $< 15R_0$, and we may be witnessing a transition from a shorter maximally growing wavelength to a longer, slower growing wavelength that reaches a larger amplitude.

## CONCLUSION

The unstable pinch and sinusoidal normal modes of a magnetized two-dimensional slab jet, and the unstable pinch, helical, and higher order normal modes of a magnetized three-dimensional cylindrical jet have been shown to be related to Alfvén, and to fast and slow magnetosonic waves strongly modified by the jet–external-medium interface. On the supermagnetosonic jet the growth rate of surface and body waves achieves a maximum at some frequency and wavelength, and the growth length scales with the magnetosonic Mach number. On transmagnetosonic jets the growth rate of the surface pinching wave and of high-order body waves is severely depressed relative to the growth rate of the sinusoidal (slab) surface wave, or of the helical and higher order (cylindrical) surface waves. When the jet is sub-Alfvénic but above the slow magnetosonic speed only the pinch surface wave can remain unstable. A series of numerical simulations was shown that agree with predictions made by the linearized stability analysis. The scaling of predicted and observed features with the magnetosonic Mach number suggests that it will be possible to model features seen in extragalactic radio jets and extract fundamental parameters.

## ACKNOWLEDGMENTS

The author acknowledges the large role played by David A. Clarke in developing ZEUS-3D. Dr. Clarke is supported in part by the Natural Sciences and Engineering Research Council of Canada. The numerical work utilized the Cray Y-MP at the National Center for Supercomputing Applications at the University of Illinois at Urbana-Champaign through Grant AST930007N, and the Cray C90 at the Pittsburgh Supercomputing Center through Grant AST930010P.

## REFERENCES

1. BURKE, T., R. MUNDT & T. P. RAY. 1988. Astron. Astrophys. **200**: 99.
2. PERLEY, R. A., A. H. BRIDLE & A. G. WILLIS. 1984. Astrophys. J., Suppl. Ser. **54**: 291.
3. PERLEY, R. A., A. G. WILLIS & J. SCOTT. 1979. Nature **281**: 437.
4. HARDEE, P. E., M. A. COOPER, M. L. NORMAN & J. M. STONE. 1992. Astrophys. J. **399**: 478.
5. FERRARI, A., E. TRUSSONI & L. ZANINETTI. 1981. Mon. Not. R. Astron. Soc. **196**: 1051.
6. BODO, G., R. ROSNER, A. FERRARI & E. KNOBLOCK. 1989. Astrophys. J. **341**: 631.
7. HARDEE, P. E. 1983. Astrophys. J. **269**: 94.
8. HARDEE, P. E., D. A. CLARKE & D. A. HOWELL. 1995. Astrophys. J. **441**: 644.
9. CLARKE, D. A., M. L. NORMAN & J. O. BURNS. 1989. Astrophys. J. **342**: 700.
10. STONE, J. M. & M. L. NORMAN. 1992. Astrophys. J., Suppl. Ser. **80**: 753.
11. ———. 1992. Astrophys. J., Suppl. Ser. **80**: 791.
12. VAN LEER, B. 1977. J. Comput. Phys. **23**: 276.
13. EVANS, C. E. & J. F. HAWLEY. 1988. Astrophys. J. **332**: 659.
14. STONE, J. M., J. F. HAWLEY, C. E. EVANS & M. L. NORMAN. 1992. Astrophys. J. **388**: 19.
15. HARDEE, P. E. & D. A. CLARKE. 1992. Astrophys. J., Lett. **400**: L9.

# Interface Instabilities in the Interstellar Medium[a]

J. H. HUNTER, JR.,[b] R. W. WHITAKER,[c] R. V. E. LOVELACE,[d]
AND C. SIOPIS[b]

[b]*Department of Astronomy*
*University of Florida*
*Gainesville, Florida 32611*

[c]*Los Alamos National Laboratory*
*EES-5, MS F-665*
*P.O. Box 1663*
*Los Alamos, New Mexico 86545*

[d]*Department of Applied and Engineering Physics*
*and Department of Astronomy*
*Cornell University*
*Ithaca, New York 14853*

## INTRODUCTION

We hold the view that star formation can be initiated by several mechanisms, and that none of these are well understood. Perhaps the most widely agreed upon mechanism is shocks associated with spiral density waves, which compress the interstellar medium (ISM) in spiral arms into Jeans mass clumps that form OB stars. Unfortunately, most of the mass of galactic disks resides in relatively inconspicuous stars of low mass and luminosity. Although conspicuous star forming regions, such as the Orion Nebula, have been studied in some detail, it is not obvious that these regions are typical. Indeed, as was emphasized by Aveni and Hunter[1-3] more than two decades ago, isolated T-Tauri stars exist, as do relatively small ($M \lesssim 100\ M_\odot$) regions of active star formation. A mechanism for forming isolated stars, proposed initially by Hunter,[4,5] is thermal instability followed by gravitational collapse. The advantage of this mechanism is that thermal collapse generates supersonic, imploding velocity fields, having maximum amplitudes $v \sim 10$ km/s, which compress a dense core of gas sufficiently to form stellar mass objects. This mechanism would work in the hot, tenuous gas well above the galactic plane, where massive OB stars exist that must have formed *in situ* because of their short lifetimes.[1] In Hunter's[5] approximate nonlinear model, the sensitivity of the minimum Jeans mass, $m_J$, to the implosion speed $\sim e^{-s(v^2/c^2)}$, where $s$ is a dimensionless constant of order unity and $c$ is the isothermal sound speed in the collapsing object. The predictions of this simple

---

[a]This work was supported in part by National Science Foundation Grant AST 9022827. One of the authors (J. H. H.) acknowledges support, in part, from National Science Foundation Grant NSFAST9022877. Another of the authors (R. V. E. L.) acknowledges partial support from NASA under Grant NAGW2293.

model were validated by numerical simulations.[5] In view of the sensitivity of $m_J$ to $v/c$, Hunter conjectured that star formation is not calculable in the same sense as is stellar evolution.

In the present communication, we reexamine two limiting cases of star-forming mechanisms involving self-gravity, thermodynamics, and velocity fields, that we believe must be ubiquitous in the ISM—the generally oblique collision of supersonic gas streams or turbulent eddies. The general case of oblique collisions has not yet been examined. However, two limiting cases have been studied in detail:

A. The head-on collision of two identical gas streams that form dense, cool accretion shocks that become unstable and may form Jeans mass clouds, which subsequently undergo collapse.
B. Linearly unstable tangential velocity discontinuities, which result in Kelvin–Helmholtz (K–H) instabilities and related phenomena.

Case A was studied in some detail by Hunter et al.,[6] denoted hereafter as Paper 1. Case B is a more difficult problem. Notwithstanding, Hunter and Whitaker have examined the linear problem in some detail, including thermodynamics and magnetic fields.[7,8] Hereafter, we denote these works as Papers 2 and 3. The compressible K–H instabilities exhibit rich and unexpected behaviors. Moreover a new thermal-dynamic (T-D) mode was discovered that arises from the coupling of the perturbed thermal behavior and the unperturbed flow. The T-D mode has the curious characteristic that it may be strongly unstable to interface modes when the global modes in either medium are absolutely thermally stable. In the present communication, additional models of case A are described and discussed in the next section, and in the following section self-gravity is added in the linear theory of tangential discontinuities, case B. We prove that self-gravity fundamentally changes the behavior of interfacial modes—*density discontinuities (or steps) are inherently unstable on roughly the free-fall timescale of the denser medium to perturbations of all wavelengths.*

## COLLIDING GAS FLOWS REVISITED

In Paper 1, the development of instabilities in highly compressed, cool, standing accretion shocks was studied numerically. The calculations were carried out in cylindrical $(r, z)$ coordinates, where $z$ is the vertical direction. In each unstable model, when a critical thickness of the accretion shock was reached, vorticity developed in dense tori of compressed gas at temperatures $T < 5$ K. The low temperatures were caused by CO cooling in the 1–0 rotational transition. Near the $r = 0$ axis, globules formed rather than tori, which became gravitationally unstable and collapsed if their mass, $m_c$, roughly exceeded the Jeans mass, $m_J$, of the local gas ($m_c \gtrsim 2m_J$ say). We emphasize that the slabs became unstable well before self-gravity became important. Indeed, in cases of low-mass slabs, the instabilities saturated, with the densest clumps being $\sim 10^3$ times denser than the gas in the individual streams. However, since $m_c < m_J$ no further collapse occurred.

Our conclusion that strongly cooling accretion shocks are unstable has been subsequently validated by Stevens et al.[9] in their studies of colliding stellar winds and by Kimura and Tosa,[10] who simulated the collisions of interstellar globules. The latter authors confirmed our conclusion that dense, strongly cooling regions collapse when their masses exceed roughly the local Jeans mass.

In our numerical experiments, we found that colliding gas flows are unstable when the dense materials cools efficiently. Moreover, isothermal colliding gas flows for which $3\,K \lesssim T \lesssim 10\,K$ are unstable also, whereas weakly cooling and adiabatic accretion shocks are relatively thick, tenuous, hot, and stable. Although we were unable to identify a new linear mode responsible for the instability, in Paper 1 we christened the behavior as Rayleigh–Taylor-like. The reasons for this appellation are: (a) matter is decelerated as it passes from the tenuous streams into the dense accretion disks, thereby mimicking the necessary condition for an R-T instability (the acceleration vectors are directed from the dense gas toward the tenuous streams), and (b) the growth rates of the instabilities approximately matched those of R-T instabilities having wavelengths and average fluid decelerations equal to those of our models. We were well aware that strong cooling was necessary for the shocks to be unstable and, consequently, the R-T explanation was only partially correct at best. More recently, Vishniac[11] has shown that our instability is inherently a nonlinear phenomenon, which he discusses further in a paper in this volume.

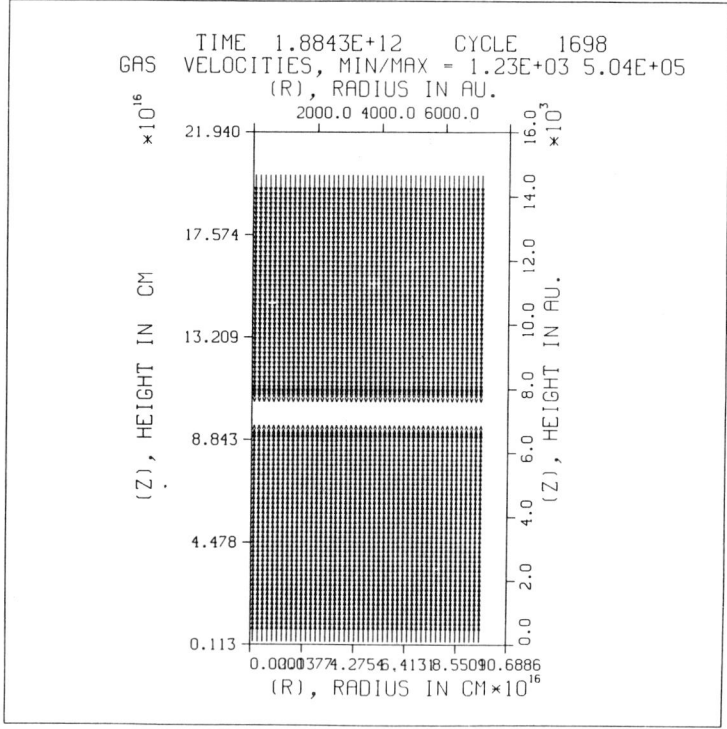

**FIGURE 1.** Velocity vectors showing a standing accretion shock at $5.97 \times 10^4$ yr after the gas streams collide. Each stream moves at 5 km s$^{-1}$.

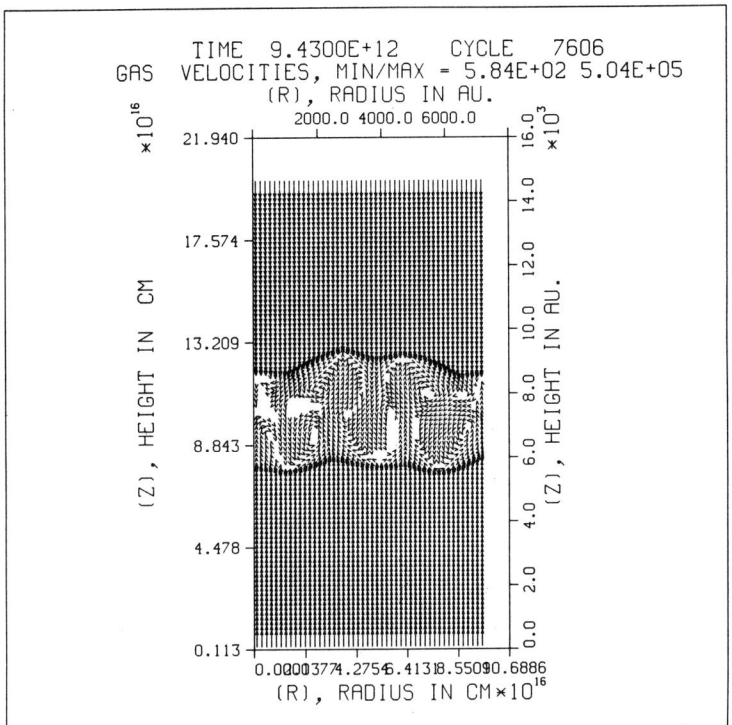

**FIGURE 2.** Velocity vectors showing the unstable accretion shock at $2.99 \times 10^5$ yr after the gas streams collide. Strong vorticity has developed in the dense material.

We comment that we are convinced that our instability is real and not a numerical artifact because:

a. The numerical code that we used, KFIX, is a standard multipurpose code that has been developed and carefully debugged at Los Alamos.

b. KFIX has a relatively large, implicit numerical viscosity, caused by its downwind differencing scheme.

We regarded the latter characteristic as a virtue—instabilities surviving such artificially viscous simulations must be robust. In addition to the models discussed in Paper 1, we have studied numerous others in which various characteristics were varied, such as the grid sizes and cooling rates. Models with finer grid sizes (and, consequently, less implicit viscosity) became unstable sooner than did the same models calculated with coarser grids. This result is in agreement with that of Stevens et al.,[9] who reported instabilities appearing so early in their simulations that they referred to them as thin-shell instabilities. FIGURES 1 through 4 show one of our heretofore unpublished calculations illustrating the development of instabilities in an accretion shock of relatively low mass. Self-gravity is relatively unimportant throughout this simulation.

In the end, our original conclusions remain unaltered; strongly cooling, colliding gas flows form unstable accretion shocks that fragment into subunits that may be

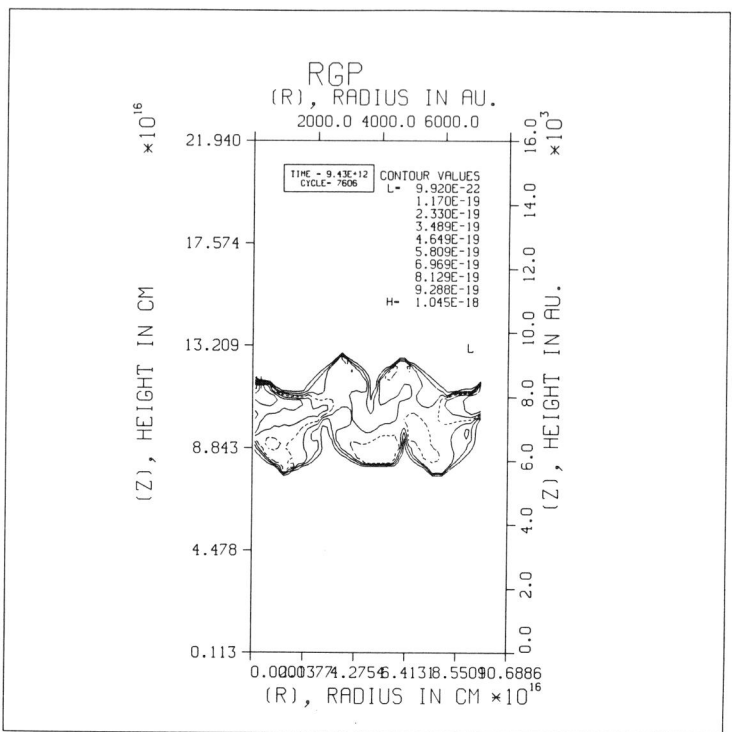

**FIGURE 3.** Density contours in the unstable accretion shock shown in FIGURE 2. The densities (gm cm$^{-3}$) of the different contours are shown on the graph.

much less massive than the Jeans mass. However, in other models with fragments that are more massive, gravitational collapse occurs if a fragment mass exceeds roughly the local Jeans mass.

## KELVIN–HELMHOLTZ AND RELATED INSTABILITIES WITH SELF-GRAVITY

In Papers 2 and 3, Hunter and Whitaker examined the stability of tangential discontinuities of velocity and density, including realistic thermodynamics and magnetic fields. Although this problem exhibits rich and varied behavior, the primary instabilities are (1) compressible K–H at moderate to low relative Mach numbers, $m$, and (2) a new, ubiquitous, supersonic instability, christened a T-D mode, because it arises from a coupling of the unperturbed flow with the heating and cooling rates of the gases on either side of the interface. The latter instability often joins continuously with the unstable K–H mode at $m \leq \sqrt{8}$, and its growth approaches an asymptotic limit at large $m$. In this present communication, we include self-gravity in the compressible hydrodynamic K–H problem. This work has been done in collaboration with Dr. R. V. E. Lovelace of Cornell. We will only outline our derivation of the fundamental dispersion relation (FDR) for this problem as a detailed derivation is being published in the *Astrophysical Journal*.

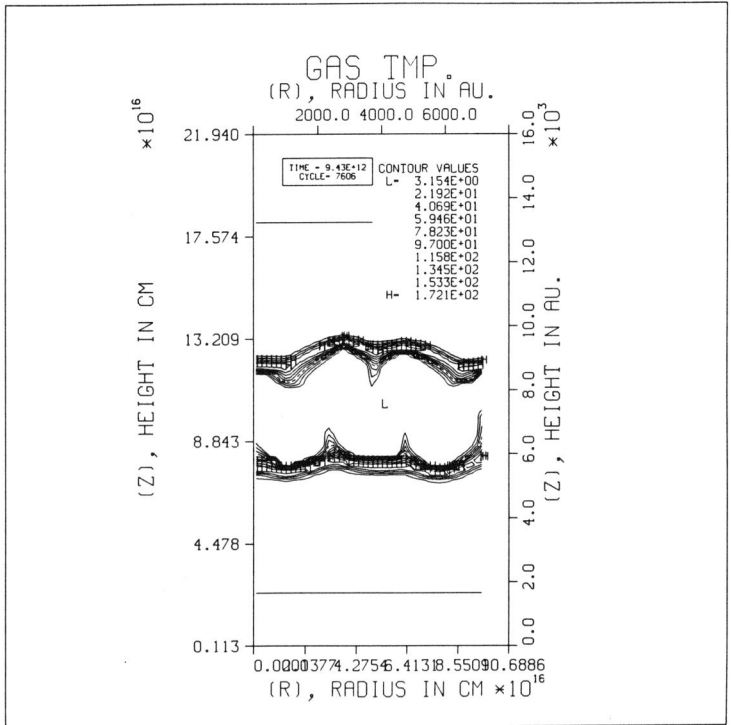

**FIGURE 4.** Temperature contours of the gas in the accretion shock shown in FIGURES 1 and 2. The temperatures of the contours are shown on the graph.

The linearized equations of continuity, motion, and energy conservation are written in a coordinate system that is at rest with respect to the (denser) fluid in the lower half-space, region 2 ($z < 0$). The unperturbed gas in the upper half-space, region 1 ($z > 0$), moves along the $x$-axis at constant speed $U$. The problem is illustrated in FIGURE 5.

In standard notation, the linear equations are

$$D\rho' = -\nabla \cdot (\rho \overline{v'}), \tag{1}$$

$$\rho D\overline{v'} + (\nabla U)\overline{v'} = -\nabla p' + \rho \nabla \psi' = -\nabla \tilde{p} - \psi' \nabla \rho, \tag{2}$$

$$(D + F_p)p' = c^2(D + F_\rho)\rho' + c^2 \overline{v'} \cdot \nabla \rho, \tag{3}$$

and

$$\nabla^2 \psi' = -4\pi G \rho'. \tag{4}$$

In these equations, $\rho$, $p$, and $c$ are, respectively, the background density, pressure, and sound speed, assumed constant for each medium, and perturbed dependant variables are adorned with primes. The velocity $\overline{V} = \overline{U} + \overline{v'} = \overline{i}U + \overline{v'}$ and the gravitational potential is denoted by $\psi$. The variable $\tilde{p} \equiv p' - \rho\psi'$. We Fourier analyze in the horizontal direction, letting each dependant variable $\zeta'(x, y, z, t) = \zeta'_0 e^{-|qz|} e^{-i(\omega t - k_1 x - k_2 y)}$, where horizontal wave number $k \equiv \sqrt{k_1^2 + k_2^2}$ and $q$ denotes

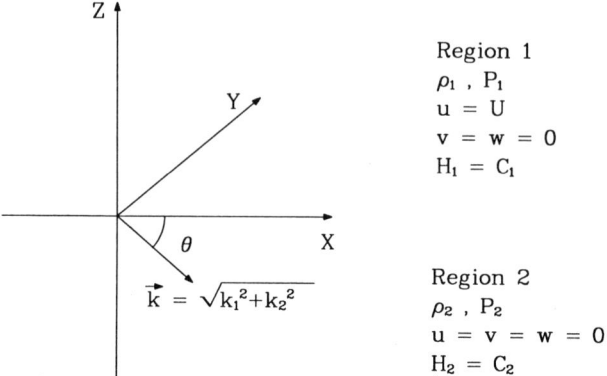

**FIGURE 5.** An illustration of the tangential problem.

the eigenfunctions in the $z$ direction. The operator $D \equiv (\partial/\partial t) + U(\partial/\partial x)$ and $F_p$ and $F_\rho$ are the thermal frequencies associated with the cooling rates $C$ and the heating rates $H/\text{cm}^3$. These frequencies are defined by

$$F_p = \frac{\gamma(\gamma - 1)}{\rho c^2} C \left( \frac{\partial \ln C}{\partial \ln T} \bigg|_\rho - \frac{\partial \ln H}{\partial \ln T} \bigg|_\rho \right), \tag{5a}$$

and

$$F_\rho \equiv \frac{(\gamma - 1)}{\rho c^2} C \left( \frac{\partial \ln C}{\partial \ln T} \bigg|_\rho - \frac{\partial \ln C}{\partial \ln \rho} \bigg|_T + \frac{\partial \ln H}{\partial \ln \rho} \bigg|_T - \frac{\partial \ln H}{\partial \ln T} \bigg|_\rho \right). \tag{5b}$$

Each medium has its own distinctive values of $F_p$ and $F_\rho$.

Our analysis is appropriate only for wavelengths that are relatively small compared with the Jeans length, which allows us to regard $\rho$, $p$, and temperature $T$ as constants in each medium. However, to preserve logical consistency, we require that the unperturbed pressure be continuous across the interface, which implies $p_1 = p_2$ and $(\rho_1 T_1/\mu_1) = (\rho_2 T_2/\mu_2)$, where $\mu$ is the mean molecular weight.

Defining $\phi_1 \equiv \omega - k_1 U$ and $\phi_2 \equiv \omega$, the solutions of the linear equations in each medium lead to the eigenfunctions,

$$q_1 = -k \left[ 1 - \frac{(\phi_1^2 + 4\pi G\rho_1)}{k^2 c_1^2} Q_1 \right]^{1/2}, \tag{6a}$$

and

$$q_2 = +k \left[ 1 - \frac{(\phi_2^2 + \pi G\rho_2)}{k^2 c_2^2} Q_2 \right]^{1/2}, \tag{6b}$$

where the thermal functions are given by

$$Q_1 \equiv \frac{(\phi_1 + iF_{p_1})}{(\phi_1 + iF_{\rho_1})} \quad \text{and} \quad Q_2 \equiv \frac{(\phi_2 + iF_{p_2})}{(\phi_2 + iF_{\rho_2})}.$$

At the new interface, the $z$ equation of motion may be written as

$$\frac{d\tilde{p}}{dz} = \rho\phi^2 z' - \psi'\frac{d\rho}{dz}, \tag{7}$$

where $z'$ is the normal displacement in the $z$ direction. Upon integrating (7) across the interface, from $z_-$ to $z_+$, we arrive at the exact result,

$$\tilde{p}_{1i} - \tilde{p}_{2i} = -\tfrac{1}{2}(\psi'_{1i} + \psi'_{2i})(\rho_1 - \rho_2), \tag{8}$$

where subscript $i$ denotes the interface. Arbitrarily close to the interface, (7) gives $\tilde{p} = (\rho\phi^2 z'/q)$.

In either medium the solution of Poisson's equation for the distributed matter yields

$$\psi' = \frac{4\pi G z'}{q}. \tag{9}$$

However, the general solution for $\psi'$ must include a contribution from the interface. Integration of (1) across the interface gives, for each Fourier component of the surface density,

$$\Sigma' = -\frac{1}{2}\left(\frac{\phi_1}{\phi_2} + \frac{\phi_2}{\phi_1}\right)(\rho_1 - \rho_2)z'_i, \tag{10}$$

where $z'_i$ is the normal displacement of the interface itself. The general solution for each Fourier component of $\psi'$ must satisfy the following two conditions:

$$\psi'_{1i} = \psi'_{2i}, \tag{11a}$$

and

$$\left.\frac{\partial\psi'}{\partial z}\right|_{z^+} - \left.\frac{\partial\psi'}{\partial z}\right|_{z^-} = -4\pi G\Sigma' \text{ (Gauss' theorem).} \tag{11b}$$

Thus, in each medium we write solutions of the form,

$$\psi' = \frac{4\pi G\rho z'}{q} + A'(x, y, t)e^{-|kz|}, \tag{12a}$$

or,

$$\psi'_{1,2} = \frac{4\pi G\rho_{1,2}}{q_{1,2}} z'_{1,2}(x, y, t)e^{-|q_{1,2}z|} + A'_{1,2}(x, y, t)e^{-|kz|}, \tag{12b}$$

where both $z'_{1,2}$ and $A'_{1,2}$ are of the forms $(z'_{o_{1,2}}, A'_{o_{1,2}})e^{-i(\omega t - k_1 x - k_2 y)}$. Consequently, the interfacial contribution satisfies Laplace's equation in both media. We must insert solutions of the preceding form into conditions (11a) and (11b) and solve for $A'_{o1}$ and $A'_{o2}$. At the interface, the solution for the perturbed potentials is,

$$\psi'_{1i} = \psi'_{2i} = \psi'_i = \left[2\pi G\left(\frac{\rho_1}{q_1} + \frac{\rho_2}{q_2}\right) - \frac{\pi G}{k}\left(\frac{k^2 U^2}{\phi_1\phi_2}\right)(\rho_1 - \rho_2)^2\right]z'_i. \tag{13}$$

Upon substituting this form for the potentials, as well as the results $\tilde{p}_{1,2i} = (\rho_{1,2}\phi_{1,2}^2/q_{1,2})$, into (8), we obtain for the FDR,

$$\frac{\rho_1\phi_1^2}{q_1} - \frac{\rho_2\phi_2^2}{q_2} + 2\pi G\left(\frac{\rho_1}{q_1} + \frac{\rho_2}{q_2}\right)(\rho_1 - \rho_2) - \frac{\pi G k_1^2 U^2}{k\phi_1\phi_2}(\rho_1 - \rho_2)^2 = 0. \qquad (14)$$

For an incompressible fluid, both $c_1$ and $c_2$ are infinite, yielding $q_1 = -k$ and $q_2 = +k$, and the FDR becomes a quartic equation with real coefficients. In the incompressible, static limit we arrive at the scale invariant result, valid for all $k$,

$$\omega^2 = \frac{-2\pi G(\rho_1 - \rho_2)^2}{(\rho_1 + \rho_2)}. \qquad (15)$$

We have independently verified this static result using the approach of Bernstein et al.[12] Thus, we predict the remarkable result that a static density discontinuity is absolutely unstable for wavelengths that are significantly *less* than the Jeans length and that the e-folding time, $t_e$, is of the same order as $t_f$, the free-fall collapse time for a uniform sphere of initial density $\rho_2$. More precisely, letting $D = \rho_1/\rho_2$,

$$t_e \equiv \omega^{-1} = \frac{4}{\sqrt{3\pi}} \frac{\sqrt{1+D}}{|1-D|} t_f \simeq \frac{0.735\sqrt{1+D}}{|1-D|} t_f. \qquad (16)$$

In the next section, we show that the growth rates of compressible analogues of this static problem are roughly the same as those of their incompressible counterparts.

When written in univariant polynomial form, the FDR becomes an equation of 24th order with complex coefficients. Thus, the challenges posed by the general problem are (a) developing accurate solutions of the polynomial, and (b) substituting each of them into the FDR. Only those roots that satisfy the FDR are acceptable solutions for our problem. As stated previously, for incompressible flows, the polynomial reduces to one of 4th order, which can be solved easily.

In the compressible case, we define a dimensionless growth rate and relative Mach number by $x \equiv -(\omega/(kc_1)) = x_r + ix_i$ and $m \equiv (k_1 U/(kc_1))$, respectively, where $x_r$ is the real (oscillatory) part and $x_i$ is the dimensionless growth rate. For unstable modes, $x_i < 0$. For incompressible fluids, $c_1$ must be regarded as a characteristic speed (rather than the speed of sound in Medium 1). Other dimensionless parameters characterizing each model are Jeans' parameter for Medium 2, $J_2 = (4\pi G\rho_2/(k^2 c_1^2))$, $D \equiv (\rho_1/\rho_2)$, $E \equiv (c_1/c_2)^2$; Jeans' parameter for Medium 1, $J_1 = DJ_2$; and dimensionless thermal frequencies $\alpha_{1,2} \equiv (F_p)_{1,2}/(kc_1)$ and $\beta_{1,2} \equiv (F_\rho)_{1,2}/(kc_1)$.

## KELVIN–HELMHOLTZ MODELS WITH SELF-GRAVITY

In FIGURE 6 the growth rates are shown for an incompressible fluid characterized by $D = \frac{1}{2}$ and $J_2 = 0.09$. At $m = 0$, we note that the unstable growth rate $x_i = -0.0866$, in agreement with (15). When $m \gtrsim 0.8$, the growth rate converges to that of an incompressible fluid in the absence of gravity. Results for an adiabatic compressible model with the same density contrast are shown in FIGURE 7. In the range

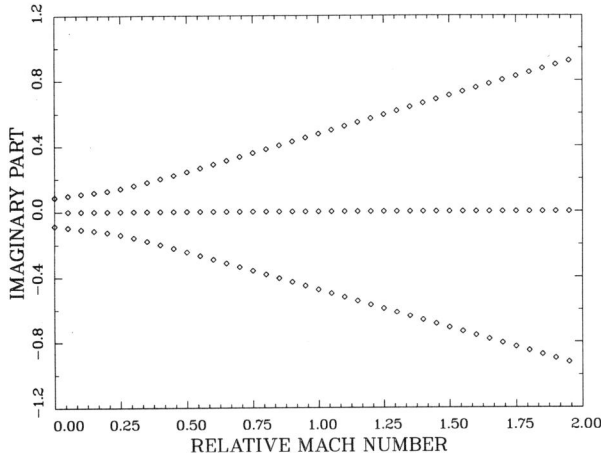

**FIGURE 6.** Incompressible Kelvin–Helmholtz instabilities including self-gravity. For this case $D = 1/2$ and the relative Mach number $= (U/c_1) \cos \theta$. The negative imaginary parts are unstable.

$0 \leq m \leq 2$, the most rapidly growing mode is the K–H solution, modified by self-gravity. At $m = 0$, $x_i = -0.09054$, and the maximum growth occurs at $m \sim 1.6$ and is scarcely affected by self-gravity.

We incorporated thermodynamics into this model ($D = 1/2$, $E = 2$, and $J_2 = 0.09$) by assuming that heating occurs at a constant rate per gram, and that both

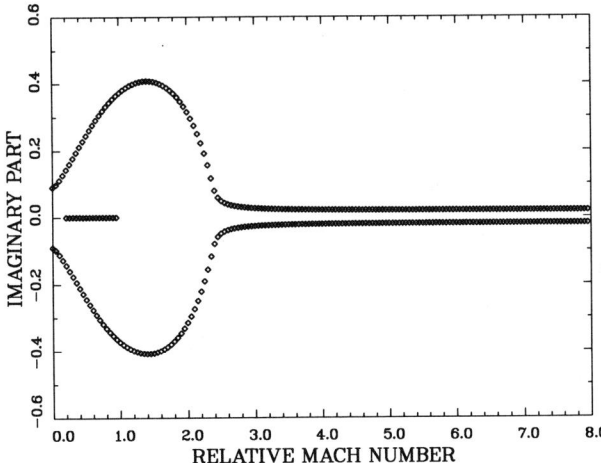

**FIGURE 7.** The adiabatic version of the problem shown in FIGURE 6 for a compressible fluid; $D = 1/2$, $E = 2$. The imaginary parts of the solutions are shown on the vertical axis and the relative Mach number on the horizontal axis. Unstable modes have negative imaginary parts, $x_i < 0$.

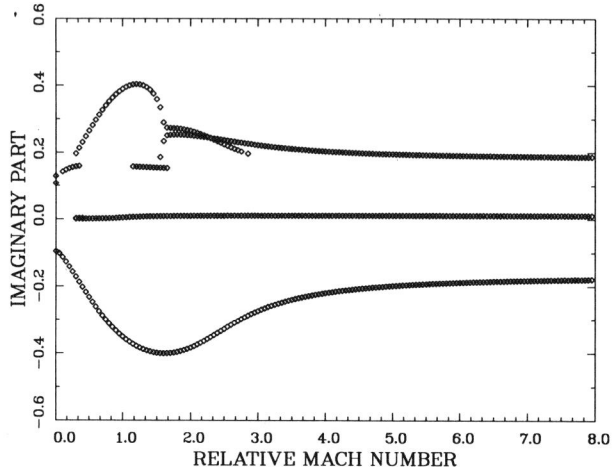

**FIGURE 8.** The anisentropic version of the problem shown in FIGURE 7.

media are molecular clouds, cooling by the $1 \to 0$ rotational transition in the CO molecule. The two molecular regions were assumed to have the following characteristics: number densities $n_1 = 100$ cm$^{-3}$, $n_2 = 200$ cm$^{-3}$, $T_1 = 20$ K, $T_2 = 10$ K, and $\gamma_1 = \gamma_2 = 1.4$. The horizontal perturbation wavelength $\lambda = 0.876$ pc, $\alpha_1 = 0.3978$, $\beta_1 = 0.1243$, $\alpha_2 = 0.4929$ and $\beta_2 = 0.1807$. Our results are shown in FIGURE 8. At $m = 0$, the unstable growth rate $x_i = -0.09561$. The maximum growth rate, $x_i \simeq -0.39$, occurs at $m \simeq 1.6$, and this unstable K–H solution joins continuously with that of the unstable T–D mode at $m \sim 2$. The $e$-folding time corresponding to $x_i = -0.39$, $t_e = 1.03 \times 10^6$ yr.

## CONCLUSIONS

In this paper we have reported our results for the limiting cases of interacting gas streams in the ISM–gas flows colliding head on, and the linear growth of instabilities in tangential velocity and density discontinuities. In both extremes, structure develops on timescales of the order of $10^6$ yr. Therefore, we think it likely that star formation takes place on timescales of $10^6$–$10^7$ years in the vicinity of accretion shock regions that form in the cases of obliquely colliding gas flows, and clouds in the ISM.

## ACKNOWLEDGMENTS

Two of the authors (J. H. H. and R. W. W.) thank Richard Gerwin for several discussions during the course of this work, and J. H. H. thanks G. Contopoulos for numerous fruitful discussions. Further, J. H. H. thanks Kenneth Eggert for his hospitality during summer visits to the Los Alamos National Laboratory.

## REFERENCES

1. AVENI, A. & J. H. HUNTER. 1967. Astron. J. **72**: 1019.
2. ———. 1969. Astron. J. **74**: 1021.
3. ———. 1972. Astron. J. **77**: 17.
4. HUNTER, J. H. 1969. Mon. Not. R. Astron. Soc. **142**: 473.
5. ———. 1979. Astrophys. J. **233**: 946.
6. HUNTER, J. H., M. T. SANDFORD II, R. W. WHITAKER & R. I. KLEIN. 1986, Astrophys. J. **305**: 309. (Paper 1)
7. HUNTER, J. H. & R. W. WHITAKER. 1987. *In* Chaotic Phenomena in Astrophysics, J. R. Buchler and H. K. Eichorn, eds.: 144. Ann. N.Y. Acad. Sci. **497**. (Paper 2)
8. ———. 1989. Astrophys. J. Suppl. **71**: 777. (Paper 3)
9. STEVENS, I. R., J. M. BLODIN & A. T. M. POLLOCK. 1992. Astrophys. J. **386**: 265.
10. KIMURA, T. & M. TOSA. 1991. Mon. Not. R. Astron. Soc. **251**: 664.
11. VISHNIAC, E. T. 1994. Astrophys. J. **428**: 186.
12. BERSTEIN, I. B., E. A. FRIEMAN, M. D. KRUSKAL & R. M. KULSRAD. 1958. Proc. R. Soc. A. **244**: 17.

# Alfvén Waves and Interstellar Turbulence

S. SRIDHAR[a]

*Canadian Institute for Theoretical Astrophysics
McLennan Labs, University of Toronto,
60 St. George Street, Toronto M5S 1A7, Canada*

## INTRODUCTION

Fluctuations of electron number density in the ionized interstellar medium (ISM) manifest themselves through the scintillation of small angular diameter radio sources.[1–3] Studies of interstellar scintillation[4–7] have established that the spectrum of electron density fluctuations is a power law on scales $\lambda$ in the range $10^9$ cm $\leq \lambda \leq 10^{15}$ cm, perhaps even extending to scales as large as 100 pc.[8] The rms density fluctuation on scale $\lambda$ is $n_\lambda \propto \lambda^\alpha$, with $\alpha \approx 1/3$, which is similar to the Kolmogorov law for velocity fluctuations in hydrodynamic turbulence. It is thought that such a power spectrum results from the action of underlying interstellar turbulence on the electrons. The elongation of the images[9,10] of strongly scattered sources suggests that the turbulence is anisotropic. This is not surprising, since the interstellar magnetic field is the dominant source of stress on the relevant length scales. Will turbulence in a predominantly ionized, magnetized plasma result in a Kolmogorov-like cascade? Recent work[11,12] strongly suggests that Alfvénic turbulence in the ionized ISM exhibits an anisotropic Kolmogorov cascade of energy to small spatial scales. The power spectrum is quasi-two-dimensional, as originally suggested by Higdon.[13] Moreover, the turbulence mixes electron density variations in precisely the right manner needed to match the interstellar spectrum.

We idealize the ionized ISM as a homogeneous, completely ionized plasma which, in the undisturbed state, supports a uniform magnetic field ($B_0 \hat{z}$). If the gas density is $\rho$, purely magnetic stresses transport disturbances at the Alfvén speed, $V_A = B_0/\sqrt{4\pi\rho}$. The sound speed is $c_s \approx 10$ km s$^{-1}$, corresponding to (mainly) ionized hydrogen at $\sim 8000$ K. The relative importance of the elasticity of the gas to the magnetic field is characterized by the plasma $\beta$, which is defined as the ratio of the gas pressure to the magnetic pressure: $\beta \approx (c_s/V_A)^2$; in the ISM, $\beta \approx 1$. We assume that the equations of magnetohydrodynamics (MHD)[14] provide a reasonble description of the medium:

$$\partial_t \rho + \nabla \cdot (\rho \mathbf{v}) = 0,$$

$$\partial_t \mathbf{B} - \nabla \times (\mathbf{v} \times \mathbf{B}) = \kappa \nabla^2 \mathbf{B},$$

$$\rho[\partial_t \mathbf{v} + (\mathbf{v} \cdot \nabla)\mathbf{v}] = \frac{1}{4\pi}(\mathbf{B} \cdot \nabla)\mathbf{B} - \nabla P + \rho\gamma\nabla^2 \mathbf{v},$$

$$\nabla \cdot \mathbf{B} = 0, \tag{1}$$

[a]Current address: Inter University Center for Astronomy and Astrophysics, Poona University Campus, Ganeshking, PUNE 411007, India.

where **v** is the velocity, **B** the magnetic field, $P$ the total (mechanical plus magnetic) pressure, $\kappa$ the magnetic diffusivity, and $\gamma$ is the kinematic viscosity. The undisturbed state is a stable solution, $\mathbf{v}_0 = 0$ and $\mathbf{B}_0 = B_0\,\hat{z}$. We imagine that this medium is stirred by a steady stochastic force that imparts rms velocities $\sim v_L$, with isotropic correlation on spatial scale $L$. By analogy with hydrodynamics, we expect the fluid to exhibit a steady, turbulent structure. Can we describe this MHD turbulence? In this paper, we shall try to give a physical picture of those aspects we understand reasonably well.

For weak disturbances, (1) support three linear modes, namely the fast and slow magnetosonic waves, and the shear Alfvén wave. The fast and slow waves suffer from collisional,[15] as well as collisionless damping,[16] both of which can be severe in a plasma with $\beta \gtrsim 1$. Collisional damping arises from the longitudinal viscosity due to proton–proton scattering, and acts most effectively on waves with wavelength of order the proton mean free path, $l_i \sim 10^{12}(\text{cm}^{-3}/n_i)\text{cm}$. Collisionless damping depends sensitively on the angle between the wave vector, **k**, and $\mathbf{B}_0$ for each of the two modes; its rate can approach the wave frequency. On the other hand, the shear Alfvén wave of small amplitude is immune from collisionless damping, and collisional damping due to the small cross-field components of the viscosity tensor occurs on spatial scales that are much smaller than those relevant for the fast and slow waves. Thus, as a first step toward simplifying the analysis, we begin with the Alfvén wave. The velocity field in the shear Alfvén wave is divergence-free, so we might learn something useful by studying the more tractable problem of incompressible MHD (IMHD) turbulence.

## INCOMPRESSIBLE MHD TURBULENCE: EARLY WORK

Incompressibility means that the continuity equation may be replaced by the simpler requirements, $\rho = $ constant, and $\nabla \cdot \mathbf{v} = 0$, so that the equations of IMHD are

$$\partial_t \mathbf{b} = \nabla \times (\mathbf{v} \times \mathbf{b}) + \kappa \nabla^2 \mathbf{b},$$
$$\partial_t \mathbf{v} = -(\mathbf{v} \cdot \nabla)\mathbf{v} + (\mathbf{b} \cdot \nabla)\mathbf{b} - \nabla p + \gamma \nabla^2 \mathbf{v}, \qquad (2)$$
$$\nabla \cdot \mathbf{v} = \nabla \cdot \mathbf{b} = 0,$$

where $\mathbf{b} = \mathbf{B}/\sqrt{4\pi\rho}$ is the magnetic field in velocity units, and $p = P/\rho$. The dissipation provided by $\kappa$ and $\gamma$ is important only on small spatial scales, so we will largely ignore the dissipative terms. *Shear* Alfvén waves and *pseudo* Alfvén waves are the two linear perturbations about this equilibrium; the latter is the incompressible limit of the slow magnetosonic wave. Both kinds of waves have the same dispersion relation, namely, $\omega = V_A |k_z|$. The perturbed velocity and magnetic field are related by $\delta\mathbf{v} = \pm\delta\mathbf{b}$, where the upper/lower signs correspond to waves traveling antiparallel/parallel to $\mathbf{B}_0$ (with $k_z < 0$ and $k_z > 0$, respectively). A miraculous cancellation of nonlinear terms permits the following exact solutions: If $\delta\mathbf{v}(\mathbf{x}) = -\delta\mathbf{b}(\mathbf{x})$ at some instant of time, $t = 0$, it can be checked that $\delta\mathbf{v}(x, y, z - V_A t) = -\delta\mathbf{b}(x, y, z - V_A t)$ for all time, irrespective of the functional form of $\delta\mathbf{v}(\mathbf{x})$ (see reference 14). This nonlinear solution (of (2), with $\gamma = \kappa = 0$) describes a wave packet of arbitrary form traveling nondispersively in the direction of $\mathbf{B}_0$. Similarly, we can also construct another class of nonlinear solutions, with $\delta \mathbf{v} = \delta \mathbf{b}$, that travels nondispersively in a direction opposite to $\mathbf{B}_0$. Both types of nonlinear solutions are stable, and the dynamics is simple so long as there is no spatial overlap ("collisions") between oppositively moving wave packets.

Iroshnikov[17] and Kraichnan[18] independently developed a theory of IMHD turbulence, which shall hereafter be referred to as the "IK theory." Kraichnan's is the deeper analysis, and the following account of the IK theory draws heavily from his paper. In the IK theory of Alfvénic turbulence, a collision between two oppositely moving wave packets creates small distortions in each of the wave packets, and successive collisions of one of these wave packets with other oppositely moving wave packets are assumed to add with random phases, until the distortions build up to an amplitude of order unity; at that point, the wave packet has lost memory of its initial state and its energy has cascaded to a smaller spatial scale. IK assume that the inertial-range energy spectrum is *isotropic* in **k**-space, and that the energy transfer is *local* in **k**-space (this is equivalent to assuming that only collisions between oppositely moving wave packets of similar spatial extent are effective in transferring energy to smaller spatial scales). Then, the collision time for packets of size $k^{-1}$ is of order $(kV_A)^{-1}$. If $E(k)$ is the three-dimensional energy spectrum, the perturbation in the velocity, $v_\lambda$, on a spatial scale $\lambda \sim k^{-1}$, is $v_\lambda \sim [E(k)k^3]^{1/2}$. IK assume [from the form of the nonlinear terms in (2)] that the fractional change in the velocity (or the magnetic) perturbation is of order $(v_\lambda/V_A) \ll 1$ during one collision. Since subsequent collisions are assumed to contribute with random phases, the number of collisions, $N$, needed for a typical wave packet to lose memory of its initial state is $N \sim (V_A/v_\lambda)^2 \gg 1$. The cascade time on scale $k^{-1}$ is $t_{cas} \sim N\omega_k^{-1}$. Assuming a $k$-independent rate of energy transfer per unit mass, $v_\lambda^2/t_{cas} \sim \varepsilon$, IK find that $E(k) \propto k^{-7/2}$ and $v_\lambda \propto \lambda^{1/4}$ (for comparison, the well-known Kolmogorov spectrum for neutral fluids has $E(k) \propto k^{-11/3}$, giving $v_\lambda \propto \lambda^{1/3}$).

Note that the theory deals with perturbations about a mean magnetic field. Kraichnan points out that, even in the absence of a mean field, an individual eddy belonging to the inertial-range sees a local, large-scale field. While a local, large-scale velocity field can be transformed away by a local Galilean transformation, a magnetic field remains unaltered by such a transformation. Thus large-scale velocity fields should not affect the interactions between two small eddies, while it would be erroneous to ignore the effect of the large-scale magnetic field. This *Alfvén effect* implies that the local structure (spatially) of IMHD turbulence resembles IMHD turbulence in the presence of a mean magnetic field.

## WHY THE IK THEORY IS INCORRECT

The preceding is a sketch of the physical arguments of the IK theory. As explained below, the apparently reasonable dimensional arguments cannot be made more quantitative, since the theory itself is incorrect. However, our salvation lies in the formalism that is developed below. It should be remembered that the Alfvén effect itself will survive, since it is independent of the details of the IK theory.

The quantity, $N \sim (V_A/v_\lambda)^2$, which is a measure of the weakness of the interactions (in the limit infinitesimal perturbations, $N \to \infty$, so that $1/N$ may be considered a small parameter), can be expressed as a function of $\lambda$ by using the form of the IK spectrum. Noting that $v_\lambda \propto \lambda^{1/4}$, we see that $N \propto 1/\lambda^{1/2}$, so that interactions weaken on small spatial scales. This is particularly encouraging, since perturbation theory can be used to put the theory on a quantitative basis. Smallness of amplitude (e.g., $v_\lambda \ll V_A$) alone is not sufficient to ensure that the perturbations may be considered weak. Let $\xi(\mathbf{x}, t)$ be the displacement vector of a fluid element that was at position $\mathbf{x}$. A necessary and sufficient condition for the perturbation to be considered weak is that the *strains* in the fluid be small:

$$\text{strain} = \left| \frac{\partial \xi_i}{\partial x_j} \right| \ll 1. \tag{3}$$

We use Newcomb's[19] action principle formulation of MHD to set up a systematic perturbation theory. The original approach has been modified to suit an incompressible fluid. The basic variable is the displacement field, $\xi(\mathbf{x}_0, t)$:

$$\mathbf{x} = \mathbf{x}_0 + \xi(\mathbf{x}_0, t). \tag{4}$$

The equations of motion—equivalent to (2)—for the fluid are obtained by varying the action ($S$) with respect to $\xi(\mathbf{x}_0, t)$, and requiring that the variation be stationary (i.e., $\delta S = 0$). The action is

$$S = \int dt \, \mathscr{L}, \tag{5}$$

and $\mathscr{L}$ is the Lagrangian, given by

$$\mathscr{L} = \int d^3x \left( \frac{\rho}{2} \left| \frac{\partial \xi}{\partial t} \right|^2 - \frac{B^2}{8\pi} \right). \tag{6}$$

Here $\mathbf{B}(\mathbf{x}, t)$ is the magnetic field at the displaced ($\mathbf{x}_0 + \xi$) position of the fluid element. For an incompressible fluid, the Jacobian of the transformation (4) should be unity:

$$J \equiv |\delta_{ij} + \xi_{ij}| = 1, \tag{7}$$

where $\xi_{ij} = \partial \xi_i/\partial x_j$. The constraint $J = 1$ must be used in the variation $\delta S$. We now use this to rewrite the magnetic energy term in the Lagrangian. If the magnetic field in the undisturbed fluid is $B_{0i}$, then $B_i = J^{-1}(\delta_{ij} + \xi_{ij})B_{0j}$. Using $J = 1$ and $B_{0j} = B_0 \delta_{j3}$, the magnetic term in the Lagrangian is

$$\int d^3x \frac{B^2}{8\pi} = \frac{B_0^2}{8\pi} \int d^3x_0 (1 + 2\xi_{33} + \xi_{i3}^2). \tag{8}$$

The $\xi_{33}$ term integrates to zero. Dropping the constant term and rewriting the Lagrangian in Fourier variables, we have

$$\mathscr{L} = \frac{\rho}{2} \int \frac{d^3k}{8\pi^3} (\dot{\xi}(\mathbf{k}) \cdot \dot{\xi}(-\mathbf{k}) - \omega^2(\mathbf{k})\xi(\mathbf{k}) \cdot \xi(-\mathbf{k})), \tag{9}$$

where $\omega(\mathbf{k}) = V_A|k_z|$. Let

$$\xi_i(\mathbf{k}) = \eta_i(\mathbf{k}) + \frac{k_i \phi(\mathbf{k})}{k}. \tag{10}$$

Here $\mathbf{\eta}$ and $\phi$ are the transverse and longitudinal parts of $\mathbf{\xi}$, respectively. We require $k_i\eta_i(\mathbf{k}) = 0$. The Lagrangian can be written as $\mathscr{L} = \mathscr{L}_\perp + \mathscr{L}_\parallel$. Then the Lagrangian is the sum of

$$\mathscr{L}_\perp = \frac{\rho}{2}\int \frac{d^3k}{8\pi^3}\left(\dot{\mathbf{\eta}}(\mathbf{k})\cdot\dot{\mathbf{\eta}}(-\mathbf{k}) - \omega^2(\mathbf{k})\mathbf{\eta}(\mathbf{k})\cdot\mathbf{\eta}(-\mathbf{k})\right), \tag{11}$$

and

$$\mathscr{L}_\parallel = -\frac{\rho}{2}\int\frac{d^3k}{8\pi^3}\left(\dot{\phi}(\mathbf{k})\dot{\phi}(-\mathbf{k}) - \omega^2(\mathbf{k})\phi(\mathbf{k})\phi(-\mathbf{k})\right). \tag{12}$$

The constraint $J = 1$ hasn't yet been completely imposed on the Lagrangian. The Jacobian has terms of up to third order in the spatial derivatives of $\xi_i$. For our purposes, an expansion of $J$ up to second order will suffice:

$$J \equiv 1 + \xi_{ii} - \tfrac{1}{2}\xi_{ij}\xi_{ji} + \tfrac{1}{2}\xi_{ii}^2 + O(\xi_{ij}^3) = 1. \tag{13}$$

To first order in $\xi_{ij}$, we have the familiar condition of incompressibility, namely $\nabla\cdot\mathbf{\xi} = 0$. This implies that $\phi(\mathbf{k}) = 0$, while there is no constraint at all on $\mathbf{\eta}$. Working to the next higher order, we find that $\phi(\mathbf{k})$ depends on $\mathbf{\eta}$:

$$\phi_k = i4\pi^3 \int \frac{d^3k_1}{8\pi^3}\frac{d^3k_2}{8\pi^3}\frac{(\mathbf{k}_1\cdot\mathbf{\eta}_2)(\mathbf{k}_2\cdot\mathbf{\eta}_1)}{|\mathbf{k}_1+\mathbf{k}_2|}\delta(\mathbf{k}_1+\mathbf{k}_2-\mathbf{k}) + O(\eta^3), \tag{14}$$

where we have used a shortened notation, $\phi_k \equiv \phi(\mathbf{k})$, $\mathbf{\eta}_i \equiv \mathbf{\eta}(\mathbf{k}_i)$, and so on. Using (14) in (12), we see that $\mathscr{L}_\parallel$ is, to lowest nonvanishing order, fourth order in $\mathbf{\eta}$. This implies that the IK theory is incorrect, as we prove below by a classical *reductio ad absurdum* argument.

Let us assume that the IK theory is correct. Then, for $\lambda$ belonging to the inertial-range, the strain $\sim(k\xi_\lambda) \sim 1/N^{1/2} \propto \lambda^{1/4}$ decreases when $\lambda$ decreases. Hence the IK inertial range has a well-behaved small parameter, and must permit a perturbative treatment in the manner attempted earlier in this section. When such an analysis is carried out, we find that third-order terms are absent in $\mathscr{L}$. This means that, when two oppositely moving wave packets collide, the perturbation experienced by each is *not* of order $(v_\lambda/V_A)$, as had been assumed in the IK theory. Rather, this change is of order $(v_\lambda/V_A)^2$, as given by the fourth-order terms. Thus the IK theory itself is incorrect.

## WEAK ALFVÉNIC TURBULENCE

In the limit of small strains, a systematic perturbative theory of interactions between waves due to nonlinear terms can be developed. This is known as *weak turbulence theory*.[20] When the nonlinear terms are ignored, the Fourier amplitudes and phases of the waves are contant in time. However, the nonlinearity will make the amplitudes change slowly, over many wave periods. It is this secular change in the amplitudes that measures energy transfer among the linear modes. A "kinetic equation" for the rate of change of energy in a mode with wave vector **k** describes how other modes in the system affect the energy in this mode. To lowest order in the nonlinearity, the kinetic equation takes account of interactions among modes taken three at a time. In the case of Alfvén waves in an incompressible fluid, it so happens that the coupling coefficients for 3-wave interactions are zero, since the Lagrangian does not have third-order terms. This is a special situation, but by no means uncommon; gravity waves in deep water also do not interact via 3-waves. The next higher order interactions involve four modes. The conservation laws that must be satisfied for resonant 4-wave interactions are

$$\mathbf{k}_1 + \mathbf{k}_2 = \mathbf{k}_3 + \mathbf{k}_4, \qquad \omega_1 + \omega_2 = \omega_3 + \omega_4. \tag{15}$$

These may be interpreted physically as momentum and energy-conservation laws governing elementary interactions between quasi particles. It must be noted that energy conservation (i.e., the equation involving $\omega$'s) is a consequence of the slow secular change of the amplitudes. The existence of the exact nonlinear solutions implies that the coupling coefficients must vanish when $\mathbf{k}_1$ and $\mathbf{k}_2$ have $z$-components of the same sign. So we need only consider cases when their $z$-components are oppositely directed. For definiteness, let $k_{1z} > 0$ and $k_{2z} < 0$. Using $\omega_j = V_A |\hat{z} \cdot \mathbf{k}_j|$, the conservation laws imply that the $z$-components of $\mathbf{k}_3$ and $\mathbf{k}_4$ must also be oppositely directed. Let $k_{3z} > 0$ and $k_{4z} < 0$. From (15) we get

$$k_{1z} + k_{2z} = k_{3z} + k_{4z}, \qquad k_{1z} - k_{2z} = k_{3z} - k_{4z}. \tag{16}$$

Therefore $k_{1z} = k_{3z}$ and $k_{2z} = k_{4z}$; the scattering process leaves the $z$-components unaltered. This implies that waves with values of $k_z$ neither present initially nor subsequently injected cannot be created by 4-wave interactions. Furthermore, since scattering conserves quasi particle numbers, *the net result of 4-wave interactions is to shuffle quasi particles around in* **k**-*space without changing their* $k_z$ *components. Energy cannot cascade along* $k_z$. The specific form of the couplings determines how quasi particles at $k'_z$ affect the shuffling of quasi particles at $k_z$.

For an isotropic dispersion relation, $\omega(k)$, and $d\omega/dk > 0$, 4-wave interactions usually result in a direct cascade of energy and an inverse cascade of quasi particles. The inverse cascade occurs because the creation of one quasi particle with large $\omega$ (and large $k$) requires input from many quasi particles with middling values of $\omega$. Conservation of total number of quasi particles implies that many of these should be created at small $\omega$ (and small $k$)—hence, the inverse cascade. For the problem being studied here, (16) implies that the $\omega$'s do not change during an elementary scattering process. Therefore the energy and quasi-particle budget are automatically satisfied without the need to create many low-energy quasi particles. *Hence there is no inverse cascade of quasi particles.*

A general incompressible perturbation is a linear combination of *shear* Alfvén waves and *pseudo* Alfvén waves. Shear Alfvén waves have $v_z = b_z = 0$, while pseudo Alfvén waves in general have nonzero $v_z$ and $b_z$. We limit our further considerations to only the shear Alfvén waves, since we expect the pseudo Alfvén wave to be heavily damped.[16] This is a consistent procedure since the coupling between shear and pseudo Alfvén waves is weak; the fractional energy loss per wave period due to the generation of pseudo Alfvén waves by shear Alfvénic turbulence is small. We now study the direct cascade of energy. Assuming *locality* of interactions in k-space, we consider the collision between an Alfvén wave packet of size $(k_z^{-1}, k_\perp^{-1})$ with another packet of comparable size that propagates in the opposite direction. During a collision time of order $(k_z V_A)^{-1}$ the change in the fluid velocity amplitude (we use a single subscript, $\lambda$, to denote $(\lambda_\perp, \lambda_z)$), $v_\lambda$, of one of the packets is

$$|\delta v_\lambda| \sim \left| \frac{d^2 v_\lambda}{dt^2} (k_z V_A)^{-2} \right|. \tag{17}$$

On dimensional grounds, from (2)

$$\frac{d^2 v_\lambda}{dt^2} \sim \frac{d}{dt}(k_\perp v_\lambda^2) \sim k_\perp v_\lambda \frac{dv_\lambda}{dt} \sim k_\perp^2 v_\lambda^3. \tag{18}$$

The quantity $k_\perp v_\lambda$ arises because, for shear Alfvén waves, $(\mathbf{v} \cdot \nabla) = \pm (\mathbf{b} \cdot \nabla) \sim v_\lambda \nabla_\perp \sim k_\perp v_\lambda$. So, in one collision the fractional change in $v_\lambda$ is

$$\left| \frac{\delta v_\lambda}{v_\lambda} \right| \sim \left( \frac{k_\perp v_\lambda}{k_z V_A} \right)^2. \tag{19}$$

When this is small, subsequent collisions contribute roughly equally with random phases. Therefore, the number of collisions needed for the packet to lose memory of its initial state is

$$N \sim \left( \frac{k_z V_A}{k_\perp v_\lambda} \right)^4. \tag{20}$$

Energy cascades only along $k_\perp$ and the cascade time is $t_c \sim N(k_z V_A)^{-1}$. Let $\varepsilon$ be the energy pumped into the system per unit time, per unit mass, per unit logarithmic interval of $|k_z|$. Since there is equipartition between kinetic and magnetic energies, the energy per unit mass is $\sim v_\lambda^2$. Therefore, $\varepsilon \sim v_\lambda^2/t_c$. The three-dimensional energy spectrum, $E$, is defined by

$$\sum v_\lambda^2 = \int E(k_z, k_\perp) \frac{d^3 k}{8\pi^3}, \tag{21}$$

where $\sum$ is a sum over wave packets of various scales. For a constant rate of cascade, we get

$$E(k_z, k_\perp) \sim \varepsilon^{1/3} V_A k_\perp^{-10/3}, \tag{22}$$

where $\varepsilon$ can depend only on $k_z$. The action principle formulation has been used (see reference 11) to develop a formal theory of weak 4-wave interactions. It can be shown that (22) is a stationary solution of the 4-wave kinetic equation that carries a positive flux of energy to large $k_\perp$.

This spectrum is valid only when the 4-wave process is the dominant interaction. A necessary condition is that the fractional change in the velocity per collision, (19), be small (equivalently, $N \gg 1$). Equation (22) implies that $v_\lambda \propto k_\perp^{-2/3}$. Using this in (20), we find that the number of collisions per cascade time is $N \propto k_\perp^{-4/3}$ at fixed $k_z$; $N$ decreases as the cascade proceeds to higher $k_\perp$. When $N$ becomes of order unity, the interactions become too strong to be described within the framework of the perturbative expansions of weak turbulence theory. Thus (22) is valid only for values of $k_\perp$ less than that for which $N \sim 1$. Writing $\varepsilon \sim v_L^3/L$, where $L$ is the spatial scale on which energy is injected, and $v_L$ is the fluid velocity amplitude on that scale, we find that

$$(k_\perp L)_{max} \sim \left(\frac{V_A}{v_L}\right)^{3/2}. \tag{23}$$

For this range in $k_\perp$ (i.e., the inertial range for the 4-wave energy cascade) to be substantial, the excitation amplitude $v_L$ must be very small compared to $V_A$. When $v_L \sim V_A$, the inertial range shrinks to zero!

## STRONG ALFVÉNIC TURBULENCE

A weak excitation of shear Alfvén waves initiates a *weak* 4-wave cascade. As energy cascades to large perpendicular wavenumbers $(k_\perp)$, the interactions between waves strengthen, ultimately invalidating the assumption of a *weak* cascade. This coincides with the strains becoming of order unity, signaling the breakdown of perturbation theory. We find it convenient to base physical arguments on the strain, defined as

$$\zeta_\lambda \sim \frac{k_\perp v_\lambda}{k_z V_A}. \tag{24}$$

A small, but nonnegligible, $\zeta_\lambda$ is associated with a finite cascade time. We can include this effect in the kinetic equation [11, eqns. (28) and (29)] by a nonlinear renormalization of the frequencies. Usually this is a higher order effect, but since the 4-wave kinetic equation conserves wave frequencies, we must include renormalization. Viewed physically, the finite lifetimes of the waves give rise to finite linewidths ("frequency–time uncertainty relationship"). So $k_z$ tends to increase from its initially small value of $L^{-1}$; the renormalized interaction permits a transfer of energy to smaller spatial scales along the parallel direction. The fractional change in $k_z$ per nonlinear interaction time is $\Delta k_z/k_z \sim N^{-1}$. As $\zeta_\lambda$ approaches unity from below, the growth rate of $k_z V_A$ approaches that of $k_\perp v_\lambda$. Let us take stock of the arguments.

(i) If $\zeta_\lambda$ is initially small, the excitation will undergo a *weak* 4-wave cascade, making $\zeta_\lambda$ grow in value.
(ii) Frequency renormalization allows for a transfer of energy to smaller spatial scales along the parallel direction.
(iii) As $\zeta_\lambda$ approaches unity from below, its growth ceases.

Each of these statements is based on weak, 4-wave couplings. Together, they lead us to conjecture that shear Alfvénic turbulence might achieve a state in which $\zeta_\lambda \sim 1$. In such a state, $k_z V_A \sim k_\perp v_\lambda$; there is a balance between the Alfvén time scale (i.e., the linear wave period) and the intrinsically nonlinear time scale at which energy is transferred to shorter scales. In this sense, the energy cascade could be called a *critically balanced* cascade. We now estimate the energy spectrum of the critically balanced cascade.

Let energy be injected into the system roughly isotropically on spatial scale $L$. If the excitation is strong, $v_L \sim V_A$, implying $\zeta_L \sim 1$. Then, the energy injected per unit mass, per unit time, on spatial scale $L$, is $\varepsilon \sim V_\lambda^3/L$. Critical balance implies that the cascade time, $t_{cas} \sim (k_z V_A)^{-1}$. Assuming a scale independent cascade rate $\varepsilon \sim v_\lambda^2/t_{cas}$, we find that

$$k_z \sim k_\perp^{2/3} L^{-1/3}, \qquad (25)$$

$$v_\lambda \sim V_A (k_\perp L)^{-1/3}. \qquad (26)$$

Equation 25 implies that the *parallel and perpendicular spatial sizes of eddies are correlated*. Since $k_\perp/k_z \sim (k_\perp L)^{1/3}$, as the cascade proceeds to larger $k_\perp$, the eddies become highly elongated along the direction of the magnetic field. The three-dimensional energy spectrum of strong Alfvénic turbulence is

$$E(k_\perp, k_z) \sim \frac{V_A^2}{k_\perp^{10/3} L^{1/3}} f\left(\frac{k_z L^{1/3}}{k_\perp^{2/3}}\right), \qquad (27)$$

where $f(u)$ is a positive, symmetric function of $u$ that becomes negligibly small when $|u| \gg 1$. We assume that $f$ has unit height and width such that $\int_{-\infty}^{\infty} du\, f(u) \sim 1$. The corresponding "one-dimensional" energy spectrum is obtained by integrating (27) over $k_z$, and over the angular coordinate in $\mathbf{k}_\perp$; it turns out to be proportional to $k_\perp^{-5/3}$, so that (27) is an anisotropic Kolmogorov spectrum.

To make connection with the action principle approach (that was so successful in dealing with weak turbulence), we note that the third-order term in (13) is of the same order as the other terms, so that this equation has to be solved nonperturbatively, if at all. Thus it would be incorrect to assert that there are no third-order terms when strains are of order unity. On the whole the action principle formalism does not appear suitable for strong Alfvénic turbulence. It is more convenient to use Elsasser's variables,[21] which may be defined as $\mathbf{U} = \mathbf{v} + \mathbf{b}$, and $\mathbf{W} = \mathbf{v} - \mathbf{b}$. Working directly with these variables, a kinetic equation for strong Alfvénic turbulence is derived in reference 12. It is demonstrated there that, when the energies in upward and downward traveling waves are equal, the energy spectrum given by (27) is a stationary solution carrying energy to large wavenumbers. Since strains are of order unity in the critically balanced cascade, perturbation theory is invalid. Interactions of all orders appear to contribute about equally, so deriving the kinetic equation with any degree of rigor is out of the question. In reference 12, a type of 3-wave interaction with frequency renormalization was used. While the closure scheme is probably state of the art, like all tractable closures for strong turbulence, it is an ugly fix.

## ELECTRON DENSITY FLUCTUATIONS

Isobaric density fluctuations in the earth's atmosphere are mixed by turbulence down to scales of order a few centimeters, where they are a well-known irritant to optical/IR astronomers. The buoyancy associated with the density variations isn't strong enough to compete with the Reynolds stresses, so that the density fluctuations act as passive contaminants in a turbulent fluid. It turns out that the power spectra of density and velocity are similar, and are given by Kolmogorov's law. Strong shear Alfvénic turbulence has some features that make it similar to atmospheric turbulence; it is noncompressive to first order in the perturbation, and the velocity amplitudes are small in comparison to the speeds of the MHD waves. The high conductivity makes the plasma stick to the field lines, although it can freely stream along them. It appears reasonable to assume isobaric density fluctuations will behave as a passive contaminant, and hence develop a power spectrum that is similar to the velocity power spectrum of the critically balanced cascade [(27)]. The corresponding temperature fluctuations could, in principle, be smoothed by thermal conduction. But the thermal diffusivity is too small, and will only be important on very small spatial scales.

The theory developed earlier naturally predicts a value for the spectral index that is close to those observed. Moreover, radio wave scattering by the electron density fluctuations will produce anisotropic images. Taking $k_\perp^{-1}$ to be the diffractive scale $\sim 3 \times 10^8$ cm, and setting the outer scale $L \sim$ pc, and using (25), yields an aspect ratio $\sim 10^3$! The observed images are much rounder than individual eddies, since they result from scattering averaged along the line of sight; passage through even one outer scale can result in considerable averaging since the magnetic field perturbations on this scale are of order unity.

## SOME COMMENTS

(i) **Damping**: Small-amplitude shear Alfvén waves are immune to collisionless damping, but suffer from *viscous dissipation* due to the small cross-field components of the viscosity tensor. In the ISM, effective scale on which this is important is close to the proton gyro radius ($\sim 10^7$ cm), so that it is very likely that the critically balanced cascade reaches these scales. *Ion-neutral damping* may, however, be important. The most vulnerable waves are those for which $k_\perp l_0 \sim 1$, where $l_0 \sim 10^{15}(\text{cm}^{-3}/n_i)$cm is the neutral mean free path. If the neutral fraction is less than a few percent, the damping for these waves will be small, so that the cascade can escape this bottleneck, and continue down to smaller scales. The coupling to the *pseudo* Alfvén wave is weak, so we expect negligible losses from the critically balanced cascade.

(ii) **Compressive medium**: The theory needs extension to the case of a compressive medium, since $\beta \sim 1$ in the ISM. However, we can be reasonably confident of some results. It is demonstrated in reference 12 that, in a $\beta \sim 1$ plasma, the critically balanced cascade suffers negligible losses due to decay into *fast* and *slow* waves. This does not imply that fast and slow waves have been shown to be unimportant. In a compressive medium, the shear Alfvén wave of large amplitude is not an exact nonlinear solution. This allows for 3-wave interactions between waves traveling in the same direction, leading to steepening of the Alfvén wave.[22,23] However, the energies in the inertial range are much smaller than energies on large scales, so that the time scale for steepening will be much longer than the cascade time. *Collisionless*, as well as *collisional damping* (due to longitudinal viscosity promoted by proton–proton scattering) of fast and slow waves is likely to restrict their amplitudes but, in the absence of a theory appropriate to a compressive medium, their role remains unclear.

## ACKNOWLEDGMENTS

I would like to thank Dr. J. H. Hunter and Dr. R. E. Wilson for inviting me to participate in this workshop. The work reported here was developed in two papers with Peter Goldreich. I take this opportunity to thank him for numerous conversations that have been immensely instructive. His comments on this manuscript were also valuable.

## REFERENCES

1. SCHEUER, P. A. G. 1968. Nature **218**: 920–922.
2. RICKETT, B. J. 1990. Annu. Rev. Astron. Astrophys. **28**: 561–605.
3. NARAYAN, R. 1992. Philos. Trans. R. Soc. London A **341**: 151–165.
4. LEE, L. C. & J. R. JOKIPII. 1976. Astrophys. J. **201**: 532–543.
5. CORDES, J. M., J. M. WEISBERG & V. BORIAKOFF. 1985. Astrophys. J. **288**: 221–247.
6. RICKETT, B. J., W. A. COLES & G. BOURGOIS. 1984. Astron. Astrophys. **134**: 390–395.
7. PHILLIPS, J. A. & A. WOLZCZAN. 1991. Astrophys. J., Lett. **382**: L27–L30.
8. ARMSTRONG, J. W., J. M. CORDES & B. J. RICKETT. 1981. Nature **291**: 561–564.
9. FRAIL, D. A., P. J. DIAMOND, J. M. CORDES & H. J. VAN LANGEVELDE. 1994. Astrophys. J., Lett. **427**: L43–L46.
10. WILKINSON, P. N., R. NARAYAN & R. E. SPENCER. 1994. Mon. Not. R. Astron. Soc. **269**: 67–88.
11. SRIDHAR, S. & P. GOLDREICH. 1994. Astrophys. J. **432**: 612–621.
12. GOLDREICH, P. & S. SRIDHAR. 1995. Astrophys. J. **438**: 763–775.
13. HIGDON, J. C. 1984. Astrophys. J. **285**: 109–123.
14. PARKER, E. N. 1979. Cosmical Magnetic Fields. Clarendon Press. Oxford.
15. BRAGINSKII, S. I. 1965. Rev. Plasma Phys. **1**: 205–311.
16. BARNES, A. 1966. Phys. Fluids **9**: 1483–1495.
17. IROSHNIKOV, P. S. 1963. Astron. Zh. **40**: 742–750.
18. KRAICHNAN, R. H. 1965. Phys. Fluids **8**: 1385–1387.
19. NEWCOMB, W. A. 1962. Nuclear Fusion Supplement. Part 2: 451–463.
20. ZAKHAROV, V. E., V. S. L'VOV & G. FALKOVICH. 1992. Kolmogorov Spectra of Turbulence I. Springer-Verlag. Berlin.
21. ELSASSER, W. M. 1950. Phys. Rev. **79**: 183.
22. COHEN, R. H. & R. M. KULSRUD. 1974. Phys. Fluids **17**: 2215–2225.
23. KENNEL, C. F., B. BUTI, T. HADA & R. PELLAT. 1988. Phys. Fluids **31**: 1949–1961.

# Sinuous Modes and Steady Warps of Polytropic Disks

N. J. BALMFORTH[a] AND E. A. SPIEGEL[b,c]

[a]*Institute for Fusion Studies*
*University of Texas*
*Austin, Texas 78712*

[b]*Astronomy Department*
*Columbia University*
*New York, New York 10027*

## INTRODUCTION

In an asymptotic development of the equations governing the equilibria and linear stability of rapidly rotating polytropes (Papers I[1] and II[2]) we employed the slender aspect of these objects to reduce the three-dimensional partial differential equations to a somewhat simpler, ordinary integrodifferential form. Specifically, the polytrope was characterized by an equatorial radius $a$ and a characteristic thickness $\varepsilon a$. The dimensionless parameter $\varepsilon = k_J a$, where $k_J$, is the Jeans wavenumber[3] based on conditions at the center of the disk. The equations were then asymptotically solved in the limit $\varepsilon \to 0$ and, once this solution is in hand, $\varepsilon$ may be expressed in terms of known global parameters of the disk, such as density, entropy, mass, and angular momentum.

The earlier calculations dealt with isolated objects that were in centrifugal balance, that is, the centrifugal acceleration of the configuration was balanced largely by self-gravity with small contributions from the pressure gradient. Another interesting situation is that in which the polytrope rotates subject to externally imposed gravitational fields. In astrophysics, this is common in the theory of galactic dynamics because disks are unlikely to be isolated objects.[4]

The dark halos associated with disks also provide one possible explanation of the apparent warping of many galaxies. If the axis of the highly flattened disk is not aligned with that of the much less flattened halo, then the resultant torque of the halo gravity on the disk might provide a nonaxisymmetric distortion or disk warp.[4]

Motivated by these possibilities we here build models of polytropic disks of small but finite thickness that are subjected to prescribed, external gravitational fields. First we estimate how a symmetrical potential distorts the structure of the disk (in the next section), then we examine its sinuous modes to confirm that they are stable, hence suggesting that a warp must be externally forced (in the third section). Finally (in the fourth section), we consider steady warps of the disk plane when the axis of the disk does not coincide with that of the halo.

[c]To whom correspondence should be addressed.

## POLYTROPES IN SYMMETRICAL POTENTIALS

We first consider polytropes whose structures are distorted under the influence of a massive ambient halo. Our purpose, in part, is to determine the leading-order effect of a halo, which we assume to be symmetrical about the disk plane. We also use this opportunity to reintroduce the notation and method developed in Paper I.

### *Equations*

Our study begins with the equations of cosmical gas dynamics, supplemented by the polytropic equation of state (with index $n$). These are given explicitly in Papers I and II, and are in each case nondimensionalized to reveal the leading-order balances when $\varepsilon$ is small. Here we write down and deal with only the dimensionless form of the equations with, in particular, the radius $a$ as unit of length. Therefore, we also stretch the axial coordinate, $z$, so that $\zeta = \varepsilon^{-1} z$ is of order unity within the disk. Axial derivatives are then of order $\varepsilon^{-1}$ and evidently large, as they must be in a slender object.

The conservation equations of momentum and mass are, in cylindrical polar coordinates $(r, \theta, \varepsilon\zeta)$,

$$\partial_t u + u\, \partial_r u + \Omega\, \partial_\theta u + \varepsilon^{-1} w\, \partial_\zeta u - r\Omega^2 = -\partial_r(\Phi + \Phi_h + H), \tag{1}$$

$$r\, \partial_t \Omega + ru\, \partial_r \Omega + r\Omega\, \partial_\theta \Omega + \varepsilon^{-1} rw\, \partial_\zeta \Omega + 2\Omega u = -\frac{1}{r}\, \partial_\theta(\Phi + \Phi_h + H), \tag{2}$$

$$\partial_t w + u\, \partial_r w + \Omega\, \partial_\theta w + \varepsilon^{-1} w\, \partial_\zeta w = -\varepsilon^{-1}\, \partial_\zeta(\Phi + \Phi_h + H), \tag{3}$$

and

$$\partial_t \rho + \frac{1}{r}\, \partial_r(r\rho u) + \partial_\theta(\rho\Omega) + \varepsilon^{-1}\, \partial_\zeta(\rho w) = 0, \tag{4}$$

where $(u, v, w)$ is the velocity vector, $\Omega = v/r$ is the angular velocity, $\Phi_h$ is the gravitational potential of the halo, $\Phi$ is that of the disk, and the specific enthalpy is given by

$$H = n\rho^{1/n}, \tag{5}$$

where $\rho$ is the density.

Poisson's equation is

$$\partial_\zeta^2 \Phi - \rho = -\varepsilon^2 \left[ \frac{1}{r} \frac{\partial}{\partial r} \left( r \frac{\partial \Phi}{\partial r} \right) + \frac{1}{r^2} \frac{\partial^2 \Phi}{\partial \theta^2} \right]. \tag{6}$$

### *Halo Models*

As a simple model of a halo, we consider a spheroidal distribution of inert material that interacts with the matter within the disk only through gravity. We presume the halo to be approximately a Maclaurin spheroid with gravitational potential,

$$\Phi_h = \tfrac{1}{2} C_h \Omega_h^2 [(1-e)z^2 + r^2], \tag{7}$$

where $\Omega_h$ is the (constant) rotation rate of the halo, $e$ is its ellipticity, and $C_h$ is a scaling factor measuring the halo mass. In scaled variables, we write this as

$$\Phi_h = \tfrac{1}{2}\beta \varepsilon^{-1} r^2 + \tfrac{1}{2}\alpha \zeta^2, \tag{8}$$

where $\alpha$ and $\beta$ are parameters. Using the external potential (8), we model the influence of externally imposed gravity on the disk's structure. More sophisticated halo structures can also be modeled with (8), since it is the general form of the Taylor expansion of $\Phi_h$ about the disk midplane.

### Steady, Symmetrical Solutions

When the basic state is axisymmetric, time-independent, and symmetric about the midplane, the derivatives with respect to $\theta$ and $t$ and the velocity components $u$ and $w$ all vanish. As in Paper I, the centrifugal potential is then an integral of the equations of motion, depending only on radius:

$$F(r) = \Phi_h + \Phi + H = \int r\Omega^2 \, dr. \tag{9}$$

To solve the equations we seek the following asymptotic series:

$$F \sim \varepsilon^{-1} F_0 + F_1 + \varepsilon F_2, \qquad \Phi \sim \varepsilon^{-1} \Phi_0 + \Phi_1 + \varepsilon \Phi_2, \tag{10a}$$

$$H \sim H_1 + \varepsilon H_2, \qquad \text{and} \qquad \rho \sim \rho_1 + \varepsilon \rho_2. \tag{10b}$$

The leading-order terms of (9) are

$$F_0 = \tfrac{1}{2}\beta r^2 + \Phi_0(r), \tag{11}$$

which also satisfies the leading-order expression resulting from Poisson's equation, (6). Thus the halo supplements the gravitational field of the disk in balancing the internal rotation, as is commonly assumed in fitting the rotation curves of galaxies to their observed visible mass distributions.

The equation determining the vertical structure is given at $O(1)$ by combining (6) and (9), and is

$$\partial_\zeta^2 H_1 + \frac{H_1^n}{n^n} = -\alpha. \tag{12}$$

Thus the density structure is

$$\rho_1(r, \zeta) = \rho_{10} \mathscr{S}_n[\zeta/\zeta_0; \alpha(n+1)/\rho_{10}], \tag{13}$$

where $\rho_{10}(r)$ is the density in the midplane, $\mathscr{S}_n(\xi; \varpi)$ is defined by the integral relation

$$\xi = \int_{\mathscr{S}_n}^1 \frac{ds}{s^{(n-1)/n}[1 - s^{(n+1)/n} + \varpi(1 - s^{1/n})]^{1/2}} \tag{14}$$

and

$$\zeta_0 = \left[\frac{2n}{(n+1)} \rho_{10}^{(n-1)/n}\right]^{-1/2}. \tag{15}$$

The disk's surface is then located at $\zeta = \Theta_1$ in leading order, where

$$\Theta_1 = \zeta_0 \int_0^1 \frac{ds}{s^{(n-1)/n}[1 - s^{(n+1)/n} + \varpi(1 - s^{1/n})]^{1/2}}. \tag{16}$$

Finally one must match the gravitational potential to the vacuum field present outside the disk, on the surface of the disk (see Paper I, and later sections). This implies that the function $\rho_{10}$ is related to the function $F_0$ through the relations

$$F_0 = -\frac{1}{2} \int_0^1 \mathcal{K}(r, s)\Sigma_1(s)s \, ds, \tag{17}$$

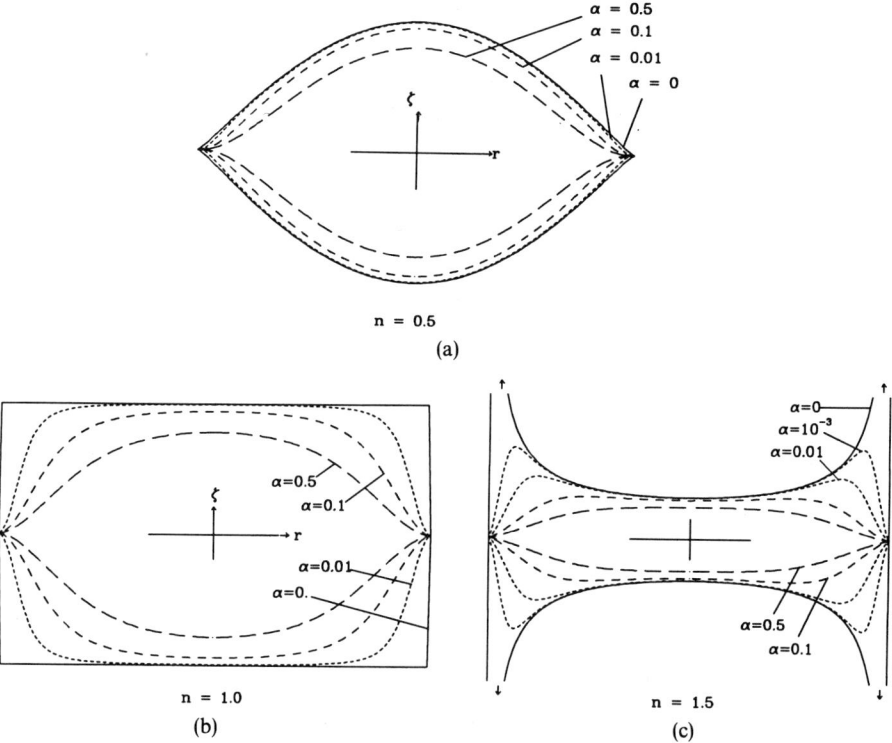

**FIGURE 1.** The shape of polytropic disks subjected to external gravitational fields. The surface density is given by $\Sigma_1 = 2x^7$. The polytropes with $n = 0.5$, 1.0, and 1.5 are shown. The curves drawn for each polytrope correspond to different values for the parameter $\alpha$. The *continuous, dotted, dashed,* and *dot-dashed curves* indicate the particular values of $\alpha$ of 0, 0.01, 0.1, and 0.5. As $\alpha$ increases the characteristic density of the surrounding halo also increases. For the polytrope with $n = 1.5$ the curve representing the shape for $\alpha = 10^{-3}$ is included.

and

$$\tfrac{1}{2}\Sigma_1 = \sqrt{\frac{2n}{(n+1)}\, \rho_{10}^{(n+1)/n}\left[1 + \frac{\alpha(n+1)}{\rho_{10}}\right]}, \qquad (18)$$

where $\Sigma_1$ is the surface density, and

$$\mathcal{K}(r,s) = \frac{\pi}{2}\sum_{k=0}^{\infty}(4k+1)\left[\frac{(2k)!}{(2^k k!)^2}\right]^2 P_k(\sqrt{1-r^2})P_k(\sqrt{1-s^2}), \qquad (19)$$

with $P_k(x)$ a Legendre polynomial.

The halo suppresses the natural tendency of the disk to flare toward its edge and compresses the structure as a whole. FIGURE 1 illustrates this distortion graphically, giving polytropic shapes for various values of $\alpha$. These disks are characterized by the surface density, $\Sigma_1 = 2(1 - r^2)^{3.5}$, and polytropic index, $n = 1.5$. Provided the halo is sufficiently massive, polytropes over wide ranges in $n$ become lenticularly shaped, and there is no need to apply boundary layer methods to correct the solution near the rim of the disk, $r = 1$, as there is for a disk with only self-gravity.[1]

## SINUOUS NORMAL MODES

The sinuous modes, which are antisymmetric about the midplane, can be treated in a very similar manner to the varicose normal modes.[2] We first split the azimuthal velocity into a mean and a perturbation, $v = r\Omega + v'$, and then, since $v$ no longer appears, we drop the prime on $v'$. The linear equations of motion and continuity are then

$$(\partial_t + \Omega\,\partial_\theta)u - 2\Omega v = -\partial_r f, \qquad (20)$$

$$(\partial_t + \Omega\,\partial_\theta)v + (2\Omega + r\,\partial_r \Omega)u = -\frac{1}{r}\partial_\theta f, \qquad (21)$$

$$(\partial_t + \Omega\,\partial_\theta)w = -\varepsilon^{-1}\partial_\zeta f, \qquad (22)$$

and

$$(\partial_t + \Omega\,\partial_\theta)\rho' + \frac{1}{r}\partial_r(r\rho u) + \frac{1}{r}\rho\,\partial_\theta v + \varepsilon^{-1}\partial_\zeta(\rho w) = 0. \qquad (23)$$

Here $f$ represents the linear, time-dependent perturbation to $F$, given by

$$f = \phi + h, \qquad (24)$$

and is no longer solely a function of $r$. Moreover, the perturbations to the thermodynamic state are described by the enthalpy change, $h$, and density change, $\rho'$. The linear version of Poisson's equation is

$$\partial_\zeta^2 \phi = \rho' - \varepsilon^2\left[\frac{1}{r}\partial_r(r\,\partial_r \phi) + \frac{1}{r^2}\partial_\theta^2 \phi\right], \qquad (25)$$

where $\phi$ is the potential perturbation.

Equations 20–22 admit normal-mode solutions whose temporal and azimuthal dependence have the form exp $(\eta t + im\theta)$, where $\eta$ is the (complex) growth rate and $m$ is the azimuthal quantum number. Hence these equations can be rearranged to give the velocity components in terms of the other variables:

$$u = -\frac{1}{\kappa^2}\left(v\, \partial_r f + \frac{2im\Omega}{r} f\right), \qquad (26)$$

$$v = \frac{1}{\kappa^2}\left(2B\, \partial_r f - \frac{imv}{r} f\right), \qquad (27)$$

and

$$w = -\frac{1}{\varepsilon v} \partial_\zeta f, \qquad (28)$$

where

$$B = \Omega + \tfrac{1}{2} r\, \partial_r \Omega, \qquad (29)$$

$$v = \eta + im\Omega, \qquad (30)$$

and

$$\kappa^2 = v^2 + 4\varepsilon\Omega B =: v^2 + \Delta. \qquad (31)$$

If we insert these expressions into the continuity equation, we obtain the dynamical perturbation equation,

$$\partial_\zeta(\rho\, \partial_\zeta f) = \varepsilon^2 v^2 \rho' - \varepsilon^2 v^2 \left\{ \frac{1}{r} \partial_r\left(\frac{r\rho}{\kappa^2} \partial_r f\right) + \left[\frac{2im}{rv} \partial_r\left(\frac{\rho\Omega}{\kappa^2}\right) - \frac{\rho m^2}{r^2 \kappa^2}\right] f \right\}. \qquad (32)$$

### Asymptotic Solution

We now develop the equations asymptotically:

$$f \sim \varepsilon^{-1} f_0 + f_1 + \varepsilon f_2, \qquad \phi \sim \varepsilon^{-1}\phi_0 + \phi_1 + \varepsilon\phi_2, \qquad (33a)$$

$$h \sim h_1 + \varepsilon h_2, \qquad \text{and} \qquad \rho' \sim \rho'_1 + \varepsilon \rho'_2, \qquad (33b)$$

with

$$\eta \sim \varepsilon^{-1/2}(\eta_0 + \varepsilon\eta_1), \qquad \Omega \sim \varepsilon^{-1/2}(\Omega_0 + \varepsilon\Omega_1), \qquad \text{and} \qquad v \sim \varepsilon^{-1/2}(v_0 + \varepsilon v_1). \qquad (33c)$$

In addition, since we consider perturbations that are *antisymmetric* about the midplane of the disk, we impose the boundary conditions:

$$\phi = 0 \qquad \text{and} \qquad \frac{1}{\rho_{10}} \partial_\zeta \phi = -A(r) \qquad \text{on} \quad \zeta = 0, \qquad (34)$$

where the function $A(r)$ is as yet unknown, and must be specified by matching the solution for the perturbation to the external potential. This yields an eigenvalue problem with $A(r)$ as the eigenfunction.

The leading-order terms of (25) and (32) are

$$\partial_\zeta(\rho_1 \, \partial_\zeta f_0) = 0 \quad \text{and} \quad \partial_\zeta^2 \phi_0 = 0. \tag{35}$$

The $O(\varepsilon^{-1})$ solution is therefore trivial: $\phi_0 = f_0 = 0$. At next order, we have

$$\partial_\zeta(\rho_1 \, \partial_\zeta f_1) = 0 \tag{36}$$

and

$$\partial_\zeta^2 \phi_1 = \rho_1'. \tag{37}$$

To avoid a logarithmic singularity in $f_1$ at the surface of the disk, we require $f_1 = 0$. Equation 37 can then be solved to give

$$\phi_1(r, \zeta) = -A_1 \int_0^\zeta \rho_1(r, \zeta') \, d\zeta' = -\tfrac{1}{2} A_1 \mu_1(r, \zeta), \tag{38}$$

where $\mu_1(r, \zeta)$ is the unperturbed surface density between $\zeta$ and $-\zeta$. The convenient relation,

$$\partial_\zeta \phi_1 = -A_1 \rho_1, \tag{39}$$

then follows.

The $O(\varepsilon)$ terms of (25) and (32) now yield

$$\partial_\zeta(\rho_1 \, \partial_\zeta f_2) = v_0^2 \, \rho_1' \tag{40}$$

and

$$\partial_\zeta^2 \phi_2 = \rho_2'. \tag{41}$$

Equation 40 can be integrated once, with the help of (37), to

$$\rho_1 \, \partial_\zeta f_2 = v_0^2 \, \partial_\zeta \phi_1. \tag{42}$$

Introduction of (39) into the right-hand side of this equation and integration yield

$$f_2(r, \zeta) = v_0^2 \, A_1 \zeta. \tag{43}$$

The amplitude of the axial velocity, $w = -v^{-1}\varepsilon^{-1/2} \, \partial_\zeta f$, is therefore constant with distance from the midplane in leading order. This means that the disturbance does not induce a true compression or rarefaction anywhere within the disk; density changes are caused only by the advection of stratified fluid.

The $O(\varepsilon)$ correction to the gravitational perturbation can be found by integrating (41). This gives

$$\phi_2(r, \zeta) = -\tfrac{1}{2}[2v_0^2 A_1(r)\zeta - 2G_0(r)A_1(r)\zeta + A_1(r)\mu_2(r, \zeta) + A_2(r)\mu_1(r, \zeta)], \tag{44}$$

where

$$G_0(r) = \frac{1}{r} \partial_r(r \, \partial_r \Phi_0) = \frac{1}{r} \partial_r(r \, \partial_r F_0) = 2\Omega_0^2 + r\frac{d\Omega_0^2}{dr}, \tag{45}$$

$\mu_2$ is the $O(\varepsilon)$ correction to $\mu_1$, namely,

$$\mu_2(r, \zeta) = \int_{-\zeta}^{+\zeta} \rho_2 \, d\zeta, \tag{46}$$

and $A_2(r)$ is another undetermined function. Accordingly,

$$[\partial_\zeta \phi_2 + \Theta_2 \rho_1']_{\zeta = \Theta_1} = [G_0(r) - v_0^2] A_1, \tag{47}$$

where $\Theta_2$ is the $O(\varepsilon)$ correction to the position of the surface of the disk.[1]

### The Eigenvalue Problem

As before, we match the internal solution

$$\phi(r, \Theta_1) = \phi_1(r, \Theta_1) + \cdots, \qquad \phi_z(r, \Theta_1) = \phi_{2\zeta}(r, \Theta_1) + \Theta_2 \phi_{1\zeta\zeta}(r_1, \Theta_1) + \cdots \tag{48}$$

(where we have used $\phi_{1\zeta}(r, \Theta_1) = 0$, following from (39)) to the external solution

$$\phi^{\text{ext}}(r, \varepsilon\Theta_1) = \phi_1^{\text{ext}}(r, 0) + \cdots, \qquad \phi_z^{\text{ext}}(r, \varepsilon\Theta_1) = \phi_{1z}^{\text{ext}}(r, 0) + \cdots. \tag{49}$$

(It is not necessary to consider the time-dependent perturbation of the surface of the disk, since the continuity of the undisturbed potential and its derivative across the surface ensure that such terms always cancel. This is explicitly seen in the fuller derivation presented in Paper II.) The resulting equation can be manipulated into the eigenvalue problem

$$[v_0^2 - G_0(x)]A_1(x) = -\frac{1}{2} \int_0^1 \mathcal{Q}_m(x, y) A_1(y) \Sigma_1(y) y \, dy, \tag{50}$$

where we have introduced the new coordinates $x^2 = 1 - r^2$ and $y^2 = 1 - s^2$. On account of the antisymmetry about the midplane, the kernel $\mathcal{Q}_m$ may be put in the convenient form[5]

$$\mathcal{Q}_m(x, y) = \frac{1}{xy} \sum_{k=1}^{\infty} \frac{(2k - 1)!}{(2k - 1 + 2m)!} (4k + 2m - 1) \Lambda_k^m P_{2k+m-1}^m(x) P_{2k+m-1}^m(y), \tag{51}$$

where $P_j^m(x)$ is a Legendre function, and

$$\Lambda_k^m = \frac{(2k-1)!(2k+m-1)!}{[2^{2k+m-1}(k-1)!(k+m-1)!]^2} \frac{\pi}{2}. \tag{52}$$

If we further set

$$A_1 = \frac{1}{\Sigma_1} d_1(x) \tag{53}$$

and

$$\varphi_k^m(x) = P_{2k+m-1}^m(x) \sqrt{(4k + 2m - 1) \frac{(2k-1)!}{(2k-1+2m!)}}, \tag{54}$$

then

$$[v_0^2 - G_0(x)] \frac{2xd_1(x)}{\Sigma_1} = -\sum_{k=1}^{\infty} \Lambda_k^m \varphi_k^m(x) \int_0^1 \varphi_k^m(y) d_1(y)\, dy. \tag{55}$$

Equation 55 is the eigenvalue problem for the sinuous modes. It is similar to that derived by Hunter and Toomre[5] for the sinuous oscillations of infinitely thin, pressureless disks. In the rigidly rotating case, the problem reduces to a much simpler form: $\Sigma_1(x) = 2x$ and the $k$th eigenvalue is

$$v_{0,k}^2 = (\eta_{0,k} + im\Omega_0)^2 = -\left(\Lambda_k^m - \frac{\pi}{2}\right). \tag{56}$$

This indicates that all of the sinuous modes are neutral in leading order for the rigid rotator.

For axisymmetrical oscillations of general disk models, $m = 0$, and we can cast the eigenvalue equation into a convenient, variational form by multiplying by $d_1(x)$ and integrating over $x$:

$$\eta_0^2 \int_0^1 \frac{2d_1^2}{\Sigma_1} x\, dx = \int_0^1 G_0(x) \frac{2d_1^2}{\Sigma_1} x\, dx - \sum_{k=1}^{\infty} \Lambda_k^0 \left[\int_0^1 \varphi_k^0 d_1\, dy\right]^2. \tag{57}$$

The form of $G_0(x)$ depends on the model considered. For highly centrally condensed models, $G_0$ is negative throughout substantial regions, whereas for models that are very nearly in uniform rotation, $G_0$ is close to the value $\pi/2$ throughout the disk. In either circumstance, the second term on the right-hand side of (57) is a strongly increasing function of radial order and dominates the first term. Axisymmetrical modes are therefore always neutral.

As illustrated by these two special examples, sinuous modes appear to be neutral at leading order. Hunter and Toomre give further examples and complementary arguments to suggest this is always the case (and that the spectrum may also be continuous). The question then arises here as to whether the effects of compressibility introduced through finite thickness can change this trend. Such effects appear at second order in ε. If, in that order, we evaluate $\phi_3$ and match the result to the analogous term in the perturbed potential outside the polytrope, we get an integral equation for the function $A_2(r)$ and the eigenvalue correction, $v_1$, as for varicose modes.[2] Then we find that the effect of compressibility leads to further stabilization of the higher-order modes. Therefore, it is not necessary to go to second order to charactrize the stability of polytropic disks of finite thickness to sinuous modes. This suggests that *warps must be externally forced rather than self-excited*. We turn next to the study of externally driven warps.

## STEADY WARPS

A common feature of observed galaxies is a warping of the disk plane and, currently, one of the most plausible explanations for this phenomenon is that the symmetry axis of the halo that shrouds the disk is not aligned with that of the disk itself.[4] Then, a weakly dipolar force acts upon the disk in addition to the leading-order, symmetrical compression discussed in the second section. For this case, we assume a form for the external potential given by

$$\Phi_h = \tfrac{1}{2}\varepsilon^{-1}\beta r^2 + \varepsilon\gamma(\theta)r\zeta, \qquad (58)$$

where $\gamma(\theta) = \tilde{\gamma}e^{i\theta}$.

Although the disk structure is steady in a suitable frame, under the influence of the external potential, it is not axisymmetric. Furthermore, the warping of the midplane introduces noncircular motions. In particular, if the axial displacement of the midplane is $O(\varepsilon\lambda)$, we expect vertical velocities of the order of

$$w \sim \frac{2}{\pi}\varepsilon^{1/2}\Omega_0\lambda \qquad (59)$$

and radial motions of similar magnitude. Such considerations provide scalings for the velocity components of a disk with a steady warp.

We also must retain the time derivatives because the external potential can provide a torque on the warped disk that forces its axis to precess.[5] In that circumstance, the disk properties can depend upon aximuthal angle and time through a term of the form $-ir\Omega_p te^{i\theta}$, where $\Omega_p$ is some constant angular precession speed. This term represents precession. There are also other nonaxisymmetrical terms that are independent of time; these represent the permanent warp.

In addition to the conventional asymptotic scalings (10), we pose

$$w = \varepsilon^{1/2}w_1 + O(\varepsilon^{3/2}), \qquad u = \varepsilon^{3/2}u_1 + O(\varepsilon^{5/2}), \qquad (60)$$

guided by the estimate (59). Moreover, we set

$$F = \varepsilon^{-1}F_0 + F_1 + \cdots \quad \text{and} \quad \Omega = \varepsilon^{-1/2}(\Omega_0 + \varepsilon\Omega_1 + \cdots), \qquad (61)$$

where both the centrifugal potential and angular momentum may now depend on time, $\theta$, and $\zeta$ in addition to radius.

### The Expansion

The leading-order asymptotic equations are

$$r\Omega_0^2 = \partial_r F_0, \qquad (62)$$

$$r^2(\partial_\vartheta \Omega_0 + rw_1\,\partial_\zeta\Omega_0) = -\partial_\theta F_0, \qquad (63)$$

$$0 = \partial_\zeta F_0 = \partial_\zeta \Phi_0 \qquad (64)$$

and

$$\partial_t\rho_1 + \partial_\theta(\rho_1\Omega_0) + \partial_\zeta(\rho_1 w_1) = 0, \qquad (65)$$

where the operator,

$$\partial_\vartheta = \partial_t + \Omega_0\,\partial_\theta. \qquad (66)$$

Equations 62–64 imply that $F_0 = F_0(r) = \Phi_0$ and $\Omega_0 = \Omega_0(r)$. This provides the leading-order solution which is, as expected, uninfluenced by the $O(1)$ dipolar force.

The second-order equations of motion are

$$2r\Omega_0 \Omega_1 = \partial_r F_1, \qquad (67)$$

$$r^2(\partial_s \Omega_1 + w_1 \partial_\zeta \Omega_1) = -\partial_\theta F_1, \qquad (68)$$

and

$$0 = \partial_\zeta F_1 = \partial_\zeta(\Phi_1 + H_1). \qquad (69)$$

Thus $F_1$ and $\Omega_1$ are again independent of $\zeta$, $\theta$, and $t$.

This information alone permits us to construct the vertical stratification. The Poisson equation can be developed to leading and second order into the relations,

$$\partial_\zeta^2 \Phi_0 = 0 \qquad (70)$$

and

$$\partial_\zeta^2 \Phi_1 = -\partial_\zeta^2 H_1 = \frac{H_1^n}{n^n}. \qquad (71)$$

Consequently,

$$\tfrac{1}{2}(\partial_\zeta H_1)^2 = \frac{(H_{1m}^{n+1} - H_1^{n+1})}{(n+1)n^n}, \qquad (72)$$

and so

$$\rho_1(r, \theta, \zeta) = \rho_{1m}(r, \theta)\mathscr{C}_n[(\zeta - \zeta_1)/\zeta_0], \qquad (73)$$

where $\partial_\zeta H_1 = 0$, $H_1 = H_{1m}$, and $\rho_1 = \rho_{1m}$ at $\zeta = \zeta_1$. Also, the integral function $\mathscr{C}_n(\xi) = \mathscr{S}_n(\xi; 0)$, and

$$\zeta_0 = \sqrt{\frac{n^{n-1}(n+1)}{2H_{1m}^{n-1}}}. \qquad (74)$$

The structure is therefore vertically symmetric about $\zeta = \zeta_1$, which denotes the location of the displaced midplane. The polytrope's surfaces are located at

$$\zeta = \zeta_1 \pm \Theta_1, \qquad (75)$$

where $\Theta_1$, as given by (16) with $\varpi = 0$, is the equilibrium half-thickness in the absence of any dipolar distortion.

### The Potential of the Warped Disk

The internal and external potential fields must now be matched at the surface of the polytrope. Proceeding as in Paper I, we derive

$$F_0(r) = \Phi_0^{\text{ext}}(r, \theta, t, 0) \quad \text{and} \quad [\Phi_{1\zeta}]_{\zeta = \Theta_1 + \zeta_1} = \Phi_{0z}^{\text{ext}}(r, \theta, t, 0), \qquad (76)$$

at the upper surface; conditions at the lower surface follow from simply changing the sign of $\Theta_1$. These relations imply that $H_{1m}$ is independent of $\theta$ and $t$, and that

$$F_0 = \frac{1}{2}\beta r^2 - \frac{1}{2}\int \mathcal{K}(r,s)\Sigma_1(s)s\,ds, \qquad (77)$$

where

$$\frac{1}{2}\Sigma_1 = \sqrt{\frac{2H_{1m}^{n+1}}{(n+1)n^n}}. \qquad (78)$$

At following order, the matching conditions are

$$F_1(r) = \Phi_1^{\text{ext}}(r,\theta,t,0) + \tfrac{1}{2}(\zeta_1 + \Theta_1)\Sigma_1 \qquad (79)$$

and

$$[\Phi_{2\zeta}]_{\zeta=\Theta_1+\zeta_1} = \Phi_{1z}^{\text{ext}}(r,\theta,t,0) + (\zeta_1 + \Theta_1)\Phi_{0zz}^{\text{ext}}(r,0). \qquad (80)$$

The $O(1)$ external potential field may now be split into dipolar and axisymmetric parts:

$$\Phi_1^{\text{ext}} = \Phi_1^{\text{ext},d} + \Phi_1^{\text{ext},s}. \qquad (81)$$

Equation 79 can then be divided into the components

$$F_1(r) = \Phi_1^{\text{ext},s}(r,0) + \tfrac{1}{2}\Theta_1(r)\Sigma_1(r) \qquad (82)$$

and

$$0 = \Phi_1^{\text{ext},d}(r,\theta,t,0) + \tfrac{1}{2}\zeta_1(r,\theta,t)\Sigma_1(r). \qquad (83)$$

Similarly, if we divide the left-hand side of (80) into pieces with different symmetry,

$$[\Phi_{2\zeta}]_{\zeta=\Theta_1+\zeta_1} = X_d(r,\theta,t) + X_s(r), \qquad (84)$$

then that equation can be divided into the two expressions,

$$X_s(r) = \Phi_{1z}^{\text{ext},s}(r,0) + \Theta_1(r)\Phi_{0zz}^{\text{ext}}(r,0) \qquad (85)$$

and

$$X_d(r,\theta,t) = \Phi_{1z}^{\text{ext},d}(r,\theta,t,0) + \zeta_1(r,\theta,t)\Phi_{0zz}^{\text{ext}}(r,0). \qquad (86)$$

The axisymmetric relations (82) and (85) can now be combined into

$$F_1 = \frac{1}{2}\Theta_1\Sigma_1 - \frac{1}{2}\int \mathcal{K}(s,r)\Sigma_2(s)s\,ds, \qquad (87)$$

where

$$\tfrac{1}{2}\Sigma_2 \equiv X_s - \Theta_1 G_0, \qquad (88)$$

and $G_0$ is defined by (45).

The asymmetric equations (83) and (86) can be combined into

$$\frac{1}{2}\int_0^1 \zeta_1(s,\theta,t)\Sigma_1(s)\mathcal{D}_1(r,s)s\,ds = X_d(r,\theta,t) - \zeta_1(r,\theta,t)G_0(r), \quad (89)$$

where $\mathcal{D}_m$ is given in (51).

## Warped Galactic Models

The leading-order density profile (73) depends on $\theta$ and $t$ only through the midplane displacement $\zeta_1(r,\theta,t)$. Hence,

$$\partial_\vartheta \rho_1 = -\partial_\vartheta \zeta_1 \partial_\zeta \rho_1. \quad (90)$$

If we insert this expression into the leading-order continuity equation, we find

$$w_1 = \partial_\vartheta \zeta_1, \quad (91)$$

which indicates that the vertical velocity is independent of $\zeta$.

The third-order vertical equation of motion is

$$\partial_\vartheta w_1 = -\partial_\zeta F_2, \quad (92)$$

which, when combined with (91), provides the relation

$$F_2 = F_{20} - (\zeta - \zeta_1)\partial_\vartheta^2 \zeta_1, \quad (93)$$

where $F_{20}$ is the value of $F_2$ on the displaced midplane. We have also the relation

$$\Phi_2 = F_2 - H_2 - \gamma r \zeta, \quad (94)$$

which yields the quantity

$$-[\Phi_{2\zeta}]_{\zeta=\zeta_1+\Theta_1} = \partial_\vartheta^2 \zeta_1 + [H_{2\zeta}]_{\zeta=\zeta_1+\Theta_1} + \gamma r. \quad (95)$$

An important feature of the second-order correction $H_2$ is that $[H_{2\zeta}]_{\zeta=\zeta_1+\Theta_1}$ is independent of both $\theta$ and $t$ if the value of $H_2$ on $\zeta = \zeta_1$ is also axisymmetric. We currently have sufficient freedom to choose $H_2(r,\theta,t,\zeta_1)$ to be axisymmetric in the scheme of asymptotic expansion. This we shall do, and so the dipolar piece of the surface gravity is given by

$$X_a = -\partial_\vartheta^2 \zeta_1 - \gamma r. \quad (96)$$

We can therefore rewrite (89) as an integral equation that determines the displacement of the midplane:

$$\frac{1}{2}\int_0^1 \zeta_1 \Sigma_1 \mathcal{D}_1(r,s)s\,ds = -(G_0 + \partial_\vartheta^2)\zeta_1 - \gamma r. \quad (97)$$

The integral expression (97) is very similar to the equation describing infinitesimal, steady warps of zero-pressure disks proposed by Hunter and Toomre.[5] It is here derived under the assumption that the warp of the galactic disk is as large as its characteristic thickness. Equation (97) is linear in the midplane displacement, $\zeta_1$,

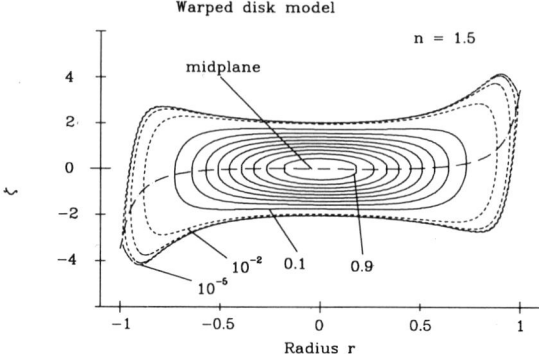

**FIGURE 2.** A model of a warped galaxy. A massive halo whose axis is not aligned with that of the disk is assumed to be responsible for the warp, and creates an external gravitational field given by (**58**). The resulting midplane distortion is computed using (**98**), and the axial stratification from (**13**) with $\varpi = 0$. In addition, the leading-order axisymmetric surface density has the form $2x^7$, and the polytropic index $n = 1.5$. Irregular behavior at the edge has been corrected by the method outlined in Paper I. The particular contour levels shown are those with values linearly spaced in increments of 0.1 between 0.1 and 0.9 (*continuous lines*), and logarithmically spaced between $10^{-2}$ and $10^{-5}$ (*dashed lines*).

because the external potential (**58**) is linear in $\zeta$. This makes (**97**) relatively straightforward to solve. We call such a solution, a *linear warp*, and we present an example below. We could allow the external potential to have an arbitrary dependence on the axial coordinate. We would then replace the final term of (**97**), $-\gamma r$, with one having general nonlinear dependence on $\zeta_1$. This would describe a *nonlinear warp*, but then the integral equation is harder to solve and we do not pursue these ideas here.

If we change variables to $x$ and $y$, the warp equation becomes

$$\frac{1}{2x} \sum_{k=0}^{\infty} \Lambda_k^1 \varphi_k^1 \int_0^1 \tilde{\zeta}_1 \Sigma_1 \varphi_k^1 \, dy = (\Omega_0^2 + D_0)\tilde{\zeta}_1 - 2i\Omega_0 \, \partial_t \tilde{\zeta}_1 - \tilde{\gamma}\sqrt{1-x^2}, \quad (98)$$

where $\zeta_1 = \tilde{\zeta}_1(r) \exp i\theta$. For models with particular surface densities, expression (**98**) can be written as a finite-dimensional matrix equation by separating $\zeta_1$ into terms representing precession (which are linear in $t$) and the permanent warp (independent of $t$), then expanding in a truncated series of the functions $\varphi_k^1$. For a rigidly rotating disk, $\Sigma_1 = 2x$ and the solution is $\tilde{\zeta}_1 \propto r$. The warp of a more centrally condensed, differentially rotating model is shown in FIGURE 2. The surface density law is given by $\Sigma_1 = 2x^7$, the polytropic index is 1.5, and the parameter $\tilde{\gamma}$ is set to unity (an edge correction has also been applied based on the method outlined in Paper I and the value for $\varepsilon$ of 0.01).

## REMARKS

In line with earlier work, we have found that in the purely gravitational case, disk warps are not self-excited. We then went on to extend previous work[5] on warped disks to develop the theory for disks with finite thickness. Our results may be of use for some further studies and we close by mentioning a few.

Although we have been concerned with disk galaxies, the asymptotic theory may also be applied to other astrophysical objects like accretion disks. These disks are normally taken to be turbulent and this raises the issue of the origin of the turbulence. Nonlinear instability is a natural process[6] that may be expected to occur in disks as it does in laboratory shears, even if overreflective instabilities[7,8] prove too mild. It is commonly believed that nonlinear instability leads to complicated two-dimensional motions that give way to fully three-dimensional turbulence through secondary instability.[9] However, in a warped disk, we face a three-dimensional flow at the outset, and this could act as a powerful catalyst favoring a primary three-dimensional instability. Hence the warp may provide a direct route to turbulence.

Moreover, magnetic fields can certainly catalyze disk turbulence.[10] On the other hand, to produce an appreciable magnetic field, we need a sufficiently complicated flow to begin with. Perhaps disks can bootstrap themselves into a situation with both magnetic fields and turbulence in mutual support. However, the solutions found here provide basic states containing nonzero helicity densities, and hence describe equilibria that may allow a dynamo of the $\alpha - \omega$ variety. It might be worthwhile to examine the dynamo action of disk warps, at least in the kinematic case. To do this, we could use the periodic Lagrangian orbits of the warped disk flows to get the dynamo growth rate.[11] In short, the complications considered here might simplify some other problems in surprising ways.

## REFERENCES

1. BALMFORTH, N. J., L. N. HOWARD & E. A. SPIEGEL. 1993. Mon. Not. R. Astron. Soc. **260**: 253 (Paper I).
2. BALMFORTH, N. J., L. N. HOWARD & E. A. SPIEGEL. 1995. SIAM J. Appl. Math. **55**: 298 (Paper II).
3. JEANS, J. H. 1928. Astronomy and Cosmogony. Cambridge Univ. Press. Cambridge, England.
4. BINNEY, J. & S. TREMAINE. 1987. Galactic Dynamics. Princeton Univ. Press. Princeton, N.J.
5. HUNTER, C. & A. TOOMRE. 1969. Astrophys. J. **155**: 747.
6. DUBRULLE, B. & J.-P. ZAHN. 1991. J. Fluid Mech. **231**: 561.
7. PAPALOIZOU, J. C. B. & J. E. PRINGLE. 1987. Mon. Not. R. Astron. Soc. **225**: 267.
8. NARAYAN, R., P. GOLDREICH & J. GOODMAN. 1987. Mon. Not. R. Astron. Soc. **228**: 1.
9. BAYLEY, B. J., S. A. ORSZAG & T. HERBERT. 1988. Ann. Rev. Fluid Mech. **20**: 359.
10. BALBUS, S. A. & J. F. HAWLEY. 1991. Astrophys. J. **376**: 214; also, ———. Astrophys. J. **392**: 662.
11. BALMFORTH, N. J., P. CVITANOVIĆ, G. R. IERLEY, E. A. SPIEGEL & G. VATTAY. 1993. Ann. N.Y. Acad. Sci. **706**: 148.

# Linear and Nonlinear Waves in Shock-Bounded Slabs[a]

ETHAN T. VISHNIAC

*Department of Astronomy*
*The University of Texas at Austin*
*Austin, Texas 78712*

## INTRODUCTION

The stability of shock waves is a rich topic with a long history. As an abstract problem in hydrodynamics, it has the attractive feature that it is an example of a solvable perturbation problem in strongly nonlinear flows, one that in certain limits can even be tested experimentally. In this paper we review theoretical work on the stability of slabs bounded on at least one side by a strong shock, in the sense that the density contrast across the shock is much larger than one. Where possible we compare these theoretical predictions to computer simulations and experimental work.

This is a problem with significant implications for astrophysics for several reasons. First, strong shocks are common in the interstellar medium (ISM), and when radiative cooling is important the density contrast may be very large. Second, the ability of shock waves to generate substructure spontaneously is important for understanding the evolution of structure in the ISM. Third, the late evolution of shock waves, when cooling has become important, affects the efficiency with which the energy associated with bulk flows in the ISM can be converted into turbulent energy and/or radiation. Fourth, the dynamical behavior of shock-bounded slabs determines whether or not gravitationally bound lumps can form inside them. Fifth, and last, the structure of strongly radiating shocks can determine what we see in terms of linewidths and line ratios. Observations are normally interpreted in terms of one-dimensional models, but if there are large postshock flows and density inhomogeneities, these results may be in error.

The earliest work on the stability of shock waves considered the case in which the shock wave can be modeled as a surface dividing two semi-infinite spaces, the shocked gas and the unshocked gas. In this case it can be shown[1] that the shock is stable under almost all circumstances. The more interesting, and realistic, problem is one in which the gas is concentrated in a slab with some characteristic thickness, $L$, and bounded by a shock on at least one face. The other face may be bounded by a shock, by thermal pressure, or by nothing at all (if the shock is decelerating). The stability of the infinite slab guarantees stability in the limit $kL \gg 1$, where $k$ is the transverse perturbation wavenumber. However, when $kL$ is of order one or larger, stability is not guaranteed. In fact, when the density contrast across the shock is

---

[a] This work was supported in part by National Science Foundation Grant 9020757.

large, meaning at least of order 10–20, it is difficult to find cases where no instabilities exist.[2] When the instability in question is purely linear and the structure of the slab is simple, then it is possible to derive exact dispersion relations that demonstrate the existence of these instabilities. On the other hand, some of the instabilities that arise are inherently nonlinear. In this case it is necessary to construct a consistent averaging procedure that will give us approximate solutions for the bulk properties of the flow in the slab.

Here we present the results of several years work on linear and nonlinear instabilities in shock-bounded slabs. For the most part we confine ourselves to quoting general results rather than presenting derivations, which can be found in the cited literature. However, we present many of the details of the averaging procedure, which allows us to find the properties of the nonlinear instabilities. A simpler version of this procedure was used in the initial discovery of the linear instabilities as well.

## DERIVING THE NONLINEAR AVERAGED EQUATIONS

The guiding philosophy of this work is to treat the shock-bounded slab as a dense sheet of matter with a small thickness and a minimal amount of substructure. Consequently, we integrate the relevant dynamical equations in the direction normal to the unperturbed shock. We define this as the $\hat{z}$ direction. In order to use this approach we have to consider a number of issues that arise in its application. A more complete discussion of these points can be found in Vishniac (1994).[2]

For example, if we consider a slab bounded on both sides by an external inward flow with density $\rho_E$ and speed $V_E$, then the integrated equation for mass conservation

$$\int_{z_1}^{z_2} (\partial_t \rho + \nabla \cdot (\rho \mathbf{v})) \, dz = 0, \tag{1}$$

becomes

$$\partial_t \Sigma + \nabla_2 \cdot \mathbf{J} = \rho_E (2V_E + \dot{z}_2 - \dot{z}_1), \tag{2}$$

where

$$\Sigma \equiv \int_{z_1}^{z_2} \rho \, dz, \tag{3}$$

$$\mathbf{J} \equiv \Sigma^{-1} \int_{z_1}^{z_2} \rho \mathbf{v} \, dz, \tag{4}$$

$\nabla_2$ is the gradient operator normal to the $\hat{z}$ direction, and the boundaries of the unperturbed slab are located at $z_1$ and $z_2$ with velocities of $\dot{z}_1$ and $\dot{z}_2$, respectively.

First, we note that in order to derive (2) from (1), it is necessary to invoke the appropriate boundary conditions at $z_1$ and $z_2$. In this case, this amounts to keeping track of the mass flux across the boundaries of the slab. This is completely obvious for mass conservation. It has a somewhat subtler appearance when we consider the force equations. In that case, the difference between the momentum accreted through the shocks and the momentum transmitted through pressure-supported boundaries becomes significant. The material accreted through the shock boundaries carries a local momentum density of $\pm \rho V_E \hat{z}$ where the sign depends on which shock we consider. Consequently, the momentum flux across the shock is purely in the $\hat{z}$ direction. The equivalent pressure tensor has only a $\hat{z}\hat{z}$ component. On the other hand, an external thermal pressure of equivalent magnitude transits a momentum flux across the boundary that is normal to the boundary. If the boundary of the slab is slightly tilted, then the external force at a shock boundary remains in the $\hat{z}$ direction, whereas an external thermal pressure bends with the boundary. An ordinary Sedov–Taylor blast wave provides an example of how important this effect can be. If we consider a small section of the shell of such a blast wave, we note that it feels an effective gravity pointing outward into the low-density external gas. Approximating the shock boundary conditions with an external thermal pressure equal to the ram pressure at the shock interface would lead to a Rayleigh–Taylor instability for all blast waves. This instability is not, in fact, ever observed for decelerating Sedov–Taylor blast waves. The shock jump conditions dictate that any pressure gradients that are not aligned with the velocity of the incoming gas are compensated by strong postshock motions parallel to the shock face. The net effect is that the transverse pressures that drive the Rayleigh–Taylor instability are neutralized for shock interfaces.

Second, treating the slab as thin, in the sense that only the grossest features of its dynamical substructure are important, implies that we will restrict ourselves to the regime where $kL \ll 1$, where $k$ is the transverse wavenumber of the perturbation and $L \equiv z_2 - z_1$ is the slab thickness. This is not terribly restrictive, since we already know that perturbations for which $kL \gg 1$ are guaranteed to be stable. The regime around $kL \sim 1$ does pose a problem, but it is possible to derive exact linear equations for this regime. Comparing the approximate nonlinear equations with the linear dispersion relation in this regime is an important check on the nonlinear equations.

Third, every equation for the evolution of the product $\rho \mathbf{vv} \ldots \mathbf{v}$ will involve a source term with an extra factor of $\mathbf{v}$. This implies that some closure scheme is necessary to end with a finite number of equations. In the work discussed here we have taken

$$T_{ij} \equiv \int_{z_1}^{z_2} \rho v_i v_j \, dz = \frac{J_i J_j}{\Sigma} + \frac{12W^2}{L^2 \Sigma} \hat{n}_i \hat{n}_j, \qquad (5)$$

where $W$ is a measure of the shearing motion within the sheet and $\hat{n}$ is defined as a unit vector perpendicular to the local normal of the shock and aligned with the local internal shearing motions. More precisely,

$$\mathbf{W} \equiv \int_{z_1}^{z_2} \rho(z - z_m)\mathbf{v} \, dz, \qquad (6)$$

where $z_m$ is the median $z$ position of material in the slab and a given $x$ and $y$ and $W \equiv \mathbf{W} \cdot \hat{n}$. This closure scheme captures the bulk motion of the slab and the effects of large-scale shearing within it. This should be sufficient to describe the transport of conserved quantities over large transverse distances. Including higher order terms would require us to make assumptions about the evolution of **vv** terms (like energy conservation, that is, no radiative cooling) as well as correlations between higher order moments. Even if some set of assumptions could be shown to be correct in some particular case, it would probably *not* be generally applicable.

Fourth, in the definition of $W$ quoted earlier, we have implicitly ignored the effects of all velocity structure within the slab except for the largest scale shearing motions. Local perturbations to the shock fronts will automatically produce such large-scale shearing motions. Smaller scale motions should be produced by a turbulent cascade. One way of treating their effects would be to introduce a hierarchy of moments depending on integrals of $(z - z_m)^n$. However, the large-scale transport properties of strong turbulence can usually be approximated by considering only the largest eddies. In an analogous way, we characterize the transport within the slab by the properties of the large-scale shear.

Fifth, the existence of a significant shear within the slab should lead to small-scale turbulence via the Kelvin–Helmholtz instability. This tends to damp the large-scale shear through the production of small-scale eddies, which in turn move momentum from one side of the slab to the other. Following this process would require a detailed model of the small-scale eddies within the slab. As a simple approximation we introduce a damping rate for $W$ given by

$$\tau_{\text{damp}}^{-1} \equiv C_d \frac{|W|}{\Sigma L}, \qquad (7)$$

where $C_d$ is a coefficient of order unity.

Sixth, the boundaries of the slab will evolve in response to internal forces that are beyond the reach of our simplified model. Nevertheless, as long as the slab evolves slowly compared to the sound crossing time, we can adopt an approximate structure based on hydrostatic equilibrium. For example, a slab bounded on both sides by shocks will have

$$z_2 \approx z_m + \frac{L}{2}, \qquad (8)$$

and

$$z_1 \approx z_m - \frac{L}{2}. \qquad (9)$$

The dynamical evolution of $z_m$ follows from this, and from its definition, that is,

$$z_m \equiv \Sigma^{-1} \int_{z_1}^{z_2} \rho z \, dz. \qquad (10)$$

The slab thickness, $L$, can be derived from hydrostatic equilibrium, the surface density $\Sigma$, the bounding ram pressure, and the average internal sound speed $c_s^2$.

Finally, we note that in practice it is impossible to treat the nonlinear effects successfully unless we retain the bending angle of the shock surfaces as a small parameter. In other words, this approximation scheme allows one to follow the evolution of a shock-bounded slab, but only as long as the displacement of the slab median position is small compared to the transverse wavelength of the perturbation.

## APPLICATIONS

The first application of vertically averaged equations to shock wave stability was aimed at the case of dense postshock layer of gas in a decelerating blast wave.[3] In this case, the slab is bounded by a shock wave on one side and thermal pressure on the other. Since the slab is decelerating, the thermal pressure is a fraction of the ram pressure at the shock boundary. This situation is a fair approximation to a Sedov–Taylor blast wave with a small adiabatic index. The initial treatment of this problem neglected internal shearing in the slab and retained only linear terms in the perturbation equations. However, it also retained the basic spherical geometry of the Sedov–Taylor blast wave solution, which provides a significant amount of stabilization for small $k$ modes. The result was the discovery of a growing oscillation with a growth rate, $\Gamma$, given by

$$\Gamma \approx \left(\frac{k^2 \dot{V}_s P_i}{\Sigma}\right)^{1/4}, \tag{11}$$

where $P_i$ is the thermal pressure on the inside surface of the slab, $\dot{V}_s$ is the acceleration of the shock surface (negative in this case), and terms due to the zeroth-order evolution of the shock and its spherical geometry have been neglected. We see from this that $\Gamma$ is complex with equal real and imaginary parts. The growing modes correspond to ripples that propagate along the slab. This expression makes it seem as though $\Gamma$ increases without limit as $k \to \infty$, but this is an artifact of neglecting terms of order $kL$. It is also possible to solve the linear problem without vertical averaging using analytic[4] or numerical techniques.[5,6] For a slab geometry, and neglecting the secular evolution of the slab, one obtains a dispersion relation of

$$\omega^4 - \omega^2 c_s^2 k^2 - \frac{k^2 \dot{V}_s P_i}{\Sigma} F = 0, \tag{12}$$

where $F$ is given by

$$F \equiv \frac{1-\beta}{2\beta}\left(\frac{1+\beta^Q}{1-\beta^Q} Q - 1\right), \tag{13}$$

and

$$Q \equiv \left(1 + 4(kL_s)^2 - \frac{4L_s^2 \omega^2}{c_s^2}\right)^{1/2}. \tag{14}$$

Here $L_s = c_s^2/|\dot{V}_s|$ is the scale height of the gas in the slab and $\beta$ is the ratio of $P_i$ to the external ram pressure. We see that $F$ goes to 1 for $kL$ small. This dispersion relation contains acoustic waves as a second branch and has a limited range of $k$ for which $\Gamma = -i\omega$ has a positive real part.

More recently this instability has been studied numerically in a two-dimensional simulation.[7] The linear growth of the perturbations was confirmed and the authors found that the instability saturated when the surface ripples were large and the bulk flows behind the shock front were slightly less than the sound speed. Since the immediate postshock flows in this state are not much less than the shock speed, this implies the existence of a series of secondary shocks that serve to diffuse the transverse momentum throughout the dense slab. These results are roughly consistent with experimental work demonstrating the existence of this instability.[8] Applying the full nonlinear formalism to this case, one can show that the nonlinear corrections are expected to be small until the ripples are large.

The source of this instability lies in the mismatch between the nature of the thermal and kinetic pressures on the opposing faces of the slab, and the effect of column density fluctuations on the slab geometry. If the slab is rippled, without otherwise altering its dynamical properties, then the thermal pressure on the back side of the slab will push material toward the lagging regions of the slab. In other words, the transverse momentum flux due to the thermal pressure will lead to an acceleration of gas toward these regions. However, as the column density in these regions rises, they will be decelerated less than neighboring regions by the accretion of stationary mass at the shock front. Consequently, a perturbation to the column density causes an effective acceleration of the denser regions. This reverses the initial rippling of the shock front, but only through a fourth-order equation in time. This would be a stabilizing effect if expressed through a second-order equation. Instead, it gives rise to growing oscillations. For the Sedov–Taylor blast wave solution, this instability appears only for $\gamma < 1.2$, that is, for density contrasts of 11 or more. This threshold is sensitive to the secular evolution of the shock. For wind-driven shocks the threshold density contrast for instability is substantially greater.

When we turn to the case of a slab confined between two shocks of equal strength, we find a curious paradox. This problem is also amenable to a simple linear perturbation analysis.[2] For a plane slab, neglecting secular evolution terms, we find two coupled sets of modes, both of which have the dispersion relation

$$\omega^2 = c_s^2 k^2. \tag{15}$$

It is not surprising to find that symmetric ripples, corresponding to acoustic perturbations of the shock-bounded slab, behave like sound waves. It is more surprising to see that the bending modes of the slab have the same dispersion relation. In spite of the lack of any tensile strength the gas slab is quite stiff. However, the most surprising aspect of this result is that it is wrong. Numerical simulations of shock-bounded stationary slabs show a powerful instability associated with bending modes.[9,10] Evidently the nonlinear terms in the evolution equations are important in the analysis. We can understand the linear result by remembering that material that passes through an oblique shock will be left with a residual velocity parallel to the shock front. This has the effect of focusing the postshock matter flows on the concave parts of the shock front. In the linear regime this is the only significant result of a bending wave and results in a rapid reversal of the local geometry. The accreted material continues piling up even after the initial concavity is filled in, resulting in a propagating bending wave.

Retaining nonlinear terms and performing the vertical averaging described earlier, one finds that the dynamics of a stationary shock-bounded slab are given by the following set of equations:[2]

$$\partial_t \, \delta\Sigma = -\partial_x J_x, \tag{16}$$

$$\partial_t J_x = -c_s^2\, \partial_x \Sigma - \partial_x\left(\frac{J_x^2}{\Sigma} + \frac{12W^2}{L^2\Sigma}\right), \tag{17}$$

$$\partial_t J_z = -\partial_x\left(\frac{J_x J_z}{\Sigma} + \frac{12W^2}{L^2\Sigma}\, \partial_x z_m\right) - \rho_E 2 V_E \dot{z}_m, \tag{18}$$

$$\Sigma\, \partial_t z_m = J_z - J_x\, \partial_x z_m - \partial_x W, \tag{19}$$

$$\partial_t W = -W \partial_x\left(\frac{J_z}{\Sigma}\right) \partial_x z_m + \frac{J_x}{\Sigma}\, \partial_x W - \partial_x\left(\frac{2WJ_x}{\Sigma}\right)$$

$$- \Sigma c_s^2\, \partial_x z_m - C_d \left|\frac{12W}{L^2\Sigma}\right| W, \tag{20}$$

and

$$\frac{\delta \Sigma}{\Sigma} = \frac{\delta L}{L}, \tag{21}$$

where we have restricted ourselves to the two-dimensional case for simplicity. It turns out that there are just two nonlinear effects that have a strong effect on the slab evolution for small bending angles. The first is the damping term in (20). A rippled slab will create a strong local shear since the gas impacting on opposing faces will move into the postshock gas in opposite directions. This term dominates the secular damping when the median position of the slab is displaced by more than the slab thickness divided by the square root of the density contrast (or equivalently, by the Mach number of the shock front relative to the postshock gas). When the displacement in $z_m$ exceeds $kL^2$ this damping term dominates over the oscillations and bending motions are strongly damped on a time scale given by

$$\tau_{\text{damp}}^{-1} \sim \left(\frac{kL^2}{C_d \langle \Delta z_m^2 \rangle^{1/2}}\right)^{1/2} c_s k. \tag{22}$$

This result does not move us materially closer to understanding the strong instability found in numerical simulations of this case. However, there is another effect described in the preceding equations that eventually dominates. Although the shearing across a rippled sheet is damped, it is not completely suppressed. The net effect is that a concave surface collects mass that has a net $\hat{z}$ momentum in the direction of the concavity, that is, that tends to depress the concavity further. Conversely, a convex surface collects mass with a net $\hat{z}$ momentum that acts to push the local bulge farther out. The effect is to increase the size of the ripples in the slab. This turns out to be the dominant effect when the median slab position is displaced by an amount comparable to the slab thickness. For $kL \sim 1$ this happens just as the bending angle becomes large so our formalism is inadequate for further analysis. However, for $kL \ll 1$ our system of nonlinear equations is still valid. In this limit we find that ripples grow at a rate given by

$$\tau_{\text{gowth}}^{-1} \sim c_s k (k \langle \Delta z_m^2 \rangle^{1/2})^{1/2}. \tag{23}$$

This leads to postshock bulk velocities of order the sound speed in the dense slab. In the simulations of Stevens et al.[10] the ripples grew until they crossed the wind sources, at which point the program crashed. It's important to note that although this nonlinear instability has certain characteristics in common with the linear instability, that is, it is most easily excited for shocks with large density contrasts, the fastest growing modes have $kL$ of order unity, the fastest growth rates are of order $c_s/L$, and the instabilities grow until the bulk velocities are of order the local sound speed, the growth rate of the nonlinear instability described here falls off much more rapidly with decreasing $k$.

It is also interesting to consider asymmetric slabs of various kinds. One of the simplest is the case where the slab is stationary, bounded by balancing thermal and ram pressures. In that case, it is possible to show that there are no bulk instabilities in the slab. Evidently, this is an exceptional case.

Another interesting case is the classic snowplow solution for supernova remnants, in which a hollow shell of dense gas plows through a uniform medium. If the interior of the shell is cold, as well as largely empty, then the shell decelerates with a radius that grows as $t^{1/4}$. Since $P_i = \beta = 0$, the linear instability described earlier does not apply. We have no numerical simulations to guide us here. The linear theory, neglecting the secular evolution of the shell and its spherical geometry, can be recovered by taking the limit $\beta \to 0$ of (12)–(14).[2] We obtain

$$(\omega^2 - c_s^2 k^2)^2 = 0. \qquad (24)$$

Superficially this resembles the linear dispersion relation for the symmetric shock-bounded slab. However, in this case the bending and compressional modes are not uncoupled and the degeneracy of the dispersion relation implies the existence of secularly growing solutions. This weak instability is driven by the fact that overdense regions are decelerated less than underdense regions, creating rippling in the slab. Consequently, an acoustic wave propagating along the slab will excite bending waves. If neither are subject to damping, then the amplitude of the bending wave will grow linearly due to repeated resonant driving by the acoustic wave. Allowing for damping due to the expansion of the shell and the accretion of additional material will moderate, but not eliminate, this instability. Nevertheless, it is not a particularly powerful instability. However, the nonlinear coupling between the bending waves and the acoustic waves allows for the appearance of a much stronger instability. The basic mechanism is evident in (17) where we see that a local shear gives rise to an effective pressure in the $\hat{x}$ momentum equation, which is proportional to $W^2$. For the snowplow solution the relevant equations are slightly different,[2] but a similar term appears in the $\hat{x}$ momentum equation and dominates the nonlinear feedback from the bending modes to the acoustic modes. The effect is that for a lateral displacement of $z_m$ exceeding the slab thickness divided by the shock density contrast (or Mach number squared), the bending waves created by an acoustic wave drive a harmonic of the original sound wave at a significant amplitude. The growth rate for this instability starts at

$$\tau_{\text{growth}}^{-1} \sim \left(\frac{\langle \Delta z_m^2 \rangle}{L^2}\right)^{1/4} \frac{c_s}{L}, \qquad (25)$$

but is partially suppressed by the internal damping of the shear when the lateral displacement exceeds $L(kL)^{5/2}$. Past this limit the growth rate is only

$$\tau_{\text{growth}}^{-1} \sim c_s k (kL)^{1/4}. \qquad (26)$$

This instability differs from the previous nonlinear instability in that the nonlinear coupling produces a cascade of waves at progressively smaller and smaller scales. The fact that the net effect of this cascade is to increase the power on the largest scales is not obvious, and emerges only from numerical integration of the relevant nonlinear equations. As before, we see that the nonlinear growth rate falls rapidly as we go to smaller and smaller $k$.

Finally, it is worth noting the implications of this work for models in which the formation of dense slabs leads to star formation through gravitational instabilities in the slabs. If one examines the gravitational instabilities of expanding shells,[3] then one finds that for wavelengths greater than a critical wavelength defined by

$$\lambda_{\text{crit}} \sim \frac{\lambda_j^2}{L}, \tag{27}$$

where $\lambda_j$ is the Jeans length for a fluid whose density and sound speed match the postshock gas, the gravitational instability growth rate is approximately

$$\tau_{\text{grav}}^{-1} \sim \left(\frac{L}{\lambda}\right)^{1/2} \frac{c_s}{\lambda_j}. \tag{28}$$

However, if the linear instability applies, then one growing and oscillating ripples whose growth rate is

$$\tau_{\text{ripple}}^{-1} \sim \left(\frac{L}{\lambda}\right)^{1/2} \frac{c_s}{L}. \tag{29}$$

We see that unless $\lambda_{\text{crit}} \sim \lambda_j \sim L$, the rippling instability will grow faster than the gravitational instabilities and it will be difficult to form bound structures, or stars.[11-13] This leads to a criterion for gravitational instability that is difficult to satisfy, that is, the age of the shock must be roughly equal to the collapse time of the unshocked gas (and the radius of the shell must exceed the Jeans length of the unshocked gas by the Mach number of the shock relative to the unshocked gas). The nonlinear instabilities discussed here are not as important in suppressing gravitational instabilities because their growth rates fall off at small $k$ much faster than $k^{1/2}$.

## SUMMARY

There are several major points that have emerged from this work. First, it is possible to treat the gross dynamics of shock-bounded slabs by integrating the equations of motion along the axis perpendicular to the unperturbed shocks. Truncating the resultant hierarchy at a low level produces a set of equations that agrees, in the appropriate limit, with more detailed, but purely linear, calculations. This system of equations suggests a rich set of nonlinear effects, some of which have been seen in computer simulations.

Second, this procedure requires a careful distinction between shock and thermal pressure boundary conditions. Substituting the latter for the former produces a different set of equations, which do not match the dynamics seen experimentally or in computer simulations.

Third, a slab bounded by two shocks of equal strength that is strongly cooled, that is, much denser than the preshock medium outside the slab, is nonlinearly unstable on scales between the slab thickness and $c_s \tau$, where $\tau$ is the time scale for evolution of the zeroth-order properties of the slab. This instability drives growing ripples in the slab so that the local center is displaced in a sinusoidal pattern. The growth rate is approximately $c_s k(k\eta)^{1/2}$, where $k$ is the transverse wavenumber of the slab perturbation and $\eta$ is the displacement amplitude. This instability is referred to as the Nonlinear Thin Shell Instability in our previous work.[2] It has a threshold of $\eta$ greater than or comparable to the slab thickness.

Fourth, a slab bounded on one side by a shock, and having no boundary at the other, that is, a decelerating slab with an exponential atmosphere, is nonlinearly unstable on scales between the slab thickness and $c_s \tau (c_s/V_s)^{1/5}$, where $V_s$ is the shock speed. The growth rate for this instability is min $[(\eta/H)^{1/2} c_s/H, c_s k(kH)^{1/4}]$. This instability is referred to as the nonlinear deceleration instability (NDI) in our previous work.[2] It has a threshold of $\eta > H(c_s^2/V_s^2)$. This result should be used with caution for $kH \ll 1$. This instability is extremely effective at generating a substructure that will grow at a rate of $c_s/H$ as $\eta$ becomes larger than $H$. Although this instability has not yet been simulated numerically, it is particularly interesting since it is likely to dominate the dynamics of a slab caught between two unequal flows.

Fifth, a slab bounded on one side by a shock and on the other by thermal pressure is Rayleigh–Taylor unstable if it is accelerating toward the shock face, and linearly unstable if it is decelerating with a large Mach number.[3] It is stable if stationary, that is, if the two pressures are equal. This is the only stable case containing a shock-bounded slab with a large density contrast.

One important implication of these instabilities is that localized injections of energy from supernovas or winds from star-forming regions will inject large amounts of vorticity into the interstellar medium as soon as radiative cooling becomes important. Another is that in cases where the linear instability is important the formation of gravitationally bound clumps from blast waves is strongly inhibited.

## REFERENCES

1. ERPENBECK, J. J. 1962. Phys. Fluids **5**: 1181–1187.
2. VISHNIAC, E. T. 1994. Astrophys. J. **428**: 186–208.
3. VISHNIAC, E. T. 1983. Astrophys. J. **274**: 152–167.
4. VISHNIAC, E. T. & D. RYU. 1989. Astrophys. J. **337**: 917–926.
5. RYU, D. & E. T. VISHNIAC. 1987. Astrophys. J. **313**: 820–841.
6. RYU, D. & E. T. VISHNIAC. 1988. Astrophys. J. **331**: 350–604.
7. MAC LOW, M. M. & M. L. NORMAN. 1993. Astrophys. J. **407**: 207–218.
8. GRUN, J., J. STAMPER, C. MANKA, J. RESNICK, R. BURRIS, J. CRAWFORD & B. H. RIPIN. 1991. Phys. Rev. Lett. **66**: 2738–2741.
9. HUNTER, J. H., M. T. SANFORD, R. W. WHITAKER & R. I. KLEIN. 1986. Astrophys. J. **305**: 309–332.
10. STEVENS, I. R., J. M. BLONDIN & A. M. T. POLLACK. 1992. Astrophys. J. **386**: 265–287.
11. ELMEGREEN, B. G. 1989. Astrophys. J. **340**: 786–811.
12. YOSHIDA, T. & A. HABE. 1992. Prog. Theor. Phys. **88**: 251–268.
13. NISHI, R. 1992. Prog. Theor. Phys. **87**: 347–365.

# Normal Modes and Continuous Spectra[a]

## N. J. BALMFORTH[b] AND P. J. MORRISON

*Department of Physics and Institute for Fusion Studies*
*University of Texas*
*Austin, TX 78712*

### INTRODUCTION

In theory of fluids, plasmas, and stellar systems, we frequently encounter the question of the stability of equilibria. The answer is provided in part on determining the evolution of an infinitesimal disturbance away from equilibrium, an approach that usually goes by way of a normal mode expansion. This approach can at times be very powerful, and amounts to solving an eigenvalue problem. It can, however, run into difficulty in circumstances for which that eigenvalue problem is, in some sense, irregular.

What we might call regular eigenvalue problems involve the solution of a set of ordinary differential equations with regular coefficients on a domain of finite size. As illustrated by the classic Sturm–Liouville problem, the eigenvalue spectrum turns out to be composed of an infinite number of distinct points. Like the characteristic frequencies of a vibrating string, these correspond to the distinct, normal modes. One might say that the set of irregular problems consists of everything that doesn't fall into this category. For many examples, the eigenvalue spectrum retains a simple form, but in general this is not the case, and the spectrum may consist of only a finite number of discrete modes or continuous intervals.

Here we are concerned with situations for which the eigenvalue problem is irregular and the resulting spectrum is at least partly continuous. This kind of a spectrum can arise as a result of solving the problem on an infinite domain, in which case there is simply no quantization condition. Of more interest are problems in which the set of ordinary differential equations is not autonomous and contains coefficients that become singular at points within the domain.

In physical situations, singularities in the equations governing the evolution of an infinitesimal disturbance can result from a variety of effects, and they do not always affect the form of the eigenspectrum. An important class of problems for which the singularity has direct repercussions on the eigenspectrum occurs in fluids, plasmas, and stellar systems. These are ideal problems in which there are wave-mean flow or wave-particle resonances that result in the creation of a continuous eigenvalue spectrum. In these circumstances, coefficients in the differential problem are formally singular at the point at which resonance occurs. Moreover, *that point is determined by the speed of a wavelike perturbation or, equivalently, the eigenvalue.*

---

[a]This research was supported by the U.S. Department of Energy under Grant DE FG05 80ET 53088.

[b]Send all correspondence to N. J. Balmforth, Institute for Fusion Studies, University of Texas, Robert Lee Moore Hall, Austin, TX 78712-1060.

The existence of a continuous spectrum for an inviscid, shearing fluid was known to Rayleigh,[1] although he was not directly interested in it. In this context, an explicit solution for the spectrum was given by Fjørtoft and Høiland in the 1940s for the special case of incompressible Couette flow.[2] The complications associated with finding the eigenvalues surround the presence of the singularity in the equations, which occurs where the advection of the perturbation exactly cancels its natural speed; a layer in the channel associated with such a singularity is commonly referred to as a critical layer.

In plasma theory we have an analogous situation at the points in phase space for which the equilibrium particle velocity matches the phase speed of the disturbance. This led to a classic problem in plasma theory that was eventually solved by Landau, leading to the celebrated phenomenon of Landau damping. That solution went by way of Laplace transforms, which are naturally tailored to the initial-value problem. The parallel procedure using a continuum variety of normal modes was proposed by Van Kampen,[3] and considered in fluid contexts by Case.[4] In this paper we follow the directions indicated by Van Kampen for more general problems than the relatively simple plasma and fluid equilibria considered by Van Kampen and Case.

In what follows, we first describe the general method (which is discussed in greater detail and applied to parallel shear flow by Balmforth and Morrison[5]). Then, in the general context, the problem of plasma oscillations is reviewed. The remaining sections on parallel shear flow, shear flow in shallow water theory, incompressible circular vortices, and differentially rotating disks, are the bulk of the paper. We conclude with a discussion of the uses of singular eigenfunctions.

## METHOD

An important feature of the solutions that compose the continuous spectrum is that they are not regular functions; they can contain kinks, discontinuities, or singularities at the critical layers. Finding the solutions with standard numerical techniques for regular ordinary differential equations is then problematic. Here we describe an alternative method to contruct the singular eigenfunctions. Related procedures have been used in neutron transport theory,[6] scattering theory,[7,8] and plasma physics.[9]

Most informally we can speak of a system governed by an equation of the form,

$$(x - x_*)\mathscr{L}_x \phi = \mathscr{M}_x \phi, \tag{1}$$

for some eigenfunction $\phi$, and differential operators $\mathscr{L}_x$ and $\mathscr{M}_x$. The point $x_*$ is contained within the domain, $\mathscr{D}$, and is really the eigenvalue. The operator $\mathscr{L}_x$ contains the leading derivatives in the problem, and consequently the equation is formally singular at the critical point $x = x_*$.

Our method follows Van Kampen's treatment of plasma oscillations in the Vlasov–Poisson equation (we give their solution in the next section). We first divide through by the coefficient $x - x_*$. Such an operation is not mathematically defined, however; the resulting equation has a right-hand side that is not a well-behaved function of position. We attach meaning to the expression by interpreting it in a distributional sense, and we use the Cauchy principal value, $\mathscr{P}$, to handle the singular term. Then,

$$\mathscr{L}_x \phi = \mathscr{P} \frac{\mathscr{M}_x \phi}{x - x_*} + \mathscr{C}(x_*)\delta(x - x_*), \qquad (2)$$

where $\mathscr{C}$ is an arbitrary amplitude and $\delta(x)$ is the delta function.

The solution of a differential equation like (2) with a delta-function inhomogeneous term is most easily found by converting that equation to an integral equation. In order to achieve this result, we introduce the Green function of the operator $\mathscr{L}_x$, which we denote by $\mathscr{K}(x, x')$. Then we can write (2) in the form

$$\phi(x) = \mathscr{P} \int_{\mathscr{D}} \mathscr{K}(x, x') \frac{\mathscr{M}_{x'} \phi(x')}{x' - x_*} dx' + \mathscr{C}(x_*)\mathscr{K}(x, x_*). \qquad (3)$$

Equation 3 is an inhomogeneous integral equation. Its kernel is singular at the critical point, and we could use the methods of singular integral equation theory[10] to solve it. However, as yet this is no clear simplification of the problem, but we have not specified $\mathscr{C}$. At our disposal is a normalization condition. If we fix the normalization of the eigenfunction, we determine $\mathscr{C}$. Certain normalizations lead to simplifications in our problem. In particular, if we require that

$$\int_{\mathscr{D}} \mathscr{L}_x \phi \, dx = \Lambda, \qquad (4)$$

we observe that

$$\mathscr{C} = \Lambda - \mathscr{P} \int_{\mathscr{D}} \frac{\mathscr{M}_x \phi}{x - x_*} dx. \qquad (5)$$

If we substitute this relation into our integral problem (3), we see that

$$\phi = \Lambda \mathscr{K}(x, x_*) + \int_{\mathscr{D}} \mathscr{F}_{x_*}(x, x')\phi(x') \, dx', \qquad (6)$$

where

$$\mathscr{F}_{x_*}(x, x') = \frac{\mathscr{K}(x, x') - \mathscr{K}(x, x_*)}{x' - x_*} \mathscr{M}_{x'} \qquad (7)$$

is a kernel with a parametric dependence on $x_*$. This is another integral equation, but, whereas (3) was singular, (6) is not. In other words, our normalizing operation (4) has regularized the integral problem. In fact (6) is a standard Fredholm equation.[11]

Fredholm theory tells us that (6) has two kinds of solutions. There are homogeneous solutions that satisfy

$$\tilde{\phi} = \lambda \int_{\mathscr{D}} \mathscr{F}_{x_*}(x, x')\tilde{\phi}(x') \, dx', \qquad (8)$$

for certain values of $\lambda$, but if there are no values of $x_*$ for which $\lambda = 1$, then there are no homogeneous solutions to (6). Fredholm theory then demonstrates that there is a unique particular solution. If homogeneous solutions do exist with $\lambda = 1$ for certain values of $x_*$, then a solution only exists if the inhomogeneous term satisfies an additional relation (the so-called Fredholm alternative), and it is not unique.

Provided we have no homogeneous solutions, then the method allows us to construct singular eigenfunctions by solving a simpler, regular problem. Moreover, it would establish the existence of a unique solution of the kind we seek. Sometimes it can be verified directly that no such homogeneous solutions exist; also, numerical techniques can be used. Should homogeneous solutions exist, precautions must be taken to assure a unique and bounded solution to our original problem. One way to do this is by suitably scaling the amplitude of the singular mode, $\Lambda$. In particular, we can select $\Lambda = \mathcal{D}(x_*)\hat{\Lambda}$, where the function $\mathcal{D}(x_*)$ vanishes at the eigenvalues for which there exists a homogeneous solution (it is the Fredholm determinant), and $\hat{\Lambda}$ is bounded. This scaling forces the inhomogeneous term to vanish whenever a homogeneous solution appears, and so we always find a unique, bounded eigenfunction.

## PLASMA OSCILLATIONS

We first apply the method to the one-dimensional, Vlasov–Poisson equation, which reproduces Van Kampen's original solution. In this problem we have an equilibrium described by a distribution function, $f_0(v)$, where $v$ is the phase-space velocity coordinate. Infinitesimal perturbations of the distribution function, $f(x, v, t)$, satisfy the linearized Vlasov equation together with the Poisson equation for the electric field, $E(x, t)$. Because the equilibrium is independent of the spatial coordinate $x$, we can Fourier transform the equations, or, equivalently, look for solutions where the perturbations of the distribution function and the electric field are, respectively, of the forms $f(v) \exp[ik(x - ut)]$ and $E \exp[ik(x - ut)]$, where $k$ is a wavenumber and $u$ is the wave speed. The governing equations are then,

$$(u - v)f + \frac{eE}{m}\frac{df_0}{dv} = 0 \tag{9}$$

and

$$k^2 E = -4\pi e \int_0^\infty f(v)\,dv, \tag{10}$$

where $e$ and $m$ are the particles' charge and mass. If we take a solution of the form (2) for $f$, by dividing (9) by $(u - v)$, we obtain

$$f = \frac{e}{m}\mathcal{P}\frac{Ef_0'}{u - v} + \mathscr{C}(u)\delta(u - v). \tag{11}$$

If we integrate this expression over $v$, and use the normalization indicated by (4), we find that

$$\mathscr{C} = \Lambda - \frac{4\pi e^2}{mk^2}\mathcal{P}\int_0^\infty \frac{f_0'}{u - v}\,dv. \tag{12}$$

In this problem, there is no dispersion relation; solutions exist for all eigenvalues, $u$. The associated eigenfunctions are given by (11) with (12). It is not necessary to solve a Fredholm problem in this case because the Poisson equation has the simple, "degenerate" kernel, $\mathscr{K} \equiv 1$. The kernel of the Fredholm equation therefore vanishes everywhere, and $\phi \equiv 1$.

## INCOMPRESSIBLE SHEARS

A slightly more complicated example is the problem considered by Rayleigh.[12] He studied an inviscid fluid configuration consisting of a shear flow contained within a channel. If we denote $x$ and $y$ as the spatial coordinates along and across the channel, then a flow with velocity profile $U(y)$ within the domain $-\infty < x < \infty$ and $-1 < y < 1$ exists as an equilibrium of the two-dimensional Euler equations. Infinitesimal perturbations about this equilibrium can be taken to be of the form $u(y) \exp ik(x - ct)$, $v(y) \exp ik(x - ct)$, and $p(y) \exp ik(x - ct)$ for the two velocity components and pressure fluctuation. The eigenvalue is $c$, and there is a critical layer at $y = y_*$, at which point $U(y) = U(y_*) = c$. The perturbations satisfy the equations

$$ik(U - c)u + U'v = -ikp, \qquad (13)$$

$$ik(U - c)v = -p' \qquad (14)$$

and

$$iku + v' = 0, \qquad (15)$$

where the equilibrium density has been set to unity. By representing the perturbation's velocity field in terms of a stream function, $\psi(y)$, we can formally manipulate these expressions into Rayleigh's equation,

$$(U - c)(\psi'' - k^2\psi) = U''\psi. \qquad (16)$$

Rayleigh's equation is a relatively well-studied equation.[2] Various integral relations can be derived from it. These indicate that there are no discrete eigenmodes unless there is an inflexion point, $U'' = 0$, somewhere within the flow. Such modes are either purely real, in which case their critical layers lie exactly at the inflexion point, or they are complex, indicating decaying/growing pairs. All other neutral modes must have critical layers that lie within the channel; they are intrinsically irregular and we expect them to make up a continuum, that is, the singular, continuous spectrum.

Rayleigh's equation is clearly of the form of (1), provided $U(y)$ is a monotonic function. If we assume this to be the case, then the generalization of the Van Kampen eigenfunction is

$$\omega(y) = \mathscr{P}\frac{U''(y)\psi(y)}{U(y) - c} + \left[\Lambda - \mathscr{P}\int_{-1}^{1}\frac{U''(y')\psi(y')}{U(y') - c}\,dy'\right]\delta(y - y_*), \qquad (17)$$

which is the vorticity fluctuation, and $\psi$ satisfies the Fredholm equation (6) (but in the variable $y$), with

$$\mathscr{F}_{y_*}(y, y') = \frac{\mathscr{K}(y, y') - \mathscr{K}(y, y_*)}{U(y') - U(y_*)} U''(y'), \tag{18}$$

and $\mathscr{K}(y, y')$ being Green's function of the two-dimensional Laplace equation, that is,

$$\mathscr{K}(y, y') = \begin{cases} -\sinh k(1 - y) \sinh k(1 + y')/k \sinh 2k & \text{for } y > y', \\ -\sinh k(1 - y') \sinh k(1 + y)/k \sinh 2k & \text{for } y < y'. \end{cases} \tag{19}$$

Some solutions to the Fredholm problem are shown in FIGURE 1. These are computed for the flow profiles, $U(y) = y + y^3/10$ and $U(y) = y + y^3$. The continuity of fluid elements requires that $\psi$ remains continuous across the channel, but it does have a discontinuity in slope. In these cases, the absence of homogeneous solutions to our Fredholm problem can be established numerically by constructing the Fredholm determinant.[11] Hence, we set $\Lambda$ to unity.

It is not necessary to assume that the profile is monotonic. If $U(y)$ is multivalued in places, we have multiple critical layers for the corresponding wave speeds. This complicates the construction of singular eigenfunctions, but it can still be done, with some modification to the method.

## SHEARS IN SHALLOW WATER

A more complicated situation than Rayleigh's problem is when the shearing fluid is compressible. An example in which the two-dimensional character of the configuration is retained is for the flow of shallow water through a channel, a physical situation of interest in an oceanographical context.[13-15]

From a physical point of view, we expect a different spectrum for the stability eigenvalue problem, because compressibility introduces an additional degree of freedom into the dynamics of the fluid. In particular, in Rayleigh's problem there are only vortical motions. For compressible fluid we also expect sound waves, or, in the shallow-water system, surface gravity waves. (The similarity between the acoustical dispersion relation of a two-dimensional compressible fluid and that of the surface gravity waves of a shallow fluid system has led to some confusion in the past.[16])

In addition to the singular modes, we therefore anticipate a new class of modes, and from the earlier studies these are expected to compose a discrete portion of the complete eigenspectrum.

The equations for perturbations to a shearing, shallow fluid of undisturbed, uniform depth and velocity profile $U(y)$ (using a coordinate system like before and assuming monotonic velocity profiles), are[13]

$$ik(U - c)u + U'v = -\frac{ik}{Fr^2} h, \tag{20}$$

$$ik(U - c)v = -\frac{1}{Fr^2} h' \tag{21}$$

and

$$ik(U - c)h + iku + v' = 0, \qquad (22)$$

where the velocity components are again given by $u$ and $v$, $h$ is the $y$-dependent piece of the depth perturbation, and the dependence $\exp ik(x - ct)$ has again been

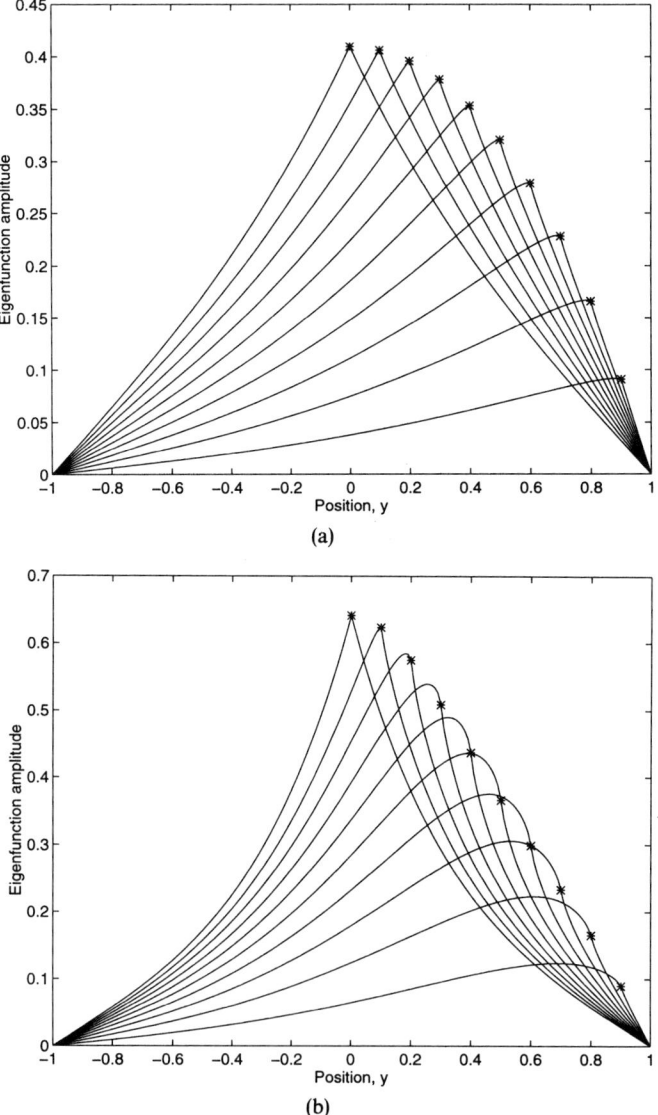

**FIGURE 1.** A selection of singular eigenfunctions, $\psi(y)$, for $U(y) = y + \alpha y^3$, with **(a)** $\alpha = 0.1$ and **(b)** $\alpha = 1$. Also, $k = 1$. Streamfunctions of various modes with different critical layers are displayed. The critical-layer amplitudes are indicated by *stars*.

introduced. These equations have been scaled to make them dimensionless; this introduces the Froude number, $Fr$, which is the ratio of the characteristic, mean flow speed to a typical surface gravity wavespeed (or a characteristic Mach number of a two-dimensional, compressible fluid).

From these expressions we can derive a second-order equation for $h$; namely,

$$(U - c)\{h'' + k^2[Fr^2(U - c)^2 - 1]h\} = 2U'h'. \qquad (23)$$

Another relation of interest comes from the vorticity and continuity equations,

$$(U - c)(v'' - k^2 v) - U''v = -ik[(U - c)^2 h]'. \qquad (24)$$

The incompressible limit, in which we should recover Rayleigh's equation, is obtained by taking $Fr \to 0$ and $h \to 0$, but with the ratio $h/Fr^2$ finite. Accordingly, (24) reduces to Rayleigh's equation since $ik\psi = v$.

The next step is to divide through by a factor of $U - c$, take a principal part, and add a delta function. In our current example, we need to be a little careful about how we should accomplish this. In analogy with Rayleigh's equation, we can clearly divide the second relation (24) by $U - c$ and proceed along the lines outlined by the method. A similar procedure for (23) does not seem to work for the following reasons.

Equation (23) contains a singular point, namely $y = y_*$. About that point, we have Frobenius expansions of the form[13]

$$h \sim (y - y_*)^3 \sum_{n=0}^{\infty} a_n (y - y_*)^n$$

and

$$h \sim -\frac{k^2 U''(y_*)}{2U'(y_*)} (y - y_*)^3 \log (y - y_*) \sum_{n=0}^{\infty} a_n (y - y_*)^n + \sum_{n=0}^{\infty} b_n (y - y_*)^n. \qquad (25)$$

If, for the moment, we consider Couette flow, for which $U'' = 0$, then we observe that the singular point in the equation for $h$ is entirely regular. In other words, it is a removable singularity (in fact the equation for $u$ in this case contains no singular terms). Moreover, the solutions of (23) form a complete basis set of regular functions (the surface gravity modes). There does not seem to be any need, then, to include singular eigenfunctions. However, in order to determine the evolution of the fluid, we need to represent both an initial height and an initial velocity field. This requires two independent sets of basis functions, and the surface gravity modes alone are in general insufficient. The singular mode spectrum is still therefore needed in order to complete the problem.

Even though there is no principal-value singularity in (23), we could nevertheless add a delta function on dividing by $U - c$. This leads to a particular solution for $h$ that might represent the singular eigenmode. Indeed, that solution generally has a discontinuity in its first derivative of $h$. However, such an eigensolution is ruled out if we use the physical requirement that the pressure gradient be continuous. Even were this objection not to preclude such solutions, we would then be forced to work with highly divergent vorticity fluctuations (in the sense that the singularity at the critical point is not just a simple pole). Moreover, these appear to be of no relation to the singular modes of Rayleigh's equation, yet Rayleigh's solutions should be recovered in the incompressible limit.

The resolution of this difficulty lies in (24) and the fact that (23) was derived from the continuity equation (22). The continuity equation contains information only about the divergence of the velocity. In deriving (23) we therefore omit crucial, singular details of the vorticity field. That field evolves according to (24). Applying our procedure to this equation gives

$$v'' - K^2 v + 2U' \frac{Fr^2 k^2}{K^2}[(U-c)v' - U'v] = \mathscr{P}\frac{U''v}{U-c} + \mathscr{C}\delta(y - y_*). \quad (26)$$

where

$$K^2 = k^2[1 - Fr^2(U-c)^2]. \quad (27)$$

This is the shallow-water version of Rayleigh's equation. In writing this equation, we have introduced another singular term, namely the term with a denominator of $K^2$. That quantity vanishes at the points $y = y_t^\pm$, for which

$$U(y_t^\pm) = c \pm \frac{1}{Fr}. \quad (28)$$

These singular terms have no counterpart in the equation for $h$, (23), reflecting how they are removable singularities (the Frobenius expansions for $v$ about these singular points are both purely regular). Formally we can write the equation for $v$ in the form of (2), thence solve it according to our method. This requires us to build a Green function for the operator on the left-hand side of (26), but then our Fredholm problem is straightforward to solve.

Buried in (26) are both the eigensolutions of the continuum and the discrete modes which correspond to the surface gravity waves. In addition, should the flow profile violate Rayleigh's criterion, there may be discrete solutions related to the vortical instabilities of the incompressible problem. When the Froude number is very small, we expect that the two types of solutions are well separated on the spectral plane. For larger values of $Fr$, the distinction may not be so clear.

This particular problem is interesting in that it provides an example where we have to be a little careful about simply writing down principal values and delta functions in order to find singular eigenfunctions. Applying the method to the equation for $h$ produces ambiguous results; the equation for $v$ seems to be the best way to go. None the less, there is a certain amount of freedom in choosing which equation to work with, or into which physical quantity we should introduce a principal-value singularity or delta function. At the end of the day, it is how well the resulting eigenfunctions behave as a unique, complete basis set that determines the optimal choice.

The problem also highlights another ambiguity. We decided not to treat the singularities occurring in (26) at the points $y = y_t^\pm$ by the method since they were removable (of zero dividing zero form), and so no principal-value piece was necessary. However, for linear shear in both compressible and incompressible fluid, $U'' = 0$, and there is no principal-value singularity in the equation for $v$ even at the critical layer. In Rayleigh's equation, the delta function piece must still be added into the equation in order to find a solution (this is the Høiland and Fjørtoft's result[2]). Similarly, in the shallow-water equation (26) we could also retain the delta function, but now there is no distinction between the importance of the critical point and the other, removable, singular points $y = y_t^{\pm}$. In principle, then, we could add alternative terms, $\mathscr{E}\delta(y - y_t^{\pm})$, to the equation. This would lead to another two sets of singular eigenfunctions.

In practice it is unlikely that the new sets of solutions are as useful as the original one because we expect continuum eigenfunctions for every wave speed that matches the mean flow. From (28) we see that this would give solutions with singularities outside the channel. From a physical point of view, we interpret $y_t^{\pm}$ to be the turning points for the surface gravity waves (the points for which these waves are reflected). There is no obvious reason why we should allow the eigenfunctions to be irregular at these points.

## INCOMPRESSIBLE VORTICES

As a prelude to discussing an astrophysical application of our method we now discuss another simple situation. This is the incompressible, two-dimensional vortex. Kelvin[17] considered such configurations with piecewise continuous vorticity distributions. These equilibria support interfacial-type discrete modes that were of interest to Kelvin, but not directly relevant to the continuum modes we derive here. More recently, the stability of a two-dimensional, incompressible vortex has regained importance, since it has become feasible to experimentally simulate the dynamics of such a configuration with an electron plasma.[18]

In polar coordinates, $(r, \theta)$, we have an equilibrium given by an angular velocity distribution, $V(r) = r\Omega(r)$. For the sake of simplicity we again take monotonic profiles for $\Omega$. Perturbations to the vortical structure can be described by a stream function, $\psi(r) \exp im(\theta - vt)$, where $m$ is the azimuthal quantum number, for which the perturbed velocity components are given by

$$u = -\frac{1}{r}\frac{\partial \psi}{\partial \theta} \quad \text{and} \quad v = \frac{\partial \psi}{\partial r}. \tag{29}$$

Then Rayleigh's equation for $\psi$ can be written in the form

$$r(\Omega - v)\left[\frac{1}{r}(r\psi')' - \frac{m^2}{r^2}\psi\right] = \zeta'\psi, \tag{30}$$

where the mean vorticity is given by

$$\zeta = \frac{1}{r}(r^2\Omega)'. \tag{31}$$

A straightforward application of our method yields the singular eigenfunctions,

$$\omega = -\mathscr{P}\frac{\zeta'\psi}{r(\Omega - v)} - \mathscr{E}\delta(r - r_*), \tag{32}$$

where $r_*$ denotes the critical ring, or the corotation radius, for which $\Omega(r_*) = v$. The stream function $\psi$ satisfies the Fredholm equation

$$\psi(r) = \Lambda \mathcal{K}(r, r_*) + \int_0^\infty \frac{\mathcal{K}(r, r') - \mathcal{K}(r, r_*)}{\Omega(r) - \Omega(r_*)} \zeta'(r')\psi(r') \, dr', \tag{33}$$

and the Green function of Laplace's equation in these coordinates is

$$\mathcal{K}(r, r') = -\int_0^\infty J_m(kr)J_m(kr') \frac{dk}{k}. \tag{34}$$

The stability criterion for the vortex is simply that $\zeta'$ does not vanish. This also excludes any discrete modes in the spectrum.

## DIFFERENTIALLY ROTATING FLUID DISKS

Compressible generalizations of Kelvin's vortices have lately prompted interest regarding noise-generation problems in aerodynamic contexts[19] and in disk theory in astrophysics.[20,21] In the latter situations, we consider slender configurations like the shallow-water shears considered earlier. These disks are essentially two-dimensional being hydrostatically stratified in the vertical, and variations in thickness provide the most important effects of compressibility. Equilibria are determined by a surface density distribution, $\Sigma(r)$, in addition to the rotation rate, $\Omega(r)$. If we consider barotropic configurations that are not self-gravitating, but rotate about some central mass, then a disturbance can be represented (to leading order in thinness), in polar coordinates, by the velocity components $u(r) \exp im(\theta - vt)$ and $v(r) \exp im(\theta - vt)$, and by perturbation in the enthalpy, $h(r) \exp im(\theta - vt)$, in the midplane of the disk. The equations of motion can be written in the form

$$im(\Omega - v)u - 2\Omega v = -h', \tag{35}$$

$$im(\Omega - v)v + \zeta u = -\frac{im}{r} h \tag{36}$$

and

$$im(\Omega - v)\sigma + \frac{1}{r}(r\Sigma u)' + \frac{im}{r} \Sigma v = 0, \tag{37}$$

where the surface density perturbation, $\sigma$, is related to the midplane enthalpy by

$$\sigma = \frac{\Sigma h}{c_s^2}, \tag{38}$$

with $c_s$ a function of the local vertical structure of the disk, or the local surface gravity wave speed. The undisturbed vorticity has again been represented by $\zeta = (r^2\Omega)'/r$.

From these equations we can derive the relation,

$$(\Omega - v)\left[\frac{1}{r}\left(\frac{r\Sigma}{D} h'\right)' - \frac{m^2\Sigma}{r^2 D} h - \frac{\Sigma}{c_s^2} h\right] + \frac{2}{r}\left(\frac{\Omega\Sigma}{D}\right)' h = 0, \tag{39}$$

where

$$D = 2\Omega\zeta - m^2(\Omega - v)^2. \tag{40}$$

This is the counterpart of (23) for the shallow-water shear. Like that equation, it is singular at the critical point, $r = r_*$ (the singularities for which $D = 0$, the so-called Lindblad resonances are removable; these are the analogues of the turning points (28) of the shallow-water problem discussed earlier), except for the case in which the potential vorticity,

$$Q = \zeta/\Sigma, \tag{41}$$

is uniform. Then there are two regular solutions at the critical ring. This case corresponds to the Couette, shallow-water shear flow example. Like that example, there would therefore appear to be no need for singular eigenfunctions, and the regularity of the pressure derivatives in (35) and (36) precludes us from dividing through by $\Omega - v$ and adding a delta function. In other words, once again we cannot straightforwardly apply the method to (39).

In order to find the continuum modes we first consider an anelastic approximation to the equations. This is obtained by taking the limit $c_s \to \infty$. Then the surface gravity waves are filtered out of the problem and the continuity equation becomes

$$\frac{1}{r}(r\Sigma u)' + \frac{im}{r}\Sigma v = 0. \tag{42}$$

We can introduce a stream function, $\psi$, to solve this equation. It is given by

$$u = \frac{im}{r\Sigma}\psi \quad \text{and} \quad v = -\frac{1}{\Sigma}\psi'. \tag{43}$$

In terms of this variable, we write the perturbed potential vorticity equation as

$$r(\Omega - v)\left[\frac{1}{r}\left(\frac{r}{\Sigma}\psi'\right)' - \frac{m^2}{r^2\Sigma}\psi\right] = Q'\psi. \tag{44}$$

This is a Rayleigh-like equation that generalizes the incompressible version, (30). Since the source of the singularity is now evident, we can apply our method to (44) and derive singular eigenfunctions for this anelastic approximation. In particular, we have

$$q = \frac{1}{r}\left(\frac{r}{\Sigma}\psi'\right)' - \frac{m^2}{r^2\Sigma}\psi = \mathscr{P}\frac{Q'\psi}{r(\Omega - v)} + \mathscr{C}\delta(r - r_*) \tag{45}$$

for the potential vorticity fluctuation, with $\psi$ determined from a suitable Fredholm equation.

To return to the full problem, we again write down the potential vorticity equation. Without approximation it is,

$$r(\Omega - v)\left[\frac{1}{r}\left(\frac{r}{\Sigma}\psi'\right)' - \frac{m^2}{r^2\Sigma}\psi\right] = Q'\psi + (\Omega - v)\mathscr{S}, \tag{46}$$

where now $\psi = r\Sigma u$, and

$$\mathcal{S} = -r\left\{\left[r^2(\Omega - \nu)\frac{\Sigma}{c_s^2} h\right]' + mr\frac{\zeta\Sigma}{c_s^2} h\right\}. \tag{47}$$

This is the generalization of the anelastic equation (**44**), and, on writing $h$ and $h'$ in terms of $\psi$ and its derivative, corresponds to (**26**) of the shallow-water example. In analogy with the anelastic equation, we can divide (**46**) by $\Omega - \nu$, treating the singular term by its principal value, and add a delta function. Eventually we solve a Fredholm problem for $\psi$.

As in the shallow-water problem, (**46**) contains two types of modes that are easily distinguished in the anelastic limit. In the disk problem, one is tempted to call the vortical modes either Rossby waves or $r$ modes, in analogy with the nomenclature of theory of geophysical fluid dynamics or stellar pulsation. These form a continuum delimited by the range of rotation speed, but, as suggested by Schutz and Verdaguer,[22] there may also be some discrete modes as a result of topographical influences (gradients in surface density), which break the Laplacian structure of the left-hand side of (**30**). Moreover, when the potential vorticity reverses sign, we violate the generalization of Rayleigh's criterion[23] and unstable/decaying mode pairs may appear.

## THE USES OF SINGULAR EIGENFUNCTIONS

In previous sections we have constructed eigenfunctions of the singular continuum for a variety of idealized problems. These eigenfunctions are characterized by irregular shapes; principal-part singularities and delta functions. As a result it is not possible to add a single continuum mode with any finite amplitude onto the basic equilibrium state without immediately leaving the linear regime and introducing singular behavior at the critical layer. An integral superposition of singular modes, however, with a distribution of amplitudes, $A(x_*)$ say, such as

$$S(x) = \mathcal{P}\int_{\mathcal{D}} \frac{A(x_*)\mathcal{M}_x \phi(x; x_*)}{x - x_*} dx_* + \left[1 - \mathcal{P}\int_{\mathcal{D}} \frac{\mathcal{M}_{x'}\phi(x'; x)}{x' - x} dx'\right] A(x), \tag{48}$$

need not be so pathological[3,4] (principal-value integrals are well-behaved functions). We have introduced a second dependence on $x_*$ into the arguments of $\phi$ in (**48**) to explicitly reveal its implicit dependence through the Fredholm problem (**6**), and set $\Lambda = 1$ to make the form of the superposition more transparent.

Integral superpositions like (**48**) can be used to represent an initial condition, which enables us to consider the initial-value problem. In fact, methods borrowed from singular integral theory[10] allow us to invert the integral relation (**48**), and to write amplitude distribution, $A(x)$, in terms of the initial condition, $S(x)$. This procedure is typically complicated by the presence of discrete modes, but often we can prove that the combination of continuum and discrete modes can represent the initial disturbance.[5] This establishes that the combination of the discrete and continuous eigenfunctions form a complete set of basis functions.

Once we have a superposition like (**48**) to represent an initial condition, we can determine the evolution for all subsequent time and show the equivalence with the solution of the problem using Laplace transforms.[4] This amounts to reinstating the temporal dependence, $\exp(-ikct)$ or $\exp(-imvt)$ within the integral superposition (**48**). Integrals of various physical quantities over the domain (such as the total vorticity across the channel) then contain factors of the form, $\exp - ikU(x)t$, within their integrands. By the Riemann–Lebesgue lemma, these integrals must vanish as $t \to \infty$ (unless there is some additional irregularity), revealing the usual phase-mixing property of an ideal system. In many situations, the integrands can be analyzed further to estimate the asymptotic temporal dependence. If this is exponential, we observe the fluid analogue of Landau damping, but in general, that phenomenon is overshadowed by algebraic decay.

Another application of a complete set of singular eigenfunctions is in perturbation theory. Superpositions like (**48**) can be posed as approximate solutions about which we can open asymptotic expansions. We can then attempt weakly nonlinear theory and investigate the ideal limit of some dissipative systems. These amount to avenues we intend to explore in the future.

A final issue that we have not mentioned until now is Hamiltonian structure. The ideal fluid or plasma equations can be recast as Hamiltonian field theories. Typically, these theories are not canonical in the sense that they do not have a standard Poisson bracket.[24] However, by defining a transformation to Hamiltonian coordinates based on the amplitudes of the singular eigenfunctions, the Poisson bracket can be transformed into a canonical form. Moreover, in these linear "normal" coordinates, the Hamiltonian itself is diagonal and of action-angle form.[5,25,26] This indicates that the singular eigenfunctions are in some sense the intrinsic degrees of freedom of the linear fluid or plasma system, like the normal coordinates that describe the modes of vibration in the classic triatomic molecule. However, the degrees of freedom are not discrete in our case; we have a continuum analogue in an infinite-dimensional Hamiltonian system.

## ACKNOWLEDGMENTS

We appreciate conversations with L. N. Howard and B. Shadwick.

## REFERENCES

1. RAYLEIGH, J. W. S. 1945. Theory of Sound, Vol. II: 399–400. Dover. New York.
2. DRAZIN, P. G. & L. N. HOWARD. 1966. Hydrodynamic stability of parallel flow of inviscid fluid. *In* Advances in Applied Mechanics, Vol. 9: 1–88. Academic Press, New York.
3. VAN KAMPEN, N. G. 1955. On the theory of stationary waves in plasmas. Physica **51**: 949–963.
4. CASE, K. M. 1960. Stability of inviscid plane Couette flow. Phys. Fluids **3**: 143–148.
5. BALMFORTH, N. J. & P. J. MORRISON. 1994. Singular eigenfunctions for shearing fluids. Submitted for publication in J. Fluid Mech.
6. SATTINGER, D. H. 1966. The eigenvalues of an integral equation in anisotropic neutron transport theory. J. Math. Phys. **45**: 188–196.
7. KOWALSKI, K. L. & D. FELDMAN. 1961. Transition matrix for nucleon-nucleon scattering I. J. Math. Phys. **2**: 499–511.

8. KOWALSKI, K. L. & D. FELDMAN. 1963. Transition matrix for nucleon-nucleon scattering II. J. Math. Phys. **4:** 507–518.
9. SEDLÁČEK, Z. 1971. Electrostatic oscillations in cold inhomogeneous plasma. Part 2: Integral equation approach. J. Plasma Phys. **9:** 187–199.
10. GAKHOV, F. D. 1990. Boundary Value Problems. Dover. New York.
11. TRICOMI, F. G. 1985. Integral Equations. Dover. New York.
12. RAYLEIGH, J. W. S. 1880. On the stability, or instability, of certain fluid motions. Proc. London Math. Soc. **9:** 57–70.
13. SATOMURA, T. 1981. An investigation of shear instability in a shallow water. J. Met. Soc. Japan **59:** 148–167.
14. HAYASHI, Y.-Y. & W. R. YOUNG. 1987. Stable and unstable shear modes of rotating parallel flows in shallow water. J. Fluid Mech. **184:** 477–504.
15. TAKEHIRO, S.-I. & Y.-Y. HAYASHI. 1992. Over-reflection and shear instability in a shallow-water model. J. Fluid Mech. **236:** 259–279.
16. QIAN, Z. S. & E. A. SPIEGEL. 1994. Autogravity waves in a polytropic layer. Geophys. Astrophys. Fluid Dyn. **74:** 225–243.
17. LORD KELVIN. 1880. On the vibrations of a columnar vortex. Phil. Mag. **10:** 155–168.
18. DRISCOLL, C. F. & K. S. FINE. 1990. Experiments on vortex dynamics in pure electron plasmas. Phys. Fluids B **2:** 1359–1366.
19. BROADBENT, E. G. & D. W. MOORE. 1979. Acoustic destabilization of vortices. Philos. Trans. R. Soc. London **290A:** 353–371.
20. DRURY, L. O'C. 1985. Acoustic amplification in discs and tori. Mon. Not. R. Astron. Soc. **217:** 821–829.
21. PAPALOIZOU, J. C. B. & J. E. PRINGLE. 1987. The dynamical stability of differentially rotating discs-III. Mon. Not. R. Astron. Soc. **225:** 267–283.
22. SCHUTZ, B. F. & E. VERDAGUER. 1983. Normal modes of Bardeen discs—II. A sequence of $n = 2$ polytropes. Mon. Not. R. Astron. Soc. **202:** 881–901.
23. RIPA, P. 1983. General stability conditions for zonal flows in a one-layer model on the $\beta$-plane or the sphere. J. Fluid Mech. **126:** 463–489.
24. MORRISON, P. J. 1982. Poisson brackets for fluids and plasmas. In Mathematical Methods in Hydrodynamics and Integrability in Dynamical Systems. M. Tabor and Y. Treve, Eds.: 13–34. American Institute of Physics Conference Proceedings, No. 88. New York.
25. MORRISON, P. J. & D. PFIRSCH. 1992. Dielectric energy versus plasma energy, and action-angle variables for the Vlasov equation. Phys. Fluids B **4:** 3038–3057.
26. MORRISON, P. J. & B. S. SHADWICK. 1994. Canonization and diagonalization of an infinite dimensional noncanonical Hamiltonian system: Linear Vlasov theory. Acta. Phys. Pol. **85:** 759–769.

# Vorticity and Mixing in Disks[a]

PHILIP YECKO

Department of Astronomy
Columbia University
New York, New York 10027
and
Department of Oceanography
Florida State University
Tallahassee, Florida 32306

## INTRODUCTION

Disks play a role in many astrophysical phenomena, typically as mixers of angular momentum. In most cases, we can think of the disk as an astro-fluid-dynamical phenomenon responsible for the redistribution of matter and angular momentum in the universe, concentrating the matter at its center and dispersing the angular momentum as far from the center as possible. How the disk is able to mix as efficiently as observations suggest is an unsolved problem, although it is unlikely that there is an all-encompassing explanation satisfactory for all disks. The redistribution of angular momentum this way in an orbital "flow" depends on there being anisotropic internal stresses (viscosity) within the disk.[1] In a simple fluid, molecular viscosity, turbulent viscosity, and wave interaction are important mechanisms for providing such stresses, although molecular viscosity is negligible for most length scales of the disk. When strong magnetic fields are present or the disk is very particulate (to cite two examples) mixing may be dominated by magnetic or collisional effects, respectively. However, even in the absence of these last two effects, the behavior of a general continuum (fluid) disk is not a completely solved problem. On the contrary, little is fully understood about the behavior of rapidly rotating strongly sheared fluid layers, even as they occur in the laboratory. This study addresses the disk in a simplified incarnation based on the premise that there are fundamental fluid dynamical questions to be answered before developing more complete models. While astronomical studies often stress the distinctions among different manifestations of disks and relevant boundary conditions (source of material infall, presence of a companion, etc.), we argue that some of the essential actions of disks are performed by dynamical processes that occur on the shortest of all disk timescales for which radial infall velocities ($u_r$) and boundary conditions are largely negligible.

## THIN DISKS

We distinguish those disks that have a dominantly massive central object (accretion disks in binary star systems) from those in which self-gravity is dominant

[a]This work was supported by U.S. Air Force Grant AFOSR F49620-93-1-0457 and National Science Foundation Grant OCE-9401977.

(galactic disks, protoplanetary disks), and concentrate on the former. In fact, we assume that the mass within the disk is small enough in comparison to that of the central object that self-gravity can be neglected entirely, and the central object is included as an external gravitational potential. Other studies have demonstrated the effects of self-gravity on the fluid dynamical behavior;[2] though it may produce instabilities, we are not concerned with those processes here. The potential of a central object can then be used:

$$\Phi_g := -\frac{GM}{R}, \tag{1}$$

where $\Phi_g$ represents the gravitational potential of a point mass $M$ at $R = 0$ ($R$ represents the three-dimensional radial coordinate) and the fluid is assumed to be barotropic:

$$p = p(\rho). \tag{2}$$

The thickness of the disk is defined by its outer surface $h(r, \phi, t)$ where we have now adopted cylindrical polar coordinates $(r, \phi, z)$. We immediately make a distinction between characteristic vertical and horizontal length scales ($H$ and $L$), and define their ratio: $\varepsilon := H/L$. In a thin disk, $\varepsilon$ is a small parameter.

Unlike standard shallow-water theory, the fluid in the disk is compressible, though the connection is made clear by introducing the vertically integrated surface density:

$$\sigma := \int_{-h}^{h} \rho \, dz, \tag{3}$$

the momentum and continuity equations can be vertically integrated to give

$$\frac{\partial \sigma}{\partial t} + \frac{\partial}{\partial r}(\sigma u) + \frac{1}{r}\frac{\partial}{\partial \phi}(\sigma v) = 0, \tag{4}$$

$$\frac{\partial u}{\partial t} + u\frac{\partial u}{\partial r} + \frac{v}{r}\frac{\partial u}{\partial \phi} - \frac{v^2}{r} = -\frac{\partial(\eta + \Phi_g(z=0))}{\partial r}, \tag{5}$$

$$\frac{\partial v}{\partial t} + u\frac{\partial v}{\partial r} + \frac{v}{r}\frac{\partial v}{\partial \phi} + \frac{uv}{r} = -\frac{1}{r}\frac{\partial(\eta + \Phi_g(z=0))}{\partial \phi}. \tag{6}$$

This description is incomplete until we relate $\sigma$ and the midplane enthalpy $\eta$ to $h$. As it appears here in dimensionless form, $\eta$ can be identified with $\frac{1}{2}h^2 r^{-3}$, so if we specify the polytropic function $p = K\rho^\gamma$, the definition of $\sigma$ can be directly integrated to give the relationships

$$\sigma = \mathcal{N} h^l r^{-3l/(\gamma+1)}, \tag{7}$$

and

$$\sigma = \mathcal{N}(2\eta)^{l/2} r^{3/2}, \tag{8}$$

where $l := (\gamma + 1)/(\gamma - 1)$ and

$$\mathcal{N} := \left[\frac{\gamma - 1}{2\gamma}\right]^{1/(\gamma-1)} \int_0^1 (1 - x^2)^{1/(\gamma-1)} \, dx. \tag{9}$$

These equations have appeared in slight variation to describe disks in other work.[3]

We now introduce a *basic state* that has only an azimuthal circulation. Recall that disks are important because they have a radial infall that carries material to their centers, but that the radial velocity is substantially smaller than the circulational velocity. If a disk is accreting centrally while receiving mass at its outer boundary, absolute equilibrium requires a balance among the central accretion, source of external material and energy dissipation. But we are interested in the dynamics nurtured by the general case of differential rotation in a gravitational potential, so we do not associate disks here with particular boundary conditions, though they are allowed to accrete.

A basic state (quantities endowed with overbars) that is time independent, nonradial, and azimuthally symmetric is then

$$\bar{v} = \left( r \frac{\partial \bar{\eta}}{\partial r} + \frac{1}{r} \right)^{1/2}. \tag{10}$$

This is simply a relationship between $\bar{v}$ and $\bar{\eta}$ such that one is determined once the other is specified. Because a disk is just an azimuthally circulating flow, we can think of it as a particular vortex whose velocity profile satisfies 2–10. The horizontal extent of the disk remains arbitrary and it will possess inner and outer "edges" for functions $\bar{\sigma}(r)$ with separated zeros. The (vertical) vorticity of the disk is:

$$\zeta = \hat{z} \cdot \nabla \times \mathbf{v} = \frac{1}{r} \left( \frac{\partial}{\partial r}(rv) - \frac{\partial u}{\partial \phi} \right), \tag{11}$$

which appears in the equation for potential vorticity $q$; this equation describes conservation of $q$ following fluid particles:

$$\frac{dq}{dt} = \frac{d}{dt}\left(\frac{\zeta}{\sigma}\right) = 0. \tag{12}$$

## STABILITY AND WAVES

By looking at small (but not in the linear sense) disturbances to steady states, we can find stability conditions that also reveal something about the mechanism of instability. Following the variational method of Arnold,[4] we find the following sufficient conditions for stability:

$$\bar{v} < \left( \frac{d}{d\bar{\sigma}} \bar{\sigma} \left[ \Phi_g(\bar{\sigma}, z=0) + \frac{l}{l+2} \bar{\eta}(\bar{\sigma}) \right] \right)^{1/2}, \tag{13}$$

and

$$\frac{\bar{v}}{d\bar{q}/dr} > 0. \tag{14}$$

Condition 13 requires that the basic state flow velocities not exceed the right-hand side of the inequality, which can be identified with the speed of gravity waves in the disk. The second condition requires that the potential vorticity gradient of the basic state not change sign except where the velocity profile itself changes sign, a relative of Fjortoft's condition for shear flow. Stronger conditions can be obtained from convexity estimates of the same global integrals of motion from which (13) and (14) were derived, but that problem is difficult (possibly impossible) to solve in shallow water.

The stability conditions also indicate that there are two mechanisms that may destabilize the disk: violation of the second condition, which corresponds to an unstable distribution of angular momentum in the disk; and violation of the first condition, which suggests the possibility of the overreflection of gravity waves in the disk.[5,6] The stability properties therefore serve to provide some insight on the behavior of waves in the disk. Another way to interpret (13), then, is that it ensures that no critical layer is contained in the disk for some part of the gravity wave spectrum. While the converse of (13) implies the appropriate wave geometry for overreflection,[7] it may not be a sufficient condition for instability since it says nothing about the resonant nature of any of the modes. While condition (14) is often attributed to the presence of a favored distribution of angular momentum, it has the alternate interpretation of disallowing the overreflection of shear-type (Rossby) waves. We look next at the behavior of linear perturbations to try to better understand the propagation and instability of waves in disks.

*Linear Waves*

To understand the behavior of wavelike disturbances, including the propagation and possible unstable growth of waves, it is essential to study the problem of linearized disturbances to some conjectured basic state. If we decompose the fields described by equations 4–6 into a time-independent basic state plus a linearized perturbation $\chi_n := \bar{\chi}_n(r) + \chi'_n$, then we get the following equations for the perturbations

$$\left(\frac{\partial \sigma'}{\partial t} + \frac{\bar{v}}{r}\frac{\partial \sigma'}{\partial \phi}\right) + \frac{\partial}{\partial r}(\bar{\sigma} u') + \frac{\bar{\sigma}}{r}\frac{\partial v'}{\partial \phi} = 0, \tag{15}$$

$$\left(\frac{\partial u'}{\partial t} + \frac{\bar{v}}{r}\frac{\partial u'}{\partial \phi}\right) - \frac{2\bar{v} v'}{r} + \frac{\partial \eta'}{\partial r} = 0, \tag{16}$$

$$\left(\frac{\partial v'}{\partial t} + \frac{\bar{v}}{r}\frac{\partial v'}{\partial \phi}\right) + \left(\frac{\bar{v}}{r} + \frac{\partial \bar{v}}{\partial r}\right)u' + \frac{1}{r}\frac{\partial \eta'}{\partial \phi} = 0, \tag{17}$$

while from (8) we can write

$$\sigma' = I(r)\eta', \tag{18}$$

where the function $I(r)$ is used to represent $(d\bar{\sigma}/d\bar{\eta})$. Notice that the coefficient of $u'$ in (17) is just the vorticity of the basic state:

$$B(r) = \frac{\bar{v}}{r} + \frac{\partial \bar{v}}{\partial r} = \frac{1}{r}\frac{\partial}{\partial r}(r\bar{v}). \tag{19}$$

Since the basic state, and therefore the coefficients of (15)–(17) are functions of $r$ only, we can consider normal mode perturbations and obtain a single second-order ordinary differential equation (ODE) for $\eta$ (after dropping primes):

$$\frac{d}{dr}\left(\frac{\bar{\sigma}}{\kappa^2}\frac{d\eta}{dr}\right) + \left(\frac{2m}{v}\frac{d}{dr}\left[\frac{\bar{\sigma}\bar{v}}{\kappa^2 r^2}\right] + I(r) - \frac{\bar{\sigma}m^2}{\kappa^2 r^2}\right)\eta = 0. \tag{20}$$

where

$$v := m\frac{\bar{v}}{r} - \omega, \tag{21}$$

and

$$\kappa^2 := v^2 - 2\frac{\bar{v}}{r}B. \tag{22}$$

This equation (with slight variation) has been examined numerically in other studies.[8] It is the eigenvalue equation for *global* modes of a shallow gas disk with a purely azimuthal circulation once we have specified boundary conditions.

The structure of the eigenfunctions is complicated by the presence of critical layers within the disk. Surface-density waves appear that are either damped or unstable. This type of instability is becoming a well-known feature of shear flows, and is that of the overreflection previously mentioned. Rossby-type modes also appear, in both a discrete and a continuous spectrum. We call these Rossby modes because the oscillation mechanism is the variation of the basic state potential vorticity. The continuous spectrum is characteristic of neutral modes in shear flows.

### Divergent Waves

Locally, compressible waves will be well approximated (away from a critical layer) by solutions over radial scales small enough that we can approximate:

$$\bar{\chi}_n \approx \text{constant} \quad \text{and} \quad \frac{d\bar{\chi}_n}{dr} \approx 0. \tag{23}$$

where the variable $\chi$ ranges over all the basic state quantities. In this case, we find the local dispersion relation and phase speed for compressible (gravity) waves in the disk:

$$c_g = \left(\frac{\bar{\eta}}{l}\right)^{1/2}. \tag{24}$$

Thus wave speeds are comparable to the mean azimuthal velocity, and the necessary instability condition 3–1 is likely to be violated. A WKBJ analysis of the modes of this generic instability problem by Knessl and Keller[9] finds the dispersion relation for the gravity waves and shows that the associated growth rate of the overreflecting mode is small. Numerical models[10] of thick disks show analagous modes. The possibility that these modes may be subcritical and therefore produce turbulence merits further investigation.

## Rossby Waves

We expect to recover Rossby waves even after simplifying this equation provided the basic state (potential) vorticity gradient is retained. So we address the simplest possible limit first. In the incompressible case, the eigenvalue relation is found to be

$$(\bar{v} - c)\left[(r\Psi')' - \frac{m^2}{r^2}\right] - \bar{\zeta}_r \Psi = 0 \tag{25}$$

where $\bar{\zeta}$ is the basic state (potential) vorticity and $\Psi$ is the stream function of the perturbation.

Apart from the cylindrical coordinate representation, this is the inviscid Rayleigh equation. For the incompressible disk, $\bar{v}$ takes the special form ($\bar{v} = r^{-1/2}$), but the well-known analyses of Heisenberg and Tollmein are applicable, indicating that there are no unstable modes and all neutral modes form a continuous spectrum of singular eigenfunctions. Balmforth has given these solutions in this volume.[11]

But we are more interested in local behavior, specifically localized wave packets. Similarly, in the atmosphere, mesoscale Rossby waves are studied independently of knowing whether they are global solutions. In fact, as the local wave propagates, we are interested in its ultimate fate—does it propagate, steepen, break, get absorbed at its critical layer, and so forth. Numerical experiments shed light on these issues.

*Parameters of Disk Flow: $\Omega$, $f$, and $\beta$*

Before beginning a localization appropriate to Rossby-type waves, we look at the global variations of the properties of a disk. There is no unique rotating frame of reference for a disk, so a localization may be performed around any radius. In an azimuthal basic state, the angular velocity varies as:

$$\bar{\Omega}(r) = \left(\frac{1}{r}\frac{d\bar{\eta}}{dr} + \frac{1}{r^3}\right)^{1/2}. \tag{26}$$

Therefore, at a particular radius $r_*$, we will have an effective Coriolis parameter $f_0 = \bar{\Omega}(r_*)$ and an effective $\beta = (d\bar{\Omega}/dr)|_{r_*}$. The local (Rossby) deformation scale is given by $L_D = (\bar{\eta}^{1/2}/f_0)$. $L_D$ is everywhere small, but is especially so at disk edges (where $\bar{\eta}$ vanishes) and inner radii (where $f_0$ is largest). Transforming to the local Cartesian coordinate system $[r, \phi] \to [y, x]$ gives the more familiar:

$$\frac{\partial u}{\partial t} + u\frac{\partial u}{\partial x} + v\frac{\partial u}{\partial y} + 2\bar{\Omega}(y)v = -\frac{\partial \eta}{\partial x}, \tag{27}$$

$$\frac{\partial v}{\partial t} + u\frac{\partial v}{\partial x} + v\frac{\partial v}{\partial y} - 2\bar{\Omega}(y)u = -\frac{\partial \eta}{\partial y}, \tag{28}$$

$$\frac{\partial \sigma}{\partial t} + \frac{\partial}{\partial x}(\sigma u) + \frac{\partial}{\partial y}(\sigma v) = 0, \tag{29}$$

with

$$\sigma = \mathcal{N}(2\eta)^{1/2} r^{3/2}. \tag{30}$$

From these equations we have reduced the near Keplerian flow of the azimuthal disk flow to the equivalent of a differentially rotating frame, and we have a system equivalent to the beta plane. The geophysical problem is recovered by introducing a local two-term Taylor expansion of the disk's basic state:

$$\bar{\Omega}(y) = \bar{\Omega}(r_*) + \frac{d\bar{\Omega}}{dr}(r_*)y + \cdots . \tag{31}$$

To be more accurate, we could include as a mean flow (in the local corotating beta plane) the higher order terms of the basic state. In the beta plane approximation, the basic state enthalpy is constant, though small amplitude variations are allowed. The conservation of potential vorticity then follows:

$$\frac{d}{dt}\left(\frac{\zeta + 2\Omega_* + \beta y}{\bar{\sigma}}\right) = 0 \tag{32}$$

where $\zeta = v_x - u_y$ and $\beta = 2[d\bar{\Omega}(r_*)/dr]$. One of the wave modes supported by (32) is the approximately nondivergent Rossby mode, having a stream function of the form $\Psi = \Psi_0 e^{ik(x-ct)}$ where the (Rossby) phase speed is given by $c = -\beta$.

## NUMERICAL MODEL

There are several features of a disk that are critically different from an ordinary shallow fluid layer. These differences are related to the finite radial extent of a typical disk, for which surface density vanishes. At such discontinuities, it is clear that a potential vorticity *front* will be present at the outer edge of the disk, and also the inner edge, should we allow such toruslike configurations.

The finite extent of the disk means that the code will have to accommodate a vanishing layer thickness at the rim. A similar problem arises in layer models of the ocean: the dynamics often forces layer interfaces to intersect horizontal boundaries, or even pinch off a layer, producing what is called an outcropping. There has been much progress recently in numerical models that effectively treat the boundaries of outcropped layers,[12] so we have adapted this code to treat a disk exactly as if it were a solitary outcropped layer.

In a purely diffusive differencing scheme, as in forward marching, there is no problem calculating the evolution of a distribution of mass that vanishes in some regions. Of course, the price is a lot of diffusion. Higher order schemes introduce the possibility of mass densities becoming *less than zero* in places. One solution to this problem is the so-called flux corrected transport (FCT) algorithm. In this scheme, a low-order (diffusive) and a high-order scheme are simultaneously calculated, and the results are used to identify the diffusive part of the low-order scheme. The density is evolved by the low-order scheme minus its diffusive part. When this subtraction is performed such that no new extrema are introduced, a scheme results that is very accurate for many flows without resorting to a complete gas-dynamical formulation.

In this way, we have adopted model equations and an algorithm that filters the highest frequency modes (sound waves) while retaining divergent effects through surface density changes and also adopted a method (FCT) of evolving layers of vanishing thickness. We are not directly interested in the details of shock structure that would be better calculated by a Boltzmann-equation-based gas dynamical treatment.

The dimensionless equations (4)–(6) are integrated using a standard finite difference prescription and grid that is energy (and enstrophy) conserving. A leapfrog timestep is used to maintain second-order accuracy. The domain is chosen to have a radius $r = 2$ such that the disk (deliminted by inner and outer radii $r_i$ and $r_o$), straddles $r = 1$, and the unit time is such that the Keplerian period at $r = 1$ is $2\pi$.

A Laplacian viscosity is included. The viscosity coefficient is chosen to damp numerical instabilities. Long time evolution of carefully chosen stable disk configurations confirm that dissipation from this viscous term is physically insignificant. Because of its Laplacian nature, the viscous term acts most strongly on the smaller scale structures, and some price is paid in the loss of very small structures. The effective Reynolds number ($Re = (UL/v)$) is approximately 1000.

At the outer edge of the computational domain, free slip tangential flow and no normal flow is imposed. The inner boundary condition plays the role of a "drain", consisting of a ring of massless points moving in purely circular Keplerian motion where mass that is advected to this point is extracted from the model. Mass and potential vorticity conservations are therefore violated at this inner boundary with the net result of global mass loss and potential vorticity gain, though we should stress that these computational boundary conditions are applied (and remain) outside of the physical disk boundaries.

Motions that deviate from the azimuthally symmetric steady flow will produe motion of the disk edges themselves. The edges of the disk may advance as propagating surface density waves, though clearly this must happen as a shock front. By analogy with the behavior of compressible vortices, we expect that the ability to radiate waves such as this will have important implications for the behavior of the disk.

## BEHAVIOR OF THE MODEL DISKS

### *The Steady Disk*

The axisymmetric steady solution given in (10) can be constructed to violate the criteria derived previously and required for stability. In fact, while it is possible to construct a disk that does not violate (14), such a disk will often violate (13). We construct one possible family of such disks, having enthalpy:

$$\bar{\eta} := \eta_0 [(r - r_i)(r_o - r)]^n. \tag{33}$$

and the associated azimuthal velocity:

$$\bar{v} = \left( r \frac{\partial \bar{\eta}}{\partial r} + \frac{1}{r} \right)^{1/2}, \tag{34}$$

which has potential vorticity

$$\bar{q} = \frac{\bar{\zeta}}{\bar{\sigma}} = \frac{1}{\mathcal{N} 2 \bar{\eta}^{1/2} r^{5/2}} \frac{\partial}{\partial r} \left( r^3 \frac{\partial \bar{\eta}}{\partial r} + r \right)^{1/2}. \tag{35}$$

For $n = 1$ the enthalpy profile is a parabolic "lens", and larger values of $n$ result in more tapered edges at $r_i$ and $r_o$. A quantitative rendering of the surface density is given in FIGURE 1 in terms of the radial mesh coordinate $\rho$ such that $r = [0, 2]$ corresponds to $\rho = [0, 40]$.

In FIGURE 1 we see the results of a control run initialized with an unperturbed disk of form (33). The curve sequence displays the evolution through $t = 32$, and consecutive curves are separated by intervals $\Delta t = 2$. There is a small amount of accretion discernable, which appears as increased surface density in the vicinity of the inner edge of the disk. This is a direct result of the viscous term of the numerical scheme. The minimal accretion is a measure of the minimal dissipation of the algorithm. Later, we calculate this directly as an *effective* $\alpha$ coefficient. Slightly less apparent is the outward spread of a small amount of material beyond the initial $r_o$. Recall that viscous stress acting on material in adjacent Keplerian orbits will result not simply in *accretion*, but in *spread*.[13]

### Perturbed Disks

Several experiments were run that included a significant nonaxisymmetric perturbation. The intent is to observe the adjustment processes. The perturbations include:

1. The gravitational potential of a companion star. A secondary of mass $M_2 = \frac{1}{8}$ is introduced at $r = 3$.
2. The presence of an accretion stream at the outer edge of the disk. A thin stream of material with a small inward velocity impinges on the outer edge ($r = r_o$) of the disk. This stream is maintained at a surface density corresponding to an enthalpy $\eta_S = \bar{\eta}/2$ and a radial velocity $u_S = -1$.

This initial analysis of the perturbed disks reveals the presence of unforeseen nonlinear processes as well as some generic global tendencies of unsteady disks.

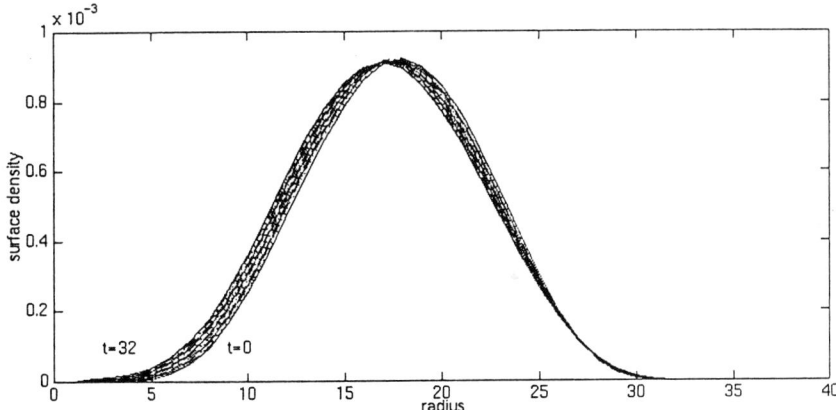

**FIGURE 1.** Time evolution of the azimuthally averaged surface density of the unperturbed disk in the time interval $t = 0$–$32$.

*Accretion*

Accretion was shown to be strongly driven by the devlopment of vortical phenomena in this study. From accretion disk theory[13] we know that the presence of viscosity induces a radial velocity:

$$u = \frac{v}{\Omega} \frac{\partial \Omega}{\partial r}. \qquad (36)$$

This velocity will carry material through the disk, producing a rate of change of mass

$$\dot{M} = 2\pi r \sigma u. \qquad (37)$$

In the $\alpha$-model, the viscosity is replaced by the mixing length parameterization

$$\alpha := \frac{v}{c_s h} \qquad (38)$$

Using the data that are represented in FIGURES 2 and 3, we evaluate $\dot{M}$ as a function of radius and time. Integrating over larger radial or temporal intervals produces more highly averaged values of $\alpha$. Calculations at $r = \frac{1}{4}$, $r = \frac{1}{2}$, and $r = 1$ ranging over $t = 8 - 40$ consistently yield $\alpha_{\text{eff}} \approx 0.1$ for all varieties of perturbation.

As a check of the numerical model and our method of calculating $\alpha_{\text{eff}}$, the process was performed on the unperturbed disk. The accretion in that case develops because of the Laplacian viscosity of the model equations. Rather than calculate $\alpha$, we find the viscosity coefficient itself $v = \frac{1}{1214}$, which nearly matches the value of $v = \frac{1}{1280}$ specified in the program.

*Potential Vorticity Smoothing*

In FIGURE 4, we see the evolution of the potential vorticity of the disk perturbed by the asymmetric potential of a binary star system. A potential function is known

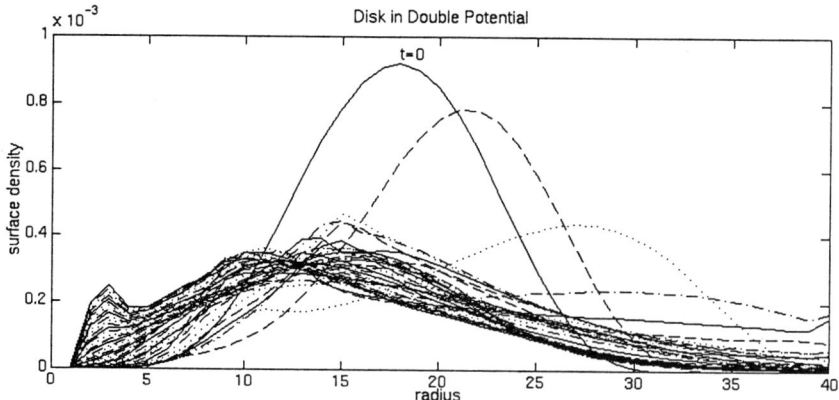

**FIGURE 2.** Time evolution of azimuthally averaged surface density of an axisymmetric disk in a binary system in the time interval $t = 0-40$.

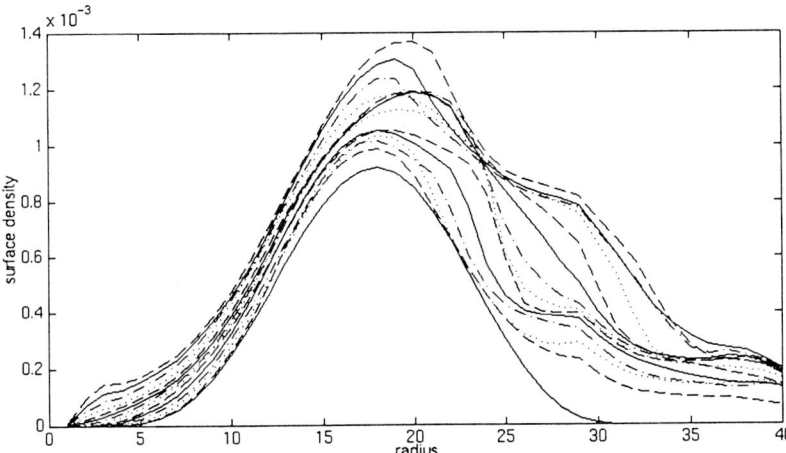

**FIGURE 3.** Time evolution of azimuthally averaged surface density of an axisymmetric disk with an accretion stream incident just beyond the outer edge in the time interval $t = 0-36$.

to preserve potential vorticity conservation, but the excitation of forced waves can and does occur. The initial condition is already rather homogeneous away from the inner and outer edges, so the effects of homogenization are most apparent in two characteristics: the steepening of the gradients near the edges, and the genericity of the end states among differently perturbed systems. Compare these characteristics to FIGURES 5 and 6, which represent the disk perturbed by an accretion stream and by an unstable mode, respectively. In FIGURE 7 four particular levels of potential vorticity are shown over a temporal period for which the smoothing of the middisk region is apparent as is the less dramatic steepening of the edge gradients.

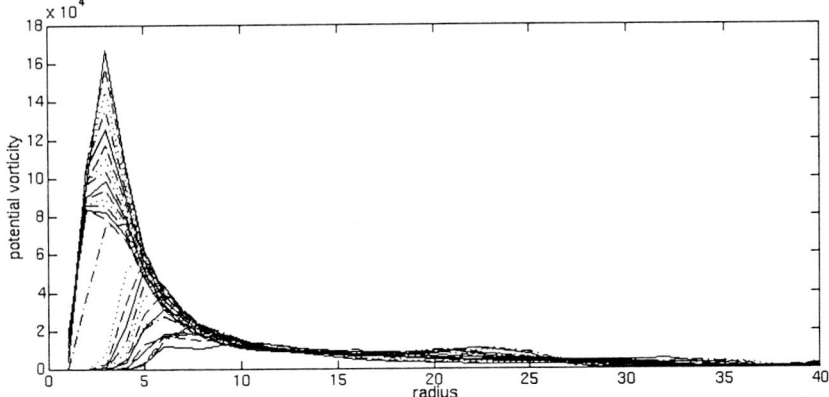

**FIGURE 4.** Time evolution of the potential vorticity of a disk in the potential of a binary star system in the time interval $t = 12-40$.

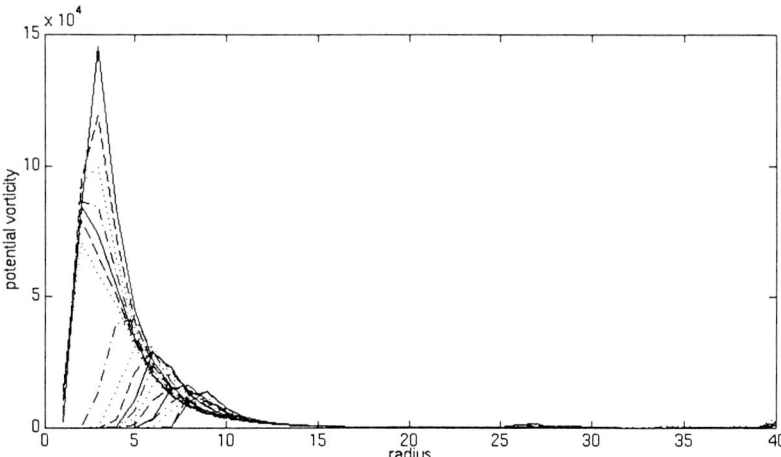

**FIGURE 5.** Time evolution of the potential vorticity in a disk with an accretion stream incident just beyond the outer edge in the time interval $t = 0$–$36$.

Batchelor[14] showed that for many closed streamline regions, potential vorticity is driven to uniformity. Since then, the so-called *Prandtl–Batchelor theorem* has been successfully modified and applied to more realistic vortices than those (nonrotating, two-dimensional, incompressible) considered by Batchelor. Rhines and Young[15] made the extension to stratified flows on a $\beta$-plane and found a related homogenization. This tendency of closed streamline regions to homogenize potential vorticity is based on the premise that eddies (and waves) will effect a down gradient flux of potential vorticity.

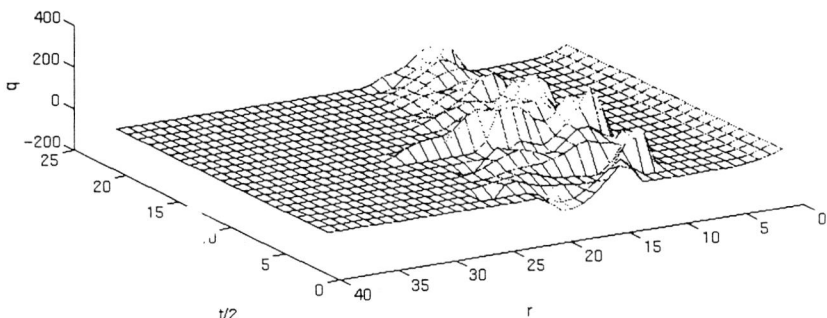

**FIGURE 6.** Time evolution of the potential vorticity of a disk with instability developing as a result of an initial potential vorticity distribution that violates the Rayleigh–Kuo stability requirement.

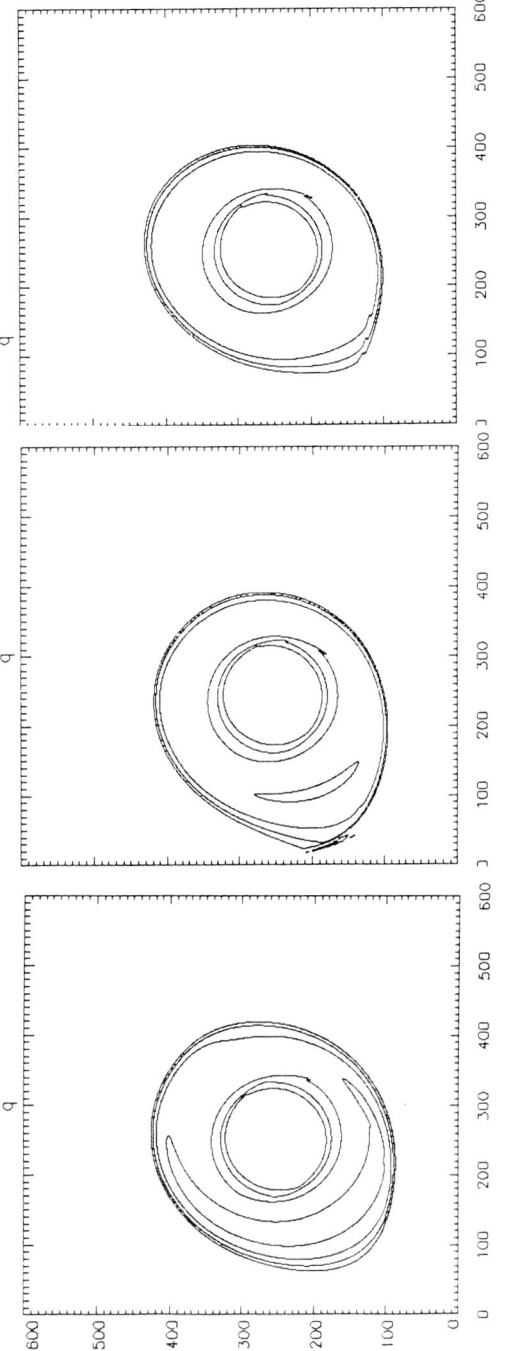

FIGURE 7. Four contours of potential vorticity at $t = 30$, $t = 40$, and $t = 50$ showing homogenization and steepening of the gradients.

*Finite Amplitude Waves and Breaking*

These disks with nonaxisymmetric influence all develop noncircular contours of (potential) vorticity and subsequently develop instabilities characteristic of nonaxisymmetric vortices. Extensive numerical and theoretical investigations of analogous models (*V-states*) have been performed by reducing the general problem to that of the evolution of the boundary between two regions of different vorticity (what is known as a vortex patch). Deem and Zabusky[16] have found instability criteria (for particular perturbation amplitude) for a variety of modes with wavenumber greater than two. What is important here is that *elliptical as well as triangular, square, and so on*, vortices are all unstable when the amplitude of asymmetry (typically expressed through the ratio of minor-to-major axes) exceeds a certain threshold. The instability develops by the steepening of the lobes of an asymmetric vortex into elongated tongues that develop into extended tendril vortices that are shed to the ambient fluid. The fate of the shed vorticity is an unsolved problem even in this simplified model, and we have not yet attempted to solve it for the complicated case of disks. In general, experiments have shown that isolated regions of vorticity tendrils are advected by the global velocity field, which is dominated by the parent vortex in most cases, resulting in advection of the tendril away from the vortex. In addition, investigations of vortex merger and equilibration of geostrophic (and two-dimensional) turbulence have observed the tendency of nonaxisymmetric vortices to axisymmetrize by shedding elongated tongues and tendrils of vorticity. The tendril ultimately becomes more sheared until its vorticity is at such small spatial scale that it falls victim to viscous dissipation (or contour surgery in cases of numerical experiments using contour dynamics). When this process happens in a disk, orbital angular momentum is placed into small packages of locally enhanced "spin" angular momentum that are advected outward and sheared to the point of viscous dissipation.

It has been demonstrated that the angular-momentum-based stability criterion for axisymmetric flows is very weak. Potential vorticity, on the other hand, is in some sense the continuum generalization of angular momentum, wherein local "degrees of freedom" are allowed because circulation is evaluated everywhere rather than with respect to a particular axis. By analogy with classical mechanics, potential vorticity encompasses *spin* as well as *orbital* angular momentum.

Vorticity in the disk varies continuously as a function of radius making quantitative comparison with $V$-state experiments impossible. But the development of asymmetric vorticity contours into lobes and into tongues that elongate into tendrils, eventually separating from their parent is observed in all asymmetrically perturbed problems. FIGURE 8 shows the process clearly for the system perturbed by the binary potential. In the eighth, ninth, and tenth frames, we see the excessively steepened, beginning to break, and breaking potential vorticity contours. In frames 11 and 12, the separated vortex is quickly elongated and absorbed by the background flow and disssipation.

In all disks, the waves developing on vorticity contours were observed to have common features. They propagate in the direction $\hat{z} \times \nabla q$ relative to the mean flow, as would be expected for localized Rossby wave packets. They must be of finite amplitude to exhibit breaking. Smaller amplitude waves that do not break also

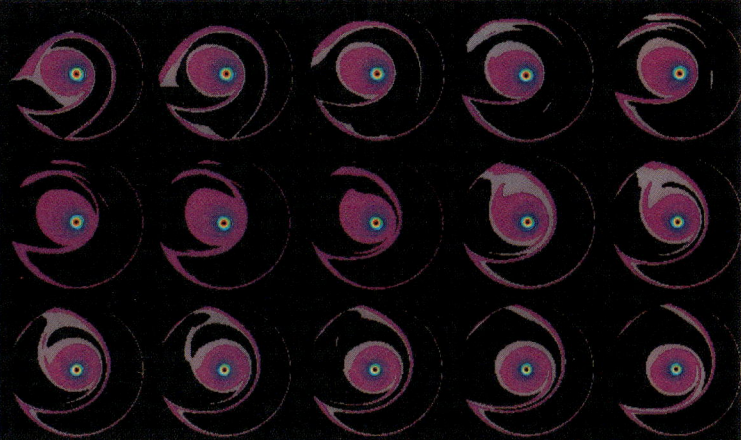

**FIGURE 8.** Steepening and breaking of a (potential) vorticity wave in a disk in a binary system.

appear, though a small amount of microbreaking may be occurring. They are local phenomena, as three or more unevenly spaced wave packets are often visible on a single $q$ contour, despite the diplar bias of the perturbations. And finally, breaking always occurs such that high potential vorticity is deposited into a lower potential vorticity region; this wave-breaking mechanism acts to transfer potential vorticity down-gradient.

## DISCUSSION

Wave breaking is, by definition, mixing. We are suggesting here that the observed accretion, potential vorticity homogenization, and waves and breaking are related and consistent processes in a disk. Enhanced angular momentum can be directly identified with shed tendrils of vorticity. But we do not yet have a predictive method for the breaking of a potential vorticity wave in a disk. Nor have we addressed the ability of nonbreaking waves to effect potential vorticity flux.

In the study of vortex dynamics, there is little interest in the fate of separated tendrils, but they have a special relevance to the accretion disk as they represent localized regions of enhanced vorticity (formerly disk angular momentum) that move outward before they deposit their vorticity into the ambient field. They are, therefore, an *unforeseen mechanism of angular momentum transport* in any accretion disk that has asymmetric nature. Additionally, the shear effectively forces isolated vortices to small scales where viscosity can dissipate them entirely, providing another *unforeseen mechanism of angular momentum transport*.

A disk in a binary systems has an asymmetric nature as a result of the gravitational influence of the secondary and as a result of the presence of an accretion stream, so asymmetric processes are prevalent. Of primary interest here is the process of adjustment, not the details of the perturbation, and we take solace in the fact that these processes are observed for a variety of perturbations. We are suggesting that these processes of vorticity axisymmetrization may be very common, and that the ability of these processes to redistribute angular momentum has not been adequately studied in the astrophysical case. We also point out that spiral waves and, in the cases of a companion star, tidal torques were also present in these experiments. The measured values of $\alpha_{eff}$ produced by these effects has been evaluated in other studies and is found to be $O(0.01)$, and therefore an insignificant contributor to the $\alpha_{eff}$ found here.

## ACKNOWLEDGMENTS

It is my great pleasure to recognize the guidance and wisdom of Ed Spiegel in this work, and the contributions of Neil Balmforth and Steve Meacham to its improvement, as well as Eric Chassignet's generosity in the use of his code. The computing was performed on the Pittsburgh Supercomputing Center C-90.

## REFERENCES

1. SHAKURA, N. I. & R. A. SUNYAYEV. 1973. Astron. Astrophys. **24:** 337–357.
2. KUMAR, S. 1991. Geophys. Astrophys. Fluid Dyn. **61:** 235–276.
3. BALMFORTH, N. J., L. N. HOWARD & E. A. SPIEGEL. 1993. Mon. Not. R. Astron. Soc. **260:** 253–272.
4. ARNOLD, V. I. 1965. Dokl. Akad. Nauk. SSSR **162:** 975–978.
5. NARAYAN, R., P. GOLDREICH & J. GOODMAN. 1987. Mon. Not. R. Astron. Soc. **288:** 1–41.
6. RIPA, P. 1983. J. Fluid Mech. **126:** 463–489.
7. LINDZEN, R. S. & J. W. BARKER. 1986. J. Fluid Mech. **151:** 189–217.
8. SCHUTZ, B. F. & E. VERDAGUER. 1983. Mon. Not. R. Astron. Soc. **202:** 881–901.
9. KNESSL, C. & J. KELLER. J. Fluid Mech. **244:** 605–614.
10. WOODWARD, J. W., J. E. TOHLINE & I. HACHISU. 1994. Astrophys. J. **420:** 247–267.
11. BALMFORTH, N. J. & P. J. MORRISON. 1995. Normal modes and continuous spectra. This issue.
12. SUN, S., R. BLECK & E. P. CHASSIGNET. 1993. J. Phys. Ocean. **23:** 1877–1884.
13. PRINGLE, J. E. 1981. Annu. Rev. Astron. Astrophys. **19:** 137–162.
14. BATCHELOR, G. K. 1956. J. Fluid Mech. **1:** 177–190.
15. RHINES, P. B. & W. R. YOUNG. 1982. J. Fluid Mech. **122:** 347–367.
16. DEEM, G. S. & N. J. ZABUSKY. 1978. Phys. Rev. Lett. **40:** 859–862.

# Bending Waves in Flattened Stellar Systems[a]

C. HUNTER

*Department of Mathematics*
*Florida State University*
*Tallahassee, Florida 32306-3027*

## 1. INTRODUCTION

Warping of disk galaxies is much more evident in the gas than in stars. The warp in the H1 layer of the Milky Way was detected from 21-cm observations in the 1950s.[1,2] Then, in the 1970s, it became apparent that warps are a common feature of the H1 layers of external galaxies.[3,4] Warps of H1 layers become strong only where they extend far beyond the optical image,[5] and the optical evidence for the occurrence of warps in stellar disks is much less strong. There are clear-cut cases such as M31,[6] but van der Kruit and Searle[7–9] detected only one very marginal warp in the seven edge-on galaxies that they observed. Although Sánchez-Saavedra et al.[10] claimed to detect warps in 42 of the 86 highly inclined spirals that they studied, they did classify 23 of the 42 as barely perceptible.

Two of the mechanisms that have been proposed for explaining warps, an intergalactic wind[11] or an infall of matter,[12] are likely to affect gaseous disks more strongly than stellar disks. Other proposed mechanisms apply equally well to stellar as to gaseous disks. It was clear to the first detectors of the warp in the Milky Way that this warp is much too large to be attributed to the tide currently being raised by any near neighbor. This explanation is similarly unsatisfactory for the warps in many external galaxies. Past close encounters could raise larger tides. Both this and other possibilities such as Lynden-Bell's[13] proposal that the warp is a free mode of oscillation, require an understanding of dynamics of bent disks and the waves that they can support. It is this dynamics, rather than the explanation of galactic warps, that is the topic of this paper. Massive dark halos may very well play a major role in galactic warps.[14–16] As Binney[17] has remarked in a recent review:

> Early theoretical work on warped galaxies, especially the classic paper by Hunter & Toomre (1969), was concerned with the dynamics of isolated disks. It is now generally supposed that even late-type galaxies, which appear to consist of a disk alone, are embedded in a dynamically dominant dark halo. This is fortunate since it is in some ways easier to understand the dynamics of a massive disk sitting in an externally generated gravitational field than to get to grips with the dynamics of an isolated disk.

I believe that the dynamics of an isolated massive disk is still a topic of astrophysical relevance, even if it is not the most astrophysically interesting one related

---

[a]This work has been supported in part by National Science Foundation Grant DMS-9304012. This paper is based on an invited talk presented at the Tenth Annual Florida Workshop in Nonlinear Astronomy, September 1994, and a revised version that was given as the Brouwer Award Lecture in April 1995.

to warps. The difficulty of detecting warps in stellar disks suggests that they are generally stable against bending perturbations. I think that an improved understanding is needed of what it takes to keep a disk stable against bending perturbations. If dark halos are indeed the major cause of the warping of the gas beyond the optical disk, it is unlikely that they would also help stabilize the stellar disk. And why should it be so difficult to come to grips with the dynamics of isolated massive disks, in view of the relative simplicity of the system with which we are dealing, one in which particles move solely under the influence of gravity?

## 2. DYNAMICS OF THE BENDING OF RAZOR-THIN DISKS

### 2.1. Equation of Bending Motion

The simplest possible dynamical model, and one that is still widely used even when coupled with much more elaborate halo models,[18,19] is that of a razor-thin flat disk in which matter rotates in circular orbits about a center, with angular velocity $\Omega(R)$. We use polar coordinates $(R, \theta)$ in the $z = 0$ plane of the unperturbed disk. Perturbations are described by the displacement $h(R, \theta, t)$ in the $z$-direction, which we suppose to be small. Taking account of the convective effect due to the circular rotation, the equation of motion in the $z$-direction is[20]

$$\left[\frac{\partial}{\partial t} + \Omega(R)\frac{\partial}{\partial \theta}\right]^2 h = F_z = F_{disk} + F_{ext}. \tag{1}$$

Here $F_z$ is the force per unit mass in the upward $z$-direction, with contributions $F_{disk}$ from the disk itself, and $F_{ext}$ from external forces.

The gravitational force due to the disk itself is

$$F_{disk} = G \iint_{\text{Disk surface } S'} \frac{\Sigma(R')[h(R', \theta', t) - h(R, \theta, t)]}{|\mathbb{R} - \mathbb{R}'|^3} dS', \tag{2}$$

where $\Sigma(R)$ is the surface density of the unperturbed disk, and $\mathbb{R}$ and $\mathbb{R}'$ are position vectors in the $z = 0$ plane. The perturbing force $F_{disk}$ can be regarded as the force at height $h$ due to an array of gravitational dipoles of spatial density $\Sigma(R)h(R, \theta, t)$. The dipole nature of this force comes from the fact that it is due to a combination of positive elements given by the perturbed disk, which is displaced from the $z = 0$ plane, and negative elements arising from the original unperturbed disk. The force $F_{disk}$ of (2) can be split into the two components given by the two numerator terms. The first component is then the vertical force caused by the dipoles, while the second component is that due to the force being evaluated at height $h$.[16] The splitting must be done carefully so as to avoid singular integrals because the improper integral (2) is convergent by virtue of the vanishing of the numerator at $\mathbb{R} = \mathbb{R}'$. The correct procedure is described in Section A.1 of the Appendix. When it is followed, the second component from the split is

$$-G \iint_{\text{Disk surface } S'} \frac{\Sigma(R')h(R, \theta, t) \, dS'}{|\mathbb{R} - \mathbb{R}'|^3} = -v^2_{disk}(R)h(R, \theta, t), \tag{3}$$

where the line through the integral signs denotes that the finite part of this singular integral, as defined in (26), is to be used. It measures the $z$-gravitational force due to the unperturbed disk at height $h$, after the discontinuous self-force due to the local density has been subtracted, and $v_{disk}(R)$ is the natural frequency of vertical oscillations at radius $R$ in this gravity field. It can be shown[20] (see also the Appendix) that

$$v_{disk}^2 = 2\Omega^2 - \kappa^2, \qquad (4)$$

where $\Omega(R)$ is the circular angular velocity and $\kappa(R)$ is the epicyclic frequency of small noncircular motions in the gravity field of the disk. Note that the positivity of $\Sigma$ is no guarantee of the positivity of the $v_{disk}^2$ given by the finite part integral of (3). For example, $v_{disk}^2$ is negative in the case of uniform rotation for which $\kappa^2 = 4\Omega^2$. If external forces, such as those due to a $z$-symmetric static halo, are also taken into account, then the natural frequency $v$ of vertical oscillations has to be modified to

$$v^2 = v_{disk}^2 + v_{ext}^2, \qquad (5)$$

where $F_{ext} = -v_{ext}^2 h$.

### 2.2. Kinematic Bending Waves

When we ignore the self-gravitational forces arising from the bent disk, so that the force $F_{disk}$ is given entirely by the component (3), the dynamics is described by the simple equation[21]

$$\left[\frac{\partial}{\partial t} + \Omega(R)\frac{\partial}{\partial \theta}\right]^2 h = -v^2(R)h. \qquad (6)$$

Assuming dependence on $e^{i(m\theta - \omega t)}$, (6) gives the dispersion relation

$$(\omega - m\Omega) = v^2, \qquad (7)$$

which allows two possible frequencies. There is a faster direct wave that propagates with angular velocity $\omega/m = \Omega + (v/m)$, and a slower, and often retrograde, wave that propagates with angular velocity $\omega/m = \Omega - (v/m)$. When either $\Omega$ or $v$ vary with $R$, the wave velocity varies with radius, and there is no single coherent wave. The winding dilemma, of a wave shape that becomes continually more tightly coiled, is avoided in the important $m = 1$ case for a disk in a spherical galaxy because then $v = \Omega$. However, the dilemma generally arises with mass distributions that are flattened in $z$ because their gravitational fields give $v > \Omega$.

### 2.3. Bending Waves with Self-Gravity

The gravitational forces induced by the bending of the disk provide one mechanism that gives coherence between different radii, and allows the possibility of global and non-winding waves. Numerical methods generally are necessary for finding such solutions of (1) with the full expression (2) for the gravitational $F_{disk}$

force. However, an approximate dispersion relation of

$$(\omega - m\Omega)^2 = v^2 + 2\pi G\Sigma k \tag{8}$$

can be obtained for waves of short and positive radial wavenumber $k$ by using a short wavelength approximation to the local gravitational forces exerted by the bent disk. This relation confirms explicitly that self-gravity exerts an extra stabilizing influence. Toomre and I[20] never found any unstable solutions in our investigations, though we were only able to prove the nonexistence of instabilities for the angular wavenumbers $m = 0$ and $m = 1$.[b] The dispersion relation (8) shows that there are still two kinds of waves, either fast with $\omega > m\Omega + v$ at each radius or slow with $\omega < m\Omega - v$, and with a wavenumber that varies with radius.

Although the modes of a finite disk that has a sharp edge form a discrete spectrum, most of these modes run together to form a continuous spectrum when the density tapers gradually to zero at the outer edge $R = R_{edge}$ of the disk.[20] A lack of clearly defined discrete modes is a difficulty for modal explanations of warps. The continuum ranges of frequencies then consist of the two discrete bands, one fast and one slow, allowed by the limiting conditions at the outer edge of the disk:

$$\omega > m\Omega(R_{edge}) + v(R_{edge}),$$
$$\omega < m\Omega(R_{edge}) - v(R_{edge}). \tag{9}$$

As (8) correctly predicts, continuum modes fluctuate extremely rapidly as the outer edge, where the surface density $\Sigma$ tends to zero, is approached. Their radial wavenumber $k$ becomes infinite while their phase velocity $\omega/k$ slows to zero in such a manner that they never reach the outer edge in finite time. Our numerical studies essentially found that no nontrivial discrete $m = 0$ or $1$ modes remain in the presence of the continuum, though some nontrivial discrete modes do so remain for $m \geq 2$.[22] However, Sparke and Casertano[16] have shown that a finite disk can have discrete $m = 1$ warping modes even when the density tapers to zero at the outer edge, provided that it does so in a region that is narrow enough so as to leave a sufficient cavity in which the discrete modes can be trapped.

### 2.4. Bending Instabilities Arising from Noncircular Motions

The dynamics is changed radically when random motions in the plane of the disk, which also add to the coherence between different radii, are present. We suppose in this model that internal forces normal to the disk are strong enough to maintain its razor-thin structure. Inertial effects now modify the left-hand side of the equation of motion (1), which becomes

$$\left[\frac{\partial}{\partial t} + \bar{\Omega}(R)\frac{\partial}{\partial \theta}\right]^2 h + \frac{1}{R\Sigma(R)}\frac{\partial}{\partial R}\left[R\Sigma(R)\sigma_R^2\frac{\partial h}{\partial R}\right] + \frac{\sigma_\theta^2}{R^2}\frac{\partial^2 h}{\partial \theta^2} = F_z. \tag{10}$$

---

[b]At the meeting, Jeremy Goodman reported that he had found numerical evidence for instabilities at $m > 1$.

Here, $\bar{\Omega}(R)$ is the mean angular velocity of the circular motion about the center, and $\sigma_R^2$ and $\sigma_\theta^2$ are the radial and azimuthal velocity dispersions, respectively. Equation (10) has been named the *membrane equation* by Polyachenko and Shukhman,[23] who first gave its full form, though Toomre[24] had earlier realized the significant effects that random motions in the plane of the disk can have. The approximate dispersion relation given by (10) for the short wavelength limit is

$$(\omega - m\bar{\Omega})^2 = v^2 + 2\pi G\Sigma k - k^2 \sigma_R^2 - m^2 \sigma_\theta^2 / R^2. \tag{11}$$

It shows that the noncircular motions are destabilizing, and increasingly so at shorter wavelengths. The reason is that this is no ordinary membrane. The random motions endow it with a pressure, not necessarily isotropic, and it is in a state of compression, rather than tension. This compression causes it to tend to buckle when it is bent, instead of pulling it back to its unbent state as a tension would.

The few known analytical solutions of the membrane equation (10) confirm the instability. There is a brief mention in Hunter and Toomre[20] of the fact that two equal superposed and oppositely circularly rotating Maclaurin disks are unstable for $m \geq 2$. The recent discovery of counterrotating stars in the disk of NGC 2550[25] has made this model seem somewhat more realistic that it did in 1969. For two equal superposed and oppositely circularly rotating disks, $\bar{\Omega} = \sigma_R = 0$ and $\sigma_\theta^2 = R^2 \Omega_c^2$, where $\Omega_c(R)$ is the angular velocity for circular motion. The dispersion relation (11) therefore reduces to

$$\omega^2 = v^2 + 2\pi G\Sigma k - m^2 \Omega_c^2, \tag{12}$$

which does indeed predict instability at large enough $m$.

Polyachenko[26] derived the exact dispersion relation for Maclaurin disks with Kalnajs[27] distribution functions. The latter have isotropic velocity dispersions and mean angular velocities that are a constant fraction of the constant $\Omega_c$. As (11) then correctly predicts, these models have instabilities at short enough wavelengths for all angular wavenumbers $m$. Polyachenko's paper ends with the suggestion that these buckling instabilities, or the hose-pipe instabilities as he called them, "might have some bearing on the notorious problem of the minimum flattening of elliptical galaxies." This problem has yet to be resolved, and Polyachenko's suggestion remains an important one.[28-30]

Malkov and his associates[31,32] used a similar model to investigate some global $m = 2$ bending modes of elliptical Freeman disks.[33] They found that many of those that are free of bar instability are prone to bending instability. The latter phenomenon is understandable in the light of the fact that self-gravity acts to stabilize, while in-plane random motions act to destabilize, the bending of a thin disk. These roles vis-à-vis stability are the reverse of what they are for density waves in the plane of a disk. This reversal of roles is apparent when the bending dispersion relation (11) is compared with the dispersion relation[21]

$$(\omega - m\Omega)^2 = \kappa^2 - 2\pi G\Sigma k + c^2 k^2, \tag{13}$$

for density waves in an isotropic one-component fluid model with velocity dispersion $c$. Here Jeans instability occurs at sufficient long wavelengths unless the velocity dispersion is large enough (i.e., meets the Toomre criterion,[34] $c > \pi G\Sigma/\kappa$).

Merritt and Sellwood[35] computed solutions of the membrane equation (10) for a class of Kuzmin–Toomre disks with two equal counterrotating stellar populations. They found infinite continuous spectra of unstable modes, the frequencies of which they could fit well with a theoretical formula. They also found a few discrete oscillatory modes for the $m = 0$ and $m = 1$ cases, though none for the $m = 2$ case.

### 2.5. N-Body Simulations

Sellwood and Merritt[36] also carried out $N$-body simulations of two initially equal counterrotating stellar populations, simulations that detected waves and instabilities in the plane as well as out of it. Their major findings concerning bending instabilities are that thin radially hot models are disrupted by $m = 0$ axisymmetric bell modes. The dominant bending instability in cooler thin models is an $m = 2$ saddle mode. They were able to detect an $m = 1$ warp, but it always grew more slowly than either the bell or the saddle modes. These instabilities are weakened, though mostly rather slowly, by increasing disk thickness. Nevertheless, they did find that models with an intermediate radial pressure become, after minor adjustments, almost axisymmetric, apparently stable, moderately thin disks. Their results therefore suggest strongly that the study of disk systems of small, though finite, thickness is a topic worth pursuing.

## 3. BENDING WAVES IN SYSTEMS OF FINITE THICKNESS

In this section, we explore how bending waves in systems of finite thickness differ from those in razor-thin disks. In particular our interest is in understanding how the destabilizing effect of noncircular motions in the plane is overcome.

### 3.1. Effect of Finite Thickness on Self-Gravity

The effectiveness of the restoring force of gravity, which counteracts bendings of razor-thin disks and causes such bendings to give rise to stable waves in the absence of noncircular motions, is diminished by a finite thickness of the disk. To justify and quantify this last statement, we consider a small wavy displacement $h_0 e^{ikx}$ in the $z$-direction of a thin layer of gravitating matter that is stratified in the $z$-direction but is uniform in other directions. Suppose that this displacement shifts, but does not otherwise alter, the internal density profile $\rho_0(z)$ of the layer. Then the total force per unit mass acting in the $z$-direction on the element at $x = x_0$ can be calculated (see Appendix) to be

$$F_z = \frac{-2\pi G k h_0 e^{ikx_0}}{\Sigma} \int_{-\infty}^{\infty} \int_{-\infty}^{\infty} \rho_0(z)\rho_0(s) e^{-k|z-s|} \, dz \, ds, \qquad (14)$$

where $\Sigma = \int_{-\infty}^{\infty} \rho_0(z) \, dz$ is now the integrated surface density of the layer. This force simplifies to $-2G\Sigma k h_0 e^{ikx_0} = -2\pi G\Sigma k h$ for a razor-thin disk for which $\rho_0(z) = \Sigma\delta(z)$, and the force appropriate to the thin disk dispersion relation (8) is then recovered. More generally, however, the force $F_z$ is described by the relation

$$F_z = -2\pi G \Sigma k \mathscr{F} h, \tag{15}$$

and contains an additional reduction factor $\mathscr{F}$. This reduction factor depends upon the wavenumber scaled with the scale height $H$ of the layer, and the explicit expression for it is

$$\mathscr{F}(kH) = \frac{1}{\Sigma^2} \int_{-\infty}^{\infty} \int_{-\infty}^{\infty} \rho_0(z)\rho_0(s) e^{-k|z-s|} \, dz \, ds. \tag{16}$$

It diminishes steadily as the scaled wavenumber $kH$ increases; that is, as the thickness of the layer increases, or as the ratio of wavelength of the bending becomes short compared with this thickness. The reduction factor can be evaluated explicitly for a top-hat density profile of width $2H$ as

$$\mathscr{F}_{TH}(kH) = \frac{2kH - 1 + e^{-2kH}}{2(kH)^2}. \tag{17}$$

It tends to 1 as the scale height $H \to 0$, but decays to zero as $(1/kH)$ as $kH \to \infty$. The latter type of decay is characteristic of other density profiles because the reduction factor (16) can be approximated asymptotically, for large $kH$ and smooth peaked density profiles, by

$$\mathscr{F}(kH) \sim \frac{2}{k\Sigma^2} \int_{-\infty}^{\infty} [\rho_0(s)]^2 \, ds. \tag{18}$$

Hence, unless the bending causes the interval density profile to be readjusted in some especially favorable manner, self-gravity cannot be the mechanism by which an unstable razor-thin disk achieves stability at finite thickness. For an isothermal density profile

$$\rho_0(z) = \frac{\Sigma}{2H} \operatorname{sech}^2\left(\frac{z}{H}\right), \tag{19}$$

the limit (18) is easily calculated to be $\mathscr{F}(kH) \sim 2/(3kH)$. The general expression for $\mathscr{F}$ has the more complicated form

$$\mathscr{F}(kH) = \tfrac{1}{2}(kH)^2 \psi'(kH/2) - 1 - kH, \tag{20}$$

where $\psi$ is the digamma function.[c]

### 3.2. Dynamics of Slabs of Finite Thickness

The altered dynamics of systems of finite thickness is the only other mechanism for achieving stability. We consider first the simpler case of gaseous systems, as dynamicists are often wont to do, because of the greater complexity of stellar systems described by the collisionless Boltzmann equation. In both instances, our

---

[c]This result is due to Jeremy Goodman (private communication).

analysis again refers to systems that are stratified only in the $z$-direction, and are uniform in the other directions, and are therefore more properly referred to as slabs, rather than disks.

### 3.2.1. Dynamics of Gaseous Slabs

Although there have been several studies of flat gaseous systems, most of them have focused on the circumstances under which instabilities of Jeans type occur. Ledoux,[37] who analyzed the stability of a static slab of isothermal gas with the density profile (**19**), did also consider bending perturbations, and showed that there were no bending instabilities. Goldreich and Lynden-Bell[38] included bending perturbations in their study of the combined effects of uniform rotation and finite thickness on the stability of gaseous slabs. They were able to show that a polytropic slab with $\gamma = 2$, and an incompressible slab with a top-hat density profile, as well as an isothermal slab, are stable against bending. For the incompressible slab they obtained full dispersion relations. That for bending waves is, in our notation,

$$\tilde{\omega}^2 = 2\pi G \Sigma k \mathscr{F}_{TH}(kH) \left[ \frac{nkH}{\tanh nkH} \right], \tag{21}$$

where $n = \tilde{\omega}/\sqrt{\tilde{\omega}^2 - 4\Omega^2}$ and $\tilde{\omega}$ is the frequency as measured in a frame that follows the rotation. This shifted frequency $\tilde{\omega}$ is akin to our $(\omega - m\Omega)$, and the dispersion relation (**21**) can be contrasted with the dispersion relation (**8**) for a razor-thin disk. Relation (**21**) has no $\nu$ term only because Goldreich and Lynden-Bell did not include any background gravitational field in the $z$-direction. It also contains the reduction factor $\mathscr{F}_{TH}(kH)$ of (**17**) by which gravitational restoring forces for the top-hat density profile are less than those for a razor-thin disk. However, the fluid nature of the slab, in this case an incompressible fluid with a sharp boundary on which there are surface waves, contributes an extra factor $(nkH/\tanh nkH)$ to the dispersion relation (**21**). This extra factor grows as $kH$ for $kH$ large and exactly counters the decay of the reduction factor $\mathscr{F}$, and adds to the stability. In fact, the dispersion relation (**21**) becomes $\tilde{\omega}^2 = ngk$ in the limit of large $kH$, where $g = 2\pi G\Sigma$ is the acceleration due to gravity felt at the surface. This is exactly the dispersion relation for surface waves on a deep rotating ocean. Although Fridman and Polyachenko[39] find a similar result for the nonbending $z$-symmetric oscillations of an incompressible layer, it is unlikely that this limit has much relevance in the present context.

### 3.2.2. Dynamics of Stellar Slabs

In view of the incompleteness of our present understanding of the dynamics of the bending of gaseous disks, it is scarcely surprising that the dynamics of the bending of stellar disks is also incompletely understood. The stellar dynamic work that is most relevant in the present context is that of Toomre[24] and Araki.[40] They discussed the stability of a stellar slab, stratified only in the $z$-direction as in Section 3.1, and with a distribution function that is Maxwellian in both $x$ and $z$, but with different velocity dispersions $\sigma_x$ and $\sigma_z$, respectively. They found that bending instabilities are suppressed provided that the ratio $\sigma_z/\sigma_x$ of the velocity dispersions exceeds 0.293. Their neutral stability curve has the form of a sheared hump in the $(\lambda/\lambda_J, \sigma_z/\sigma_x)$-plane, and is plotted in FIGURE 6 of Merritt and Sellwood.[35] Here $\lambda = 2\pi/k$ is the radial wavelength, and $\lambda_J = \sigma_x^2/G\Sigma$ is the Jeans length for instabilities in the plane. A band of unstable wavelengths, intermediate between short and long and lying within the hump, occurs for values of $\sigma_z/\sigma_x$ less than 0.293. A remarkably similar neutral stability curve is obtained from Polyachenko and Shukhman's[39,41] analysis of bending waves in a uniformly dense stellar slab of finite thickness, and with a Maxwellian distribution of $x$-velocities. Polyachenko and Shukhman's neutral stability is a somewhat inflated version of Toomre and Araki's sheared hump, and peaks at $\sigma_z/\sigma_x = 0.370$, rather than 0.293. The difference between the two minimum values of $\sigma_z/\sigma_x$ needed to avoid instability is relatively modest in view of the extreme differences between the $z$-components of the distribution functions in the two cases, one a decaying exponential in the partial energy $E_z$ and the other with a singular $(E_0 - E_z)^{-1/2}$ dependence, where $E_0$ is the partial energy of the most energetic particles that are marginally confined within the slab. In both cases, the results are consistent with the instability at short wavelengths ($\lambda < \lambda_J$) of razor-thin disks with random motions in the plane of the disk only, for which $\sigma_z/\sigma_x = 0$. They are also consistent with the stability of the gaseous disks discussed in the previous section that, to some extent, approximate the $\sigma_z/\sigma_x = 1$ case of isotropic velocity dispersions, which is is well clear of the unstable regions. The longer wavelength branches of the two neutral stability curves both bend initially to the right with increasing $\sigma_z/\sigma_x$, indicating that a small amount of $z$-velocity dispersion is initially destabilizing. This feature is barely present in Merritt and Sellwood's approximation [their eq. (26)] to this curve, which is also shown in their FIGURE 6. The form $(\omega - m\Omega)^2 = 2\pi G\Sigma k + (\sigma_z^2 - \sigma_x^2)k^2 = (4\pi^2\sigma_x^2/\lambda^2)(\sigma_z^2/\sigma_z^2 - 1 + \lambda/\lambda_J)$ proposed by Bertin and Mark[18] for the dispersion relation for bending waves in a stellar disk [their eq. (D.1)] does not come close to giving even a qualitatively correct neutral stability curve.

The dynamics of even one-dimensional $z$-motions of a collision stellar slab is complicated. Weinberg[42] found that such a slab can undergo purely oscillatory and undamped modes. These modes have frequencies in the gaps between the bands $[2n\pi/T_{max}, 2n\pi/T_{min}]$, formed by the orbital frequencies ($n = 1$) and their harmonics ($n > 1$), where $T_{max}$ and $T_{min}$ are the longest and shortest orbital periods. What is of greatest interest to the present discussion is that Weinberg also considered bending disturbances with wavenumber $k$ in the $x$-direction perpendicular to the stratification, also assuming a Maxwellian stellar distribution function in $x$-velocities. He found such bending disturbances to be damped, and that the damping is strong except in the case of the fundamental *sloshing* mode, for which it is much weaker. The fundamental sloshing mode, in which the slab moves up and down as a whole, is precisely the one that is of most interest in the present context.

Louis[43] was able to reproduce some of Weinberg's low-order one-dimensional modes using a simple truncated systems of moment equations. He found that discrete oscillation modes could be tracked into the continuum bands, where they are damped, initially only slightly. He was also able to identify the modes in $N$-body simulations. He did not study bending modes, remarking only that "the lowest order sloshing mode is somewhat boring . . . unless one takes into account the horizontal structure of the galactic disk which then leads to the study of bending modes." Of course it is precisely that structure that makes the sloshing mode, rather than higher-order modes with more complex z-structure, the most interesting of all and not the least bit boring. The most important implication of Louis's work is that simple truncated systems of moment equations may be capable of giving good descriptions of the bending oscillations of stellar disks, and allow one to avoid dealing directly with the collisionless Boltzmann equation.

## 4. DISCUSSION

The understanding of the dynamics of the bending of flattened stellar systems is significantly incomplete. It is most complete for the razor-thin disks discussed in Section 2. However, the severe instabilities predicted by Polyachenko and Shukhman's[23] membrane equation (10) when there are random velocities in the plane, as there surely are in flat stellar systems, shows that that model can have only a limited range of validity. Nature, and the $N$-body simulations of Sellwood and Merritt,[36] suggest that it is not difficult to build rather flat rotating stellar systems that are free of bending instabilities. Presumably the consequence of the bending instability that results if a system becomes too flat is that it thickens. The work of Toomre,[24] Polyachenko and Shukhman,[41] and Araki[40] predicts that a relatively low level of random velocities perpendicular to the plane of the disk is sufficient to achieve stability. Since gravity is too weak to cure bending instabilities of a thin disk in the presence of random motions, and since gravity's restoring force weakens as the disk thickens, it must be inertial effects that induce the stability. Merritt and Sellwood[35] suggest that the most important mechanism by which a finite thickening succeeds in quenching bending instabilities, is the out-of-phase response of stars when forced at a frequency greater than their natural frequency. Although promising, a fully convincing dynamical demonstration of their suggestion is lacking, and their application of it illustrated in their Figure 6, which also incorporates the reduction factor $\mathscr{F}$ of equation (20), is not quite right.

Another important suggestion that Merritt and Sellwood make is that inhomogeneities play a major role, and that infinite slab analyses, on which all of the discussion of Section 3 is based, are deceptive in overestimating the stabilizing effect of both self-gravity and z-velocity dispersion. They found that bending modes in inhomogeneous disks can remain unstable at very long wavelengths. In $N$-body experiments, they found instability to bending even when the velocity anisotropy is much less extreme than the critical value for instability in an infinite slab. These phenomena have yet to be explained, and deserve an explanation.

Finally, there remains the question of whether it is possible to construct a simplified model of a flattened stellar system that includes the major phenomena of ordered and random motions, small but finite thickness, and inhomogeneities. Louis[43] showed, in a simpler context, that it is possible to construct relatively simple, yet useful, models of low-order modes of oscillation of stellar slabs using simple truncated systems of moment equations, and our interest here is also in modeling low-order modes. If it isn't, then this would appear to be another field that theoreticians may have to abandon to the $N$-body computationalists.

## APPENDIX

### A.1. The Finite Part Integral for the Gravitational Force Due to a Dipole Array

The gravitational potential at a general point of space arising from a distribution of dipoles of strength $\mu(x, y)$ distributed on a region $S$ of the $(x, y)$-plane is given by

$$\Phi_d(x, y, z) = -G \iint_{S'} \frac{z\mu(x', y')\, dx'\, dy'}{[(x - x')^2 + (y - y')^2 + z^2]^{3/2}}. \qquad (22)$$

This potential, which is odd in $z$, is discontinuous at the plane in which the dipoles lie, where it has the one-sided limits[44]

$$\lim_{z \to 0^+} \Phi_d(x, y, z) = -2\pi G\mu(x, y), \qquad \lim_{z \to 0^-} \Phi_d(x, y, z) = +2\pi G\mu(x, y). \qquad (23)$$

The $z$-derivative of this potential, which we need for the upward force exerted by the dipoles, must be evaluated with some care. It is an even function of $z$ and its two one-sided limits $z \to \pm 0$ are identical. However, the direct differentiation of (22), with its discontinuity, gives rise to a delta-function component.

To isolate this singular behavior, we first split (22) into the three parts:

$$\begin{aligned}\Phi_d(x, y, z) = & -G \iint_{S'-S_\delta} \frac{z\mu(x', y')\, dx'\, dy'}{[(x - x')^2 + (y - y')^2 + z^2]^{3/2}} \\ & - G \iint_{S_\delta} \frac{z[\mu(x', y') - \mu(x, y)]\, dx'\, dy'}{[(x - x')^2 + (y - y')^2 + z^2]^{3/2}} \\ & - G \iint_{S_\delta} \frac{z\mu(x, y)\, dx'\, dy'}{[(x - x')^2 + (y - y')^2 + z^2]^{3/2}}, \end{aligned} \qquad (24)$$

where $S_\delta$ is a small circle in the $(x, y)$-plane, center $(x, y, 0)$, and radius $\delta$. The third integral is elementary and can be evaluated as

$$2\pi G\mu(x, y)\left\{\frac{z}{(\delta^2 + z^2)^{1/2}} - \mathrm{sgn}\,(z)\right\}. \qquad (25)$$

This is the component that gives rise to the jump discontinuity (23). We now differentiate with respect to $z$, evaluate at $z = \pm 0$ so as to avoid the delta function, and take the limit $\delta \to 0$, which makes the second component of (24) negligible provided that the dipole density $\mu$ is sufficiently smooth, for example, twice continuously differentiable.[44] We then obtain the upward force exerted by the dipole distribution as

$$F_z = -\left.\frac{\partial \Phi_d}{\partial z}\right|_{z=\pm 0}$$

$$= \lim_{\delta \to 0}\left\{G \iint_{S'-S_\delta} \frac{\mu(x', y')\,dx'\,dy'}{[(x-x')^2 + (y-y')^2]^{3/2}} - \frac{2\pi G\mu(x, y)}{\delta}\right\}. \quad (26)$$

The right-hand side here defines the finite part integral that is used in (3) of the text. It is this finite part integral that is computed by the expansion methods described in sect. II(d) of Hunter and Toomre,[20] or the Fourier–Bessel methods of Sparke and Casertano.[16]

Equation (4) is a consequence of the fact that the dipole potential (22), with $\mu$ replaced by $\Sigma$, is the z-derivative of the simple source potential

$$\Phi_s(x, y, z) = -G \iint_{S'} \frac{\Sigma(x', y')\,dz'\,dy'}{\sqrt{(x-x')^2 + (y-y')^2 + z^2}}. \quad (27)$$

It follows from our earlier discussion that the $v_{\text{disk}}^2$ of (3) is simply $\partial^2\Phi_s/\partial z^2|_{z=\pm 0}$. Because the potential $\Phi_s$ satisfies Laplace's equation away from the disk, it follows that

$$v_{\text{disk}}^2 = -\left(\frac{\partial^2}{\partial x^2} + \frac{\partial^2}{\partial y^2}\right)\Phi_s(x, y, 0). \quad (28)$$

Equation (4) follows when the derivatives of $\Phi_s$ are related to the circular velocity via the equation $\Omega^2(R) = \partial\Phi_s/R\,\partial R$.

### A.2. The Reduction Factor for a Disk of Finite Thickness

The small wavy displacement considered in Section 3.1 induces a density field $\rho_0(z - h_0 e^{ikx})$, and hence a density perturbation $-h_0 e^{ikx}\rho_0'(z)$ and, by Poisson's equation, a gravitational potential $\Phi_1(z)e^{ikx}$, such that

$$\Phi_1''(z) - k^2\Phi_1(z) = -4\pi Gh_0\,\rho_0'(z). \quad (29)$$

The relevant solution of this equation is

$$\Phi_1(z) = (2\pi Gh_0/k)\int_{-\infty}^{\infty} \rho_0'(s)e^{-k|z-s|}\,ds, \quad (30)$$

which gives

$$\Phi_1'(z) = 2\pi Gh_0\left\{-2\rho_0(z) + k\int_{-\infty}^{\infty} \rho_0(s)e^{-k|z-s|}\,ds\right\}. \quad (31)$$

The total $z$-force per unit length that acts on the surface density is

$$\Sigma F_z = \int_{-\infty}^{\infty} \rho_0(z - h_0 e^{ikx})[-\Phi_0'(z) - \Phi_1'(z)e^{ikx_0}] \, dz,$$

$$\simeq e^{ikx_0} \int_{-\infty}^{\infty} [h_0 \Phi_0'(z)\rho_0'(z) - \Phi_1'(z)\rho_0(z)] \, dz, \qquad (32)$$

where $\Phi_0(z)$ is the gravitational potential of the perturbed state. Integration by parts of the $\rho_0'(z)$ of the first component of (32), followed by the use of Poisson's equation for $\Phi_0''(z)$ and equation (31) for $\Phi_1'(z)$, then gives equation (14).

## ACKNOWLEDGMENTS

I am grateful to Jeremy Goodman, the referee, for a careful reading of my original manuscript, and many helpful suggestions for its improvement.

## REFERENCES

1. BURKE, B. F. 1957. Astron. J. **62**: 90.
2. KERR, F. J. 1957. Astron. J. **62**: 93.
3. SANCISI, R. 1976. Astron. Astrophys. **53**: 159–161.
4. BOSMA, A. 1978. Ph.D. thesis, Groningen Univ., Groningen, The Netherlands.
5. BRIGGS, F. 1990. Astrophys. J. **352**: 15–29.
6. INNANEN, K. A., K. W. KAMPER, K. A. PAPP & S. VAN DEN BERGH. 1982. Astrophys J. **254**: 515–516.
7. VAN DER KRUIT, P. C. & L. SEARLE. 1981. Astron. Astrophys. **95**: 105–115.
8. VAN DER KRUIT, P. C. & L. SEARLE. 1982. Astron. Astrophys. **110**: 61–78.
9. VAN DER KRUIT, P. C. & L. SEARLE. 1982. Astron. Astrophys. **110**: 79–94.
10. SÁNCHEZ-SAAVEDRA, M. L., E. BATTANER & E. FLORIDO. 1990. Mon. Not. R. Astron. Soc. **246**: 458–462.
11. KAHN, F. D. & L. WOLTJER. 1959. Astrophys. J. **130**: 705–717.
12. OSTRIKER, E. C. & J. J. BINNEY. 1989. Mon. Not. R. Astron. Soc. **237**: 785–798.
13. LYNDEN-BELL, D. 1965. Mon. Not. R. Astron. Soc. **129**: 299–307.
14. TUBBS, A. D. & R. H. SANDERS. 1979. Astrophys. J. **230**: 736–741.
15. SPARKE, L. S. 1984. Astrophys. J. **280**: 117–125.
16. SPARKE, L. S. & S. CASERTANO. 1988. Mon. Not. R. Astron. Soc. **234**: 873–898.
17. BINNEY, J. 1990. *In* Dynamics and Interactions of Galaxies, R. Weilen, Ed: 328–337. Springer-Verlag. Berlin.
18. BERTIN, G. & J. W.-K. MARK. 1980. Astron. Astrophys. **88**: 289–297.
19. NELSON, R. W. & S. TREMAINE. 1995. Mon. Not. R. Astron. Soc. **275**: 897–920.
20. HUNTER, C. & A. TOOMRE. 1969. Astrophys. J. **155**: 747–776.
21. BINNEY, J. & S. TREMAINE. 1987. Galactic Dynamics, Chapt. 6. Princeton Univ. Press. Princeton, N.J.
22. HUNTER, C. 1969. Stud. Appl. Math. **48**: 55–76.
23. POLYACHENKO, V. L. & I. G. SHUKHMAN. 1979. Astron. Zh. **56**: 724–731. (Transl. in Sov. Astron. **23**: 407–411.)
24. TOOMRE, A. 1966. Geophysical Fluid Dynamics, Notes on the 1966 Summer Study Program at the Woods Hole Oceanographic Institution, Ref. No. 66-46, 111.
25. RUBIN, V. C., J. A. GRAHAM & J. D. P. KENNEY. 1992. Astrophys. J. **394**: L9–L12.

26. POLYACHENKO, V. L. 1977. Pis'ma Astron. Zh. **3**: 99–103. (Transl. in Sov. Astron. Lett. **3**: 51–53.)
27. KALNAJS, A. J. 1972. Astrophys. J. **175**: 63–76.
28. FRIDMAN, A. M. & V. L. POLYACHENKO. 1984. Physics of Gravitating Systems, Vol. 2, Chapt. X. Springer-Verlag. New York.
29. MERRITT, D. & L. HERNQUIST. 1991. Astrophys. J. **376**: 439–457.
30. VANDERVOORT, P. O. 1991. Astrophys. J. **377**: 49–71.
31. MALKOV, E. A. 1989. Astron. Zh. **66**: 1189–1197. (Transl. in Sov. Astron. **33**: 614–618.)
32. MALKOV, E. A., T. N. NUZHNOVA & B. S. SAGINTAEV. 1991. Pis'ma Astron. Zh. **17**: 469–473. (Transl. in Sov. Astron. Lett. **17**: 200–202.)
33. FREEMAN, K. C. 1966. Mon. Not. R. Astron. Soc. **134**: 15–23.
34. TOOMRE, A. 1964. Astrophys. J. **139**: 1217–1238.
35. MERRITT, D. & J. A. SELLWOOD. 1994. Astrophys. J. **425**: 551–567.
36. SELLWOOD, J. A. & D. MERRITT. 1994. Astrophys. J. **425**: 530–550.
37. LEDOUX, P. 1951. Ann. Astrophys. **14**: 438–447.
38. GOLDREICH, P. & D. LYNDEN-BELL. 1965. Mon. Not. R. Astron. Soc. **130**: 97–124.
39. FRIDMAN, A. M. & V. L. POLYACHENKO. 1984. Physics of Gravitating Systems, Vol. 1, Chapt. I. Springer-Verlag. New York.
40. ARAKI, S. 1985. Ph.D. thesis, MIT, Cambridge, Mass.
41. POLYACHENKO, V. L. & I. G. SHUKHMAN. 1977. Pis'ma Astron. Zh. **3**: 254–257. (Transl. in Sov. Astron. Lett. **3**: 134–136.)
42. WEINBERG, M. D. 1991. Astrophys. J. **368**: 66–78.
43. LOUIS, P. D. 1992. Mon. Not. R. Astron. Soc. **258**: 552–570.
44. COURANT, R. & D. HILBERT. 1962. Methods of Mathematical Physics, Vol. 2, Chapt. IV. Wiley Interscience. New York.

# On Global Wave Patterns in Galaxies: Their Generation and Maintenance[a]

C. C. LIN[b] and G. BERTIN[c]

[b]Massachusetts Institute of Technology
Cambridge, MA 02139
and
Florida State University
Tallahassee, FL 32306

[c]Scuola Normale Superiore
I-56126 Pisa, Italy

## 1. INTRODUCTION

The objective of this paper is to discuss the formation and maintenance of global structures observed in galactic disks in normal (nonbarred) spirals with emphasis on the generation of density waves. Both internal dynamical mechanisms and interaction with external galaxies are considered. Such a discussion contains four important aspects: (1) the discription of the physical process of evolution that a galactic disk undergoes from the time of its formation; (2) the study of the relevant dynamical mechanisms; (3) the examination of the relevant physical contexts in which the global structure is formed; and (4) the examination of the likelihood and the degree of participation of the various dynamical mechanisms in the process of initiation of density waves and the formation and maintenance of the observed global wave patterns.

After studying all possible types of dynamical scenarios for the generation of global structures of density waves, we aim at identifying, if possible, an appropriate approach applicable to the *statistical majority* of galaxies that are observed to have regular global structures. This is a practical goal, for the result, as we shall see, would be useful for the detailed analysis of observational data and therefore to check the theoretical predictions against observational data.

Before we go into the detailed discussions, let us point at the difference between the physical perspective with emphasis on observational data, and the dynamical perspective with emphasis on mechanisms. For example, in the mathematical models considered for the study of mechanisms, one usually adopts an axially symmetric basic state of equilibrium as the initial condition. But did a galaxy ever exist in such a state as it evolves from its formation during primordial times? Did density waves begin to propagate, though not necessarily as a global pattern, over a galactic disk as soon as the disk took shape? The chaotic conditions in a gas-dominated protogalaxy would certainly provide sufficient disturbances for the initiation of such

---

[a]Invited paper at the Tenth Florida Workshop in Nonlinear Astronomy, September 22–24, 1994.

waves. But the formation of global spiral structures may come at a later stage of development when the stellar component and the gaseous component reach a proper balance and distribution.

Alternatively, copious formation of stars beginning in the early galaxy may lead to a galactic disk in which the older stellar population with large velocity dispersion would dominate the dynamics and keep the disk globally stable, featureless, and approximately axisymmetric. A global wave pattern cannot appear unless the disk is externally excited. But would it be easy to excite such a stable galactic disk into the formation of a long-lasting global pattern? Indeed, would it be easy to excite it into any significant global pattern at all? Clearly, we are now dealing with a physical situation different from that just discussed, and there may be galaxies belonging to either category.

There may therefore be considerable differences when one approaches the problem of generation from the observer's perspective with emphasis on the present conditions, or the dynamicist's perspective, with emphasis on past processes that lead to the present state. The failure to recognize different categories of galaxies—"apples and oranges"—has sometimes been the source of controversy among theorists. In the rest of this section, we attempt to clarify the nature of the issues that naturally arise.

### 1.1. The Observational Perspective

From the physical point of view, the study of the generation of density waves in one particular normal spiral is not likely to lead to unique answers as to what initiated the observed spiral structure, for we are seeking plausible *past* physical scenarios that are compatible with the *present* observational data. It is thus clear that one should keep an open mind and be prepared to accept a number of plausible *alternative* theoretical scenarios for comparison with observational data for individual objects. However, for the totality of all galaxies, there are clear *statistical* characteristics such as the well-known Hubble classification. It is therefore desirable that we should also be able to provide a theoretical framework of the same general nature.

The existence of apparently long-lasting global structures was emphasized by Oort.[1] Lindblad[2] suggested that such structures might be *quasi stationary*. Lin and Shu followed up this idea with the formulation of a quantitative working hypothesis of quasi-stationary spiral structure (QSSS hypothesis) and a study of its implications (see review by Lin[3]). The first challenging problem is thus how these structures might be sustained, perhaps as intrinsic characteristics for each galaxy. If so, we should understand the dynamical basis for these characteristics. A deeper issue on global structure of galaxies was also discussed by Oort[4] almost exactly 25 years ago. At the 38th symposium of the International Astronomical Union (IAU), he raised the following issue of *regularity*: "A more serious problem seems that of the long-term permanence of the spiral wave—can they continue to run around during 50 revolutions without fatal damage to their regularity?" In response to this issue, he stated that "in most cases, we cannot invoke interaction. Moreover, if there had been interaction in the past it might well have contributed to *disturb* the regular wave pattern, but less likely to rebuild it." He continued by pointing out that a statement on the existence of the global strrructure should be "supplemented by two essential additions. First, that in about half of the spirals, the structure is either unclear, or there are more than two arms. Second, that even in the half that can be classed among the two-armed spirals there are invariably important additional features *between* the two principal arms."

These two supplementary points were also stressed by Lin[3] in terms of the *coexistence* of all kinds of spiral features. Three-armed features and interarm bridges are now usually attributed to contributions from the gaseous component; the dominant contribution to the two-armed global spiral gravitational field is believed to be associated with the "older" evolved stars. This point receives direct observational support from recent infrared observations (see, e.g., Block *et al.*[5]), where the two-armed grand design stands out clearly despite the irregularities present in optical and radio observations.

## 1.2. The Dynamical Perspective

It is the obligation of the theorist to propose dynamical scenarios that would be compatible with a wide range of observational data. The prominence of grand designs is obviously one outstanding feature to be addressed.

In Oort's perception described earlier, as well as in the Lin–Shu hypothesis of quasi-stationary spiral structure, the global spiral structure is regarded as intrinsic to a galaxy, although it may be disturbed by interaction among galaxies. In contrast, Toomre and Toomre[6] suggested that the density wave pattern might have been recently generated (e.g., in the case of NGC 5194/5) through interaction among galaxies. But it is clear that these authors are emphasizing *different physical contexts.* The basic states considered are assumed to be stable (cf. Section 1). *A priori*, the formation of global spiral structures may be achieved through intrinsic mechanisms alone, or with the help of external forcing. The issue we should address is thus this: *Which are the physical contexts and dynamical scenarios that apply to the statistical majority of the observed global spiral structures?*

One can approach this problem with two different perspectives. In one approach, the emphasis is on the present. One begins with the observational data—in the present context, the apparent existence of regular global structures—and seek the dynamical basis for the *maintenance* of such modelike global structures. The search for the dynamic evolutionary scenario that would lead to such structures is given second priority. This is the *modal approach*. This approach has been well developed and successfully applied by Bertin, Lin, and their collaborators[7–12] for describing a wide range of astrophysical phenomena, qualitative and quantitative (see Section 3 below). An alternative approach focuses on evolution: various *evolutionary* scenarios are considered in order to understand the mechanisms and then one attempts to make use of this understanding to identify dynamical scenarios to be applied to specific physical situations. This approach is especially needed in the study of interacting galaxies. In this second approach, it is essential that all plausible alternative models relevant to the physical context are actually included in the theoretical scenarios studied. For it is now well known that, in a nonlinear dynamical system, two different dynamical scenarios could lead to quite similar configurations at one particular instant (see Section 3). In general, it is natural that both approaches should be kept in mind and explored quantitatively.

In either case, it is important to keep in mind that the global structure evolves in time. The question is whether the structure is (either in general or in individual cases) long lasting and *slowly* evolving or whether it is *fast* evolving for a limited period of time and then settles down to a slowly evolving state, or even decays away. (See Appendix A.1 for more detailed discussions.) For this would decide how the observational data should be analyzed. Indeed, in the observational context, it is of secondary importance whether the generation process depends on intrinsic mechanisms, external forcing, or a combination of both.

The generation and maintenance of global patterns with a high degree of regularity, will now be discussed for isolated galaxies and for interacting galaxies, with emphasis on the discussion of long-lived characteristics, and on different physical contexts.

## 2. THE MODAL AND NONMODAL APPROACHES

The observation of regular global structures in both barred and nonbarred field galaxies and other essentially isolated galaxies naturally suggests the modal approach; that is, we visualize the possibility for standing wave patterns or (nonlinear) natural global modes to be generated through *intrinsic* gravitational instability, which is most likely to be associated with the gaseous component in the gas-rich *outer* galactic disk. The detailed dynamical mechanisms for the formation of such natural modes have been quite fully worked out and described.[7,8,12]

Selective amplification of the modes would lead to the limitation of important modes to a small number, often as small as one or two. An immediate requirement is that the theoretical global patterns must resemble that *observed*. A typical scenario of a long-lasting but slowly evolving global pattern is shown in FIGURE 1, reproduced from a paper by Lin and Lowe.[11]

It is to be emphasized that the possibility for the support of long-lasting global structures depends critically on *the nature of the basic state*. Some basic states do not support long-lasting global structures, but may support fairly fast-evolving spiral structures on a large scale as a response to a suitable external excitation; these will eventually decay away. Statistically, it is obvious that the long-lasting global structures would be more frequently observed. Examples of three types of basic states, with different modal characteristics, have been described.[11] These points are discussed further in the following sections (3 and 4). It is argued that the statistical majority of the global spiral structures observed are slowly evolving and long-lasting, however they were generated.

The *implications* of the modal theory of spiral structure has been well worked out in recent years.[12] In general terms, the modal approach has succeeded in providing a viable framework for interpreting the variety of observed morphologies for spiral galaxies, including grand-design normal spirals, bars, multiarmed galaxies, and flocculent galaxies. Such a variety of morphologies is traced back to the variety of basic states that may involve heavy disks, light disks, self-regulated disks of stars

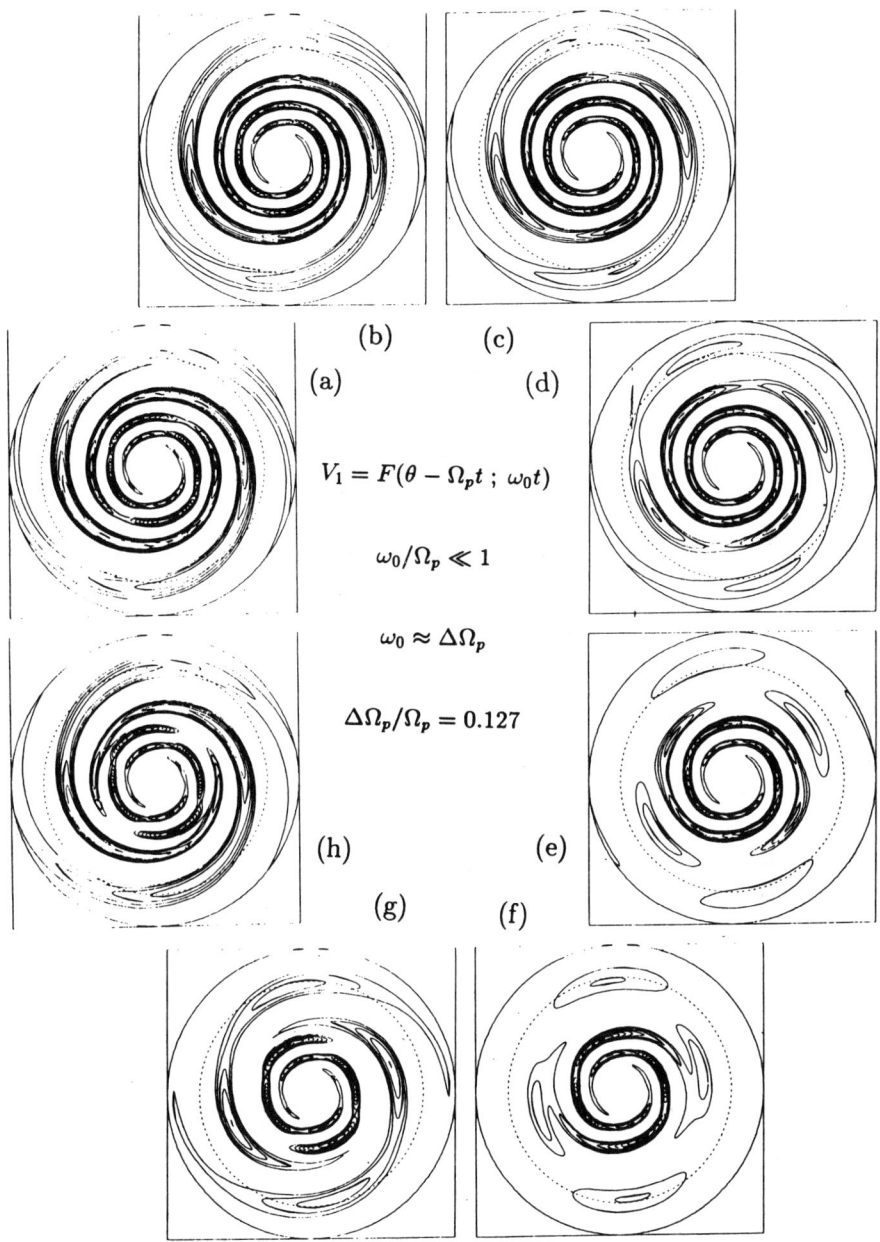

**FIGURE 1.** A simple example illustrating the concepts of quasi-stationary and quasi-periodic evolutions. For a basic state with two important modes, the successive frames, (**a**) to (**h**), show long-lasting slowly evolving global structures with *regularity*. There are changes of shape over several periods of revolution. Possible changes of basic state over the Hubble time scale are not considered here. The two modes are assigned equal amplitudes for simplicity of calculation. Actually, the *lower mode* has the higher growth rate and is expected to dominate during the long evolutionary process from the protogalaxy to early galaxy and then to the present stage. The degree of regularity would be expected to be greater than shown here. (After Lin and Lowe.[11])

and gas, or disks where the hot stellar disk is essentially decoupled from the gas and remains stable. However, as just mentioned, if we consider all the "typical grand design" normal (nonbarred) spirals, or if we focus our attention on specific systems (such as M51 and M81), where signs of interaction are clearly present, we should include the consideration of various options involving external excitation in an *ad hoc* manner. Examples of such interacting galaxies are discussed in Section 3.

The modal approach is clearly not suitable for describing fast evolutionary scenarios that can be generated by the close encounter of a galaxy with its satellite. Here a different dynamical perspective with emphasis on the evolutionary process is desired. Here we also have to face the issue of the nature of the basic state (which should be capable of supporting the wave patterns), and furthermore, we have to add another dimension to the complexity of the problem beyond that of the stability or instability of the basic state: the nature of the external forcing and its consequent *effectiveness* in the generation of global wave patterns. There is the contrast between persistent forcing (e.g., when a satellite is bound) and one-time forcing (e.g., when the satellite passes by in close encounter in a hyperbolic orbit). There is also the contrast between orbits in the plane of the galactic disk and overhead orbits in a plane at right angles. Finally, there is obviously the very important factor of the magnitude of tidal forcing, or the closeness of the satellite to the main galaxy. In one extreme, the interaction could end up as a merger of the two galaxies; in the other extreme, we have the case of essentially isolated galaxies discussed earlier, or the case of a limited "disturbance," as mentioned by Oort.

In view of this diversity of possible scenarios, we examine (in Section 3) two examples that have been studied in some detail: the case of NGC 3031/3077 (M81 system) and NGC 5194/5 (M51 system) in order to get a feel for the complexity of the issues. Some discussion of the generally relevant dynamical mechanisms may be found in the Appendix.

In Section 4, we address the problem mentioned in Section 1.2, namely: Which scenarios are most likely to be applicable to the majority of observed global spiral structures? This is bound to be somewhat speculative since we do not have, as yet, a systematic quantitative study of all types of dynamical scenarios of interacting galaxies in the context for generation of global wave patterns. Indeed, the perception visualized by Oort of the disturbance of a preexisting spiral structure has not yet been studied in detail, and a recent review[13] on interacting galaxies dealt mostly with peculiar galaxies, with rather limited comment on the topics of this paper. In order to focus our attention on *observed* global structures, we emphasize empirical confirmation of theoretical predictions (Secs. 4 and 5). In this empirical perception, we shall see that the essential issue is maintenance rather than the process of generation.

## 3. TWO CASE STUDIES OF GLOBAL STRUCTURES

As mentioned earlier, generation of density waves and formation of global wave patterns over a galactic disk involve quite complex mechanisms, both intrinsic and external. We, therefore, first consider two well-known specific cases that have been studied in great detail both theoretically and observationally. These are the cases of the M51 system (NGC 5194/5) and the M81 system (NGC 3031/3077).

We wish to make it clear that our discussions below usually refer to the behavior of the physical system. The $N$-body simulation may differ from the physical system in some aspects. Specifically, one should note that a *featureless initial condition* is usually assumed in those simulations, and this may be unrealistic for many galaxies, as noted before. There are also intrinsic difficulties of "faithfulness" with the modeling process.[14] It is unclear, for example, whether a single "softening parameter" could embody all the complications in different parts of the galactic disk.

### 3.1. The M81 System

The case of M81 has been studied in detail in both perceptions. The tidal forcing scenario, starting with a featureless initial condition, was carried out with a hyperbolic orbit by Thomasson and Donner,[15] who decided on a plausible path of the orbiting satellite through studies of the distribution and motion of neutral hydrogen. The general features of the inner spiral structure was successfully reproduced, at a certain instant in the evolutionary process, but the observed amplitude *modulation* along the spiral arm was not. This feature was reproduced together with the overall spiral pattern, in the model approach carried out by Lowe et al.,[9] showing that a self-sustained spiral pattern is viable (FIGS. 2 and 3). This difference is discussed further below.

Another independent support for the modal approach comes from the data on the distribution and motion of neutral hydrogen observed by using the Very Large Array (VLA) telescope system and analyzed by Westpfahl.[16] He found,[d] on the basis

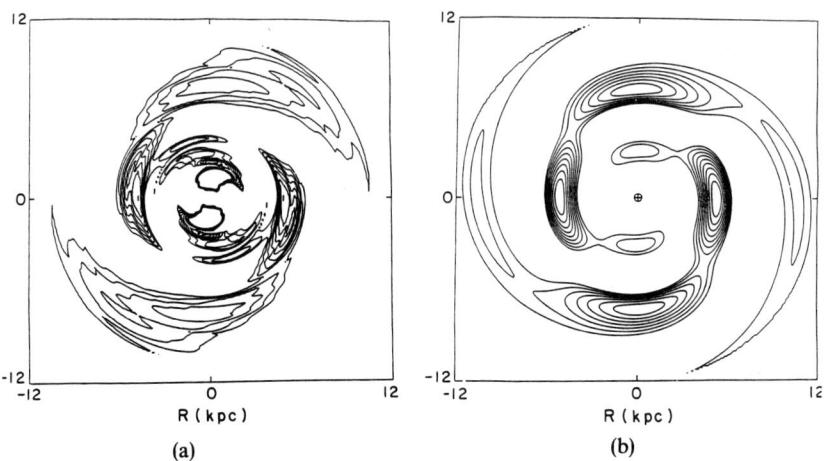

**FIGURE 2.** Density distribution the galaxy M81. (**a**) *Observed*: The two-armed Fourier component of the data given by Elmgreen et al. (**b**) *Theoretical*: Amplitude distribution of the modal pattern computed by Lowe et al.

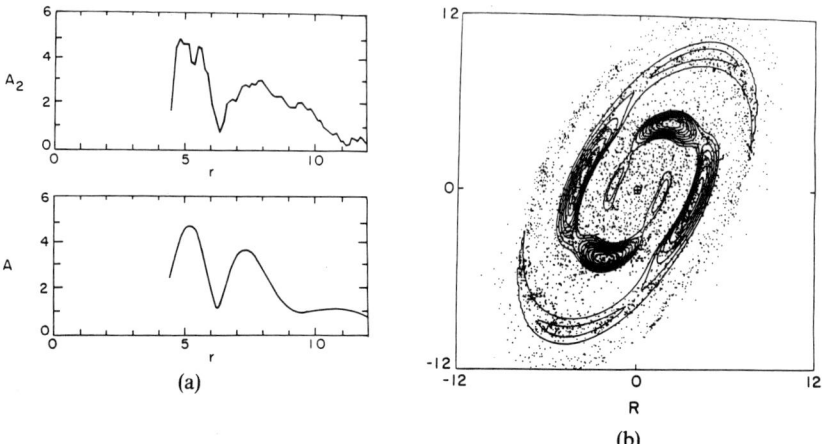

**FIGURE 3.** The galaxy M81 (after Lowe et al.[9]), further comparison with observational data. (**a**) Comparison of amplitude distribution along a radial direction from observational data (*upper panel*) and from theoretical prediction (*lower panel*). Note the location of the dip in amplitude slightly with $r = 5$ kpc. (**b**) Distribution of the interstellar medium compared with the gravitational potential in the theoretical model. Note the concentration of the interstellar medium around the minimum of the gravitational potential.

of the QSSS hypothesis, that the data are compatible with a pattern rotating at approximately the same angular velocity as that predicted by Lowe *et al.*

We note that in this case, a fairly long time elapsed between the instant of closest encounter and the present state so that transient effects might have subsided, except for some details. We can therefore conceive the following scenario. Before the passage of the satellite galaxy NGC 3077, the galaxy NGC 3031 had an existing spiral structure in a dynamically preferred state that may be approximately represented by the mode calculated. After the passage of the satellite, the outer spiral arms are slightly though noticeably more open than those expected from the mode calculated, suggesting that indeed there is still some residual effect from tidal interaction. The spiral structure in NGC 3031 (observed at the present) might have returned approximately to the *same* natural preferred state of a modelike global structure, even though the satellite NGC 3077 did disturb it during its passage. Note that even the magnitude of the amplitude distribution in the final state might have returned to that determined by intrinsic mechanisms; only the orientation might have changed.

We now go on to discuss a rather subtle point: the lack of success for the nonmodal approach to produce the observed amplitude modulation. This suggests that there might be as yet insufficient time for the waves refracted from the central regions to reach an adequate magnitude. Thus one would be led to adopt Oort's point of view that external excitation would mainly disturb a *preexisting* spiral

---

[d]In effect, Westpfahl was simply applying the original working hypothesis of quasi-stationary spiral structure, even though he referred to a procedure developed for a barred spiral; but M81 is clearly nonbarred.

pattern. One should perhaps replace the initial condition of a featureless disk with one carrying a spiral structure of the kind calculated. This is not as simple as it appears, for there is a general issue, as yet not fully unresolved, concerning a subtle point of $N$-particle simulation; that is, how to construct an $N$-body simulation of long-lasting global spiral structures.

Thomasson and Donner did test the initial *stability* of the disk (private communication) before the calculation of interaction began, claiming global stability, that is, absence of a preexisting global structure, even though the nominal value for Toomre's $Q$-parameter has the value of unity (cf. Romeo[14] regarding the criterion for stability). Note also that their $N$-body model does not include a gaseous component. Some further work would thus be desirable to accommodate the possible presence of a preexisting spiral pattern.

Note that, in the $N$-body simulation of M81, the trailing wave pattern is still wrapping up tighter and tighter during the late stages of evolution. This is, however, not to be identified with swing *amplification*, which is a proces limited in space and in time (cf. third frame of [17, Fig. 8], also reproduced as [11, Fig. 2]) and where the amplification (of density or wave action) only occurs inthe early leading phase.

### 3.2. The M51 System

Tully[18] showed, a long time ago, that the spiral pattern in NGC 5194 can be described by assuming a standing wave pattern rotating at a fixed angular speed, according to the scenario described in the Lin-Shu hypothesis of QSSS. Indeed, that scenario receives further support from the studies of synchrotron radiation observed by Mathewson *et al.*[19] (see Roberts and Yuan[20] for the theoretical studies).

The strong outer arm (pointing toward NGC 5195) obviously does not fit into a smooth global pattern and is almost certainly related to a recent passage of the smaller galaxy NGC 5195. This point was addressed by Toomre and Toomre,[6] by Hernquist,[21] and by Salo and Byrd.[22] (See also Byrd's paper at this conference.) While this outer spiral arm can be reproduced in all the studies cited, the details of the inner spiral structure was reasonably reproduced only in the recent work of Salo and Byrd.

In the simulation by Salo and Byrd, the forcing is periodic (at least approximately) and the initial condition of the disk is taken to be featureless rather than a preexisting global structure, as discussed in the case of the M81 system. In this case, we have definitive observational data of the mass distribution in infrared. Rix and Rieki[23] mapped the stellar mass distribution in M51. They found that "the spiral arm amplitudes clearly show smooth, strong radial variations ... These variations may arise from interference of a pre-existing spiral pattern with the tidally induced spiral arms." Indeed, if the preexisting spiral pattern were essentially that maintained by internal mechanisms, the following scenario appears plausible. The intrinsic spiral pattern could, by itself, supply the primary contribution to the observed smooth, strong radial variations (cf. the case of M81). If so, the primary role of tidal forcing would be the shaping of the outer spiral arm, a modification of a preexisting spiral structure, in the manner suggested by Oort (Section 1.1). The calculation of a highly nonlinear mode for the case of M51 (beginning with a linear approximation) would be very useful.

### 3.3. Some Lessons Learned

One lesson we learned from the preceding case studies is the uncertainty of "predicting the past," that is, it is difficult to determine which past dynamical processes led to the present observed state. We must therefore make a special effort to examine all the *signatures* of past scenarios, such as the amplitude modulation observed in both cases. Another lesson we learn is that for the global mode to be well settled, the feedback cycle[7,8] for the maintenance of the wave pattern must be sufficiently *short* so that it becomes comparable to the period of revolution of the wave pattern, since the signal propagation in the radial and angular directions must be at comparable rates to form a coherent pattern. In the case of M81, this can be easily checked, for the group velocity of wave propagation in the radial direction is on the order of the radial velocity dispersion that is available from the model constructed. Note that the implication is that we must have a *two-component* system to form a global pattern through the intrinsic mechanism of gravitational instability. In a galaxy recently formed from a gas-dominated protogalaxy, even if there is a galactic disk, it is *unlikely* that a global spiral pattern can be formed through intrinsic mechanisms alone, since the speed of wave propagation would be too slow to complete an effective feedback cycle.

In the context of the Lin–Shu hypothesis of QSSS, the spiral structure observed is assumed to be dynamically maintained in a state preferred by the (angle-averaged) basic state of the galactic disk; the generation mechanism may be purely internal, but contribution from external forcing is not excluded. This existence of a preferred state could be in the nature of an *attractor*, as is known to occur in many nonlinear dynamical systems with dissipative mechanisms.[24] It finds analogy in the ringing of church bells and the playing of other musical instruments, where a pure mode is approached by a process of selective amplification and decay (cf. Chandrasekhar[e]). In contrast to these familiar systems characterized by damped modes, however, realistic galaxy disks have been shown to be able to support long-lasting or self-excited global modes. The properties of global modes in stable galaxy disks are not well known. This point is discussed further below and in Appendix A.1.

## 4. FREQUENCY OF OCCURRENCE

We now turn to a discussion of the main problems posed in Section 1.2: the frequency of occurrence of the various scenarios. Since there is, as yet, no systematic study of interacting galaxies, in contrast to the case of essentially isolated galaxies, the present discussion begins with the latter. An examination of three broad prototypes of dynamical scenarios follows.

---

[e]As quoted by K. C. Woli.[25] The topic there is the quasi-normal modes of black holes.

### 4.1. Essentially Isolated Galaxies

We first discuss the case of essentially isolated galaxies briefly given in Section 2. Here, the driving mechanism for nonbarred structures is traced to the gravitational instability of the gaseous component in the outer disk. The likely formation of global structure is traced to the relative shortness of the feedback cycle, which is estimated to be on the order of magnitude of one period of revolution, for example, $\sim \frac{1}{4}$ billion years in many cases. This is also amply demonstrated through the calculation of global modes on highly plausible basic states. The regularity of the spiral structure is traced to the fact that there are eventually very few important modes, perhaps only one or two. The detailed discussions of these points and of the frequency of occurrence may be found in the monograph by Bertin and Lin.[12] Only some relevant highlights are sketched here.

In contrast to external excitation, to be discussed below, the intrinsic mechanism of gravitational instability is ever present as long as there is a sufficient amount of gas present in the outer galactic disk (see also Section 4.3 below). It is important to note that our confidence in the likelihood of the modal scenario is not only based on dynamical studies, but also on the provision of a framework for interpretation of a wide range or major categories of observational data:

1. The modal scenario provides a framework for interpretation of observational data in several broad *categories*: (i) the Hubble classification, including both normal spirals and certain barred galaxies (e.g., NGC 3359);[26] (ii) the luminosity classes; (iii) the regularity classes; and (iv) the regular infrared structure underlying the rather complex optical and radio spiral features.
2. There is compatibility with a number of detailed quantitative observational data; indeed, there has been no obvious incompatibility with observed phenomena in essentially isolated galaxies. (There are certain rare situations, such as smooth arm spirals, that require *ad hoc* discussions.)
3. The basic states adopted—with a disk, a bulge, and a halo—have mass distributions that are in general agreement with those obtained from independent systematic model studies carried out by Kent[27,f] on the basis of detailed studies of the rotation curve without reference to the discussions of spiral structure. In the case of M81, even a detailed agreement has been found.[9] Beyond this agreement, the velocity dispersion obtained in this case has a reasonably smooth variation without sudden changes. The basic states adopted are therefore quite plausible.

### 4.2. Satellite in Hyperbolic Orbit

This is the type of scenario discussed in Section 3 in connection with the M81 and the M51 systems. In this type of scenario, it is often implicitly assumed that the basic state is devoid of global spiral structure before the encounter occurs. Actually, obviously there are also cases where the galaxy in question has a preexisting global

---

[f] Note that the models for M81 determined by Kent without consideration of the density waves agrees with our determination on the basis of the spiral wave pattern.

structure, as discussed by Oort in terms of a disturbance. The frequency of occurrence and the effectiveness of this type of scenario depend on the frequency of close passage of a satellite in a proper orbital orientation and on the responsiveness of the galactic disk. As can be easily verified (cf. Appendix A.4), a satellite in the plane of the disk would be close enough to exert a tidal forcing comparable to $m/M$ times the binding force of a galactic disk at radius $R$ only when it comes to within a galactic-centric distance $r = 4^{1/3}R \sim 1.6R$ (where $m/M$ is the ratio of the masses of the satellite and the principal galaxy). A number of questions would naturally come to mind. Except when the galaxy is in a rather dense cluster, would such a close passage be easily realized? For the generation process to be most effective, the best orbit would probably be in the plane of the galaxy. How effective would a random close encounter be? On the nature of the basic sate, there are also certain questions to be answered. First, what role would be played by the galactic halo? Second, if the galactic disk were very stable, would the global spiral pattern be generated effectively and how long would it last? The longer it lasts, the more likely it will be observed. How close to marginal instability would the disk have to be so that the global pattern would be excited and long lasting?

To summarize, there is no doubt that we are dealing with a dynamically impressive scenario. But there is no clear answer to the following crucial question: What fraction of galaxies would fit into this type of dynamical scenario? Because of the several requirements to be fulfilled for effective tidal forcing, the impression is that this scenario might not occur as frequently as the other dynamical scenarios, such as that discussed previously and that to be considered below. (For M51, the alternative scenario of a preexisting spiral structure was discussed earlier. For M81, this point was discussed in the paper by Lowe et al.[9])

### 4.3. Interacting Galaxies, Bound Systems

This case was considered in connection with the discussion of the M51 system. A more thorough investigation of interaction of this type has now been carried out by Donner and Thomasson,[28] who reported "that the patterns can sometimes survive with an almost constant amplitude for five revolutions or more, and tend to be regenerated after disappearing temporarily." The authors did not include the gaseous component in their model. Thus, it is likely that this is a case of "forced oscillation" at finite amplitude (cf. Appendix A.2). Note that in such cases, the working hypothesis of quasi-stationary spiral structure would still be applicable.

Scenarios of generation and maintenance of long-lasting global spiral structures through persistent forcing should be pursued further and compared with observational data. In the meantime, such work has been carried out by Shu and his collaborators in the case of planetary rings (see Appendix A.5) with theoretical predictions in agreement with observational data. One important difference is the presence of the halo in the galactic case. A bound satellite orbiting repeatedly through the halo would likely be captured.

The frequency of occurrence of such scenarios depends naturally on the establishment of the proper orbiting configuration. When an appropriate orbiting scenario is established, in which there is continual but not necessarily steady forcing, long-lasting global spiral structures are expected to be maintained as well. Thus, the probability for a long-lasting global structure to occur through this type of mechanism is quite significant, perhaps even more so than in the previous case.

It would indeed be most interesting to extend this study to galactic disks with gas and to examine how the two mechanisms, gravitational instability and external excitation, would share in the generation process. It would be dynamically very interesting to compare the results obtained with other studies of forced motions in nonlinear dynamical systems, for example, the van der Pol oscillator[13] (see Appendix A.2). Indeed, with the gaseous component included, which should be rich in the outer disk, tidal forcing could, in some cases, impart energy to this more responsive outer disk in a general manner. This would in turn support the usual intrinsic mechanisms (which includes gravitational instability in the outer disk) for the formation and maintenance of the global structure in the nature of standing wave patterns.

### 4.4. Tentative Conclusion

Although we have not yet discussed all the possible cases of interacting galaxies as listed in Section 3 (see also reference 13 for studies of interacting galaxies related to peculiar galaxies, etc.), the type of arguments used here can be applied to other contexts to help to decide whether a significant fraction of observed galactic spirals would fit into those contexts. A cursory examination of these more complex cases shows that the general conclusions given earlier are not likely to be changed. Further exploration of those contexts by careful modeling should be continued, so that a clearer comprehensive picture can be reached regarding the roles of galactic interaction. Especially of interest would be the tidal forcing of galaxies with preexisting spiral structures. From an observational perspective, however, one need not pay special attention to the past dynamical scenario that led to the present state. The central issue is whether one can demonstrate the likelihood for the long-lasting global modes to be maintained and observed, and whether one could apply the QSSS hypothesis to the analysis of observational data. This we have well demonstrated with the modal theory, with the support of infrared observations in the Whirlpool,[23] and in other galaxies.[g] Block et al.[5] concluded that "the underlying mass distributions observed in the infrared are exceptionally *regular*, suggesting that large scale spiral structure is principally *intrinsic*, as argued by the modal theory..... The absence of infrared multiple-armed structure is attributed to the efficiency of Inner Lindblad Resonance absorption in the evolved Population II disk." Until a significant modification of basic concepts is indicated by serious discrepancies between theory and observation, the working hypothesis of a quasi-stationary structure may be applied with high expectancy of success.

## 5. FUTURE OUTLOOK

Attempts to quantify Bertil Lindblad's density wave theory began in the early 1960s as radio observatories furnished abundant data on the distribution and

---

[g]To be sure, alternative scenarios must be considered. See Note Added in Proof at the end of the article.

motion of Pop I objects.[3] The high-speed stream outside of the Sagittarius arm and other spiral arms in the Milky Way and the approximate coincidence of the peak of the synchrotron emission with the dust lane in the Whirlpool are some of the highlights. We now have infrared data to provide direct information on Pop II objects from which we can determine the dominant spiral gravitational field. The self-consistency[12] of the Newtonian potential inferred from observed spiral mass distribution and the gravitational field required to provide the dynamical forcing that yields the mass distribution in Pop II objects would provide a direct quantitative check on the concept of slowly evolving standing wave patterns (QSSS hypothesis). This observational approach to density wave patterns should be attempted with various methods of calculation available.

In the meantime, infrared and optical observations of amplitude modulation in galactocentric distance—or along spiral arms—furnishes convincing evidence that the principal spiral structures observed in most cases are indeed associated primarily with a standing wave pattern that is slowly evolving in the nature of a (nonlinear) characteristic mode. Many galaxies appear to have spiral arms similar in structure to that shown in FIGURE 1 for M81 (cf. the spiral arms of NGC 1300 and the recent infrared data on a number of galaxies[5]). Had the spiral structure been generated very recently through external excitation from an originally stable and featureless state, it would most likely have a different variation of amplitude.

Detailed quantitative confirmation of this kind naturally depends on the full development of the nonlinear theory of density waves for a two-component system of stars and gas. This is as yet not available. Such development would be highly desirable, but not easy. One should note, however, that there are clear general features in the observational data—the observed highly regular amplitude modulation in the infrared (and in the optical)—would clearly support the concept of standing wave patterns, while much of the discussion in this paper—for example, how much of a role does intrinsic mechanisms or external excitation play in the generation process?—cannot be, by itself, subject to observational testing. The linear theory of spiral modes provides the general concepts—such as the processes of feedback cycle and overreflection at corotation zone (WASER)—that allows a clear description of essential dynamical processes that support nonlinear *characteristic* modes to be identified with the observed slowly evolving standing wave patterns. Even though much work remains to be done, the general perception described in the introductory sections (1 and 2) of this paper appears to be essentially supported by observational studies.

## 6. SUMMARY

Formation of density waves in galactic disks is studied with emphasis on the various dynamical scenarios for the generation and the maintenance of regular global structures and on their frequency of occurrence. Both internal dynamical mechanisms and external excitation are considered. Emphasis is placed on the maintenance of long-lasting normal (nonbarred) global spiral structures and their time rate of evolution. The dynamical mechanisms that favor these standing wave patterns are gravitational instability in the gas-rich outer galactic disk and persistent tidal forcing, if present, from external galaxies. The perception of a standing wave pattern is supported by the observed amplitude variation along the spiral arms in galaxy M81. It is suggested that the alternative scenario of a fast-evolving spiral structure generated from an initially featureless galactic disk through recent tidal interaction is likely to have only limited applicability.

Understanding of the dynamical mechanisms is helped by referring to the general theory of nonlinear dynamical systems for which there is a well-developed literature. The working hypothesis of *quasi-stationary* spiral structure may be considered in analogy with the existence of preferred states of "attractors" in certain nonlinear dynamical systems. It is also noted that different *past* dynamical scenarios may lead to approximately the same wave pattern observed at *present*—a point already brought out in previous studies of galaxy M81.

Recent infrared observations support the long-standing perception that the global spiral structure in the Pop II stellar disk in most galaxies is regular and *intrinsic* in nature, and that this structure is long-lasting, slowly evolving, and *maintained* primarily through internal dynamic mechanisms. This perception is also supported by the successes of the modal theory in providing a comprehensive framework for the categorical classification of galaxies in checking against specific observational data, as documented in the extensive literature on the density wave theory. Besides the specifics already mentioned, the list of these successes included the correlation of the dust lane with the peak of the synchrotron emission in M51, the Hubble morphological classification, the luminosity classes, and the regularity classes. The same modal approach is also successful in explaining why and how some galaxies are subject to flocculent and others to barred spiral structure, in contrast to approaches emphasizing interactions where these issues are not addressed in a coherent framework.

## APPENDIX: DYNAMIC MECHANISMS

The dynamic mechanisms in the galactic disk are quite complicated. They are treated in many papers in the published literature from various points of view. The reader is referred to a forthcoming monography,[12] which is written from the point of view of combining observational and dynamical approaches. Here we just mention some relevant highlights in the dynamical approach, with references to the literature. Detailed studies of some of the topics are available to the reader upon request.

### A.1. Slowly Evolving Global Structures

As mentioned in Section 1, B. Lindblad[2] first suggested that the global spiral structure observed might be evolving in a quasi-stationary manner; that is, even though there is little change in the spiral pattern on a short time scale, there could be a variation of the structure on a long time scale, as was indeed suggested earlier by P.O. Lindblad[29] and described by him as "quasi-periodic." Mathematically, this variation in time may be described by the equation[30,h] (cf. formulas at the center of FIG. 1):

$$q = F\left(r, \theta - \Omega_p t; \frac{2\pi t}{\tau_0}\right), \qquad \text{(A.1.1)}$$

where $\Omega_p$ is the angular velocity of the wave pattern and $\tau_0$ is a time scale much longer than the period of revolution $2\pi/\Omega_p$.

In a complex system like a galaxy, it is not natural to expect that the global structure would remain stationary over many (or even several) orbital periods of the stars. Indeed, even the "basic state" is expected to be slowly evolving. The development of a full nonlinear modal theory for the galactic context is therefore not expected to be easy or even rewarding for understanding the mechanisms or comparison with observational data. We shall therefore adopt a semi-empirical approach with emphasis on the empirical verification of the "hypothesis of slowly evolving global spiral structures" described by (A.1.1), whose working form is the conventional QSSS hypothesis. As demonstrated in this paper, this approach has so far not met with any serious difficulties, but has derived support from a wide range of observational information. For an understanding of the general dynamical mechanisms involved, we rely on analogy with simpler nonlinear dynamical systems, for which there is a well-developed literature.[24]

For a complex system like a galaxy to develop and to maintain a global pattern, there must be some special characteristics in the intrinsic dynamic mechanisms that allows the system to *prefer* the existence of a global structure. This preference is likely to be dominated by energy considerations, as in the case of classical ellipsoids,[31] but the dissipative mechanism would probably play a significant role in leading the system to develop toward that state. Such development in the preferred states is well known to occur in many *nonlinear dynamical systems*. The preferred state is called the *attractor*.[24] In simple dynamical systems, one can develop a precise mathematical theory for them. The term is suggestive of the fact that, the mathematical solution would eventually approach ("be attracted to") that state no matter what the initial condition may be (within limits). A well-known example of an attractor is the limit cycle in the van der Pol oscillator (cf. [24, pp. 75 and 171]). Its free and forced oscillation will give us a good idea how a nonlinear dynamical system will behave.

In such theories, it has been shown that in certain classes of physical situations, organized structures will develop while in others, they do not. (These are often referred to as order versus chaos). This is exactly the situation in galaxies. Here we focus our attention primarily on order since we are interested in cases in which grand designs are *observed* or inferred to exist in galaxies. One often used approach is to begin the study on the nonlinear system by starting with the linear approximation when the amplitude of the oscillation is known to be small. In the case of galaxies, early studies of observational data in the Milky Way already allows us to estimate the order of magnitude of the spiral gravitational fields, and indeed it is established that they are often only a small fraction of the axisymmetric field in the

---

[h]See Eq. (1) on p. 384 of reference 30. The term "grand design" appears to be first used here.

galaxy. Thus, it is reasonable to start from the linear approximation and follow up the development of the quasi-linear theory of *equilibration* that has been developed extensively in the theory of hydrodynamic stability.[32] To be sure, this perception does not apply to all cases [cf. Section 3.2].

### A.2. The van der Pol Oscillator: Free and Forced Oscillations

The best known nonlinear dynamical system with a preferred state, or an *attractor*, is the van der Pol oscillator defined by the differential equation [32, p. 190]

$$\frac{d^2s}{dt^2} + \varepsilon(x^2 - 1)\frac{dx}{dt} + x = 0. \quad \text{(A.2.1)}$$

We consider the free oscillation of this system and its forced oscillation described by Drazin [32, p. 217]

$$\frac{d^2x}{dt^2} + x = \varepsilon\left\{\gamma \cos \omega t + (1 - x^2)\frac{dx}{dt}\right\} \quad \text{(A.2.2)}$$

for $\gamma, \varepsilon, \omega > 0$, $\omega = 1$. Resonance occurs at $\omega = 1$.

The solution of (1) is well known to have a unique limit cycle (see [32, p. 192] for references to the mathematical proof). Thus, the solution of (1) with initial conditions within rather broad limits is expected to lead to the preferred state of the limit cycle. It is natural to refer to it as an attractor because solutions are generally attracted toward this preferred state.

The solution for a forced oscillator can also be easily obtained by the perturbation theory when the parameter $\varepsilon$ is small or moderate. When $\omega$ is distinct from unity, the solution is found to approach "the sum of the stable limit cycle of van der Pol's equation without forcing and a forced oscillation." However, for $\omega \approx 1$ "the forced oscillation of period $2\pi/\omega$ dominates the free oscillation of period $2\pi$"—this is called *entrainment*. This leads to the need to use a somewhat different method of applying the perturbation theory, but the analysis can be carried through as well without serious difficulties (see [32, pp. 219–225]).

In the galactic context, resonance can only occur rarely. Also note that the forcing is usually not exactly periodic. Thus, in general, the resonance scenario is not likely to play an important role in the galactic disk (cf. discussion of mechanism (c) in the previous section).

### A.3. Tidal Forcing

Because of universal gravitation, there is always a tidal forcing mechanism due to differential gravitational forcing from all the galaxies in the universe, far and near. The classic method of studying stellar encounters[33] can be easily adapted for this purpose (no divergence is found here). The cumulative effect of distant galaxies can be shown to be not negligible. It is proportional to the square root of the parameter $nd^3$, where $n$ is the number density of galaxies and $d$ is a typical value of the diameter of the galaxy under consideration. Thus it is usually small except in dense clusters.[34,35] Nearby galaxies obviously have the largest impact.

The effects of tidal forcing on the galactic disk may be roughly described in three forms: (a) tearing, (b) shaking and ringing, and (c) resonant excitation. In general, mechanism (c) will not be effective because the orbit frequency is generally too low to be appropriate to excite waves in the principal part of the galactic disk. Under special circumstances, for example, when the principal galaxy has a bound (or approximately bound) satellite, it is possible for such resonance to occur, if one of the higher harmonics of the orbiting frequency coincides essentially with that of a mode. This can also happen only infrequently.

Mechanism (a) will in general be quite effective in the generation of evolving spiral features in the outer realms of the galactic disk, such as bridges and tails, and outer spiral arms in the galactic disk if there was a *recent* close passage of a satellite; for example, in the case of the galaxies NGC 5194/5. These features are usually fast evolving, not suitable for description by the modal approach, but they are generally not long lasting (cf. below).

Mechanism (b) is the most effective in the initiation of density waves, and it is the most common in terms of frequency of occurrence. The origin of the tidal forcing could come from nearby galaxies, but it could also come from the cumulative effect of all the galaxies, far and near. Either way, the wave generation through the shaking of the galactic disk would be analogous to the ringing of a church bell.

Persistent tidal forcing, such as in the case of the bound system of two galaxies (e.g., the M51 system considered in Section 3) will likely lead to a long-lasting global structure. In $N$-particle simulation, the calculations are often carried out for a finite amount of time, starting with a featureless condition. This yields a pattern in the "early" stages. Thus, the pattern is likely to be still evolving. Long-term calculations may, however, involve changes of the basis state, which may or may not be a faithful representation of the dynamical process in real physical systems.

### A.4. Magnitude of Tidal Forcing Estimated

For close encounters, one can estimate the order of magnitude of external forcing in relation to the "binding" forces. At a distance $R$ from the center of the galactic disk, the binding acceleration is $g_0 = GM/R^2 = R\Omega^2$, where $\Omega$ is the local angular velocity, which in turn defines the mass $M$. For a mass point $m$ in the plane of the galactic disk at a distance $r$ from the galactic center, the difference of its gravitational attraction at the two ends of a diameter that passes through this point, that is, the tidal forcing, is

$$\Delta g = \frac{Gm}{r_1^2} - \frac{Gm}{r_2^2}, \tag{A.4.1}$$

where $r_1$ and $r_2$ are the distances from $m$ to the end points of the diameter. Thus,

$$\frac{\Delta g/g_0}{m/M} = \frac{r^2}{r_1^2} - \frac{R^2}{r_2^2} \cong 4\left(\frac{R}{r}\right)^3, \tag{A.4.2}$$

and it decreases rapidly with increasing $r/R$. The complete formula includes an extra factor $\{1 - (R/r)^2\}^{-2}$ on the right-hand side, implying a larger tidal forcing.

## A.5. Galactic Spirals Versus Planetary Rings

The nonlinear dynamical mechanisms of external excitation in planetary rings and circumstellar disks have been fully worked out by Shu and collaborators,[36,37] with observational confirmation. These will serve as a good basis for judging the role of external excitation in the case of galactic disks, since the dynamical similarity between galactic spirals and planetary rings is easily noticeable. However, there are important differences. Planetary rings do not have the intrinsic local gravitational instability that is present in the two-component galactic disk, where the high dispersion speed of the stars also facilitate the formation of standing global patterns by shortening the feedback cycle. Without this mechanism, persistent external forcing through a shepherding satellite plays the dominant role in the formation of planetary rings. In the case of galactic disks, this driving mechanism also has much less importance because of the stability of the inner disk and the presence of the halo. A bound satellite galaxy that comes too close to the principal galaxy would lead to a merger. Since tidal forcing decreases as the inverse third power of the distance, only weak persistent forcing is possible. Furthermore, the response in the disk would be stronger in the less stable outer disk. Thus, the patterns excited would tend to be of the same general nature as that normally excited by intrinsic gravitational instability of the gaseous component. In the nonlinear context, the galactic disk would be expected to behave in a manner similar to that of the forced van der Pol oscillator discussed in Section A.2.

NOTE ADDED IN PROOF: Recently, H. Salo sent to one of us (C. C. L.) the preprint of a paper[38] giving the full details of a model for the interacting M51 system (cited in reference 22 in the text). The interaction is found to be very strong. Thus the nature of the spiral in NGC 5194 may be dependent on external interaction to an unusual extent, more so than in other interacting systems, for example, the system NGC 7752/3 [Arp86].[39] However, it should be noted that in all the interacting systems, the internal dynamics must be playing a role in the determination of the nature and the charateristics of the resultant spiral structure, and that this internal dynamics depends crucially on the nature of the basic state specified as part of the modeling process (cf. Section 2). Of all the parameters, the distributions of the gas content and of the velcocity dispersion in the stars stand out as very important, for they decide whether a standing wave pattern can be internally generated and maintained. A preexisting standing wave system (which might even be a set of modes) prior to the latest encounter, supported by a favorable basic state, may be more likely to leave its imprint after an encounter than any other wave system. Unfortunately, it is difficult to model the evolutionary process reliably over the whole history since primordial times in the $N$-body simulation process (cf. comments in Sections 3.1, 4.3 and 6).

## ACKNOWLEDGMENT

The authors wish to thank Drs. George Contopoulos, Frank Shu, Christopher Hunter, and Louis Howard for helpful discussions and for supplying important references.

## REFERENCES

1. OORT, J. H. 1962. *In* The Distribution and Motion of Interstellar Matter in Galaxies, L. Woltjer, Ed.: 234–244. Benjamin. New York.
2. LINDBLAD, B. 1963. Stockholm Obs. Ann. **22**(5). (See also reference 29 cited below.)
3. LIN, C. C. 1971. IAU Gen. Assem. Proc., pp. 88–121. (See also reference 30 for the mathematical specification of the QSSS hypothesis.)
4. OORT, J. H. 1970. IAU Symp. Proc., No. 38, p. 1.
5. BLOCK, D. L., G. BERTIN, A. STOCKTON, P. GROSBOL, A. F. M. MOORWOOD & R. F. PELETIER. 1994. Astron. Astrophys. **288**: 365–382.
6. TOOMRE, A. & J. TOOMRE, Astrophys. J., **178**: 623.
7. BERTIN, G., C. C. LIN, S. A. LOWE & R. P. THURSTANS. 1989. Astrophys. J., Part I **338**: 78.
8. BERTIN, G., C. C. LIN, S. A. LOWE & R. P. THURSTANS. 1989. Astrophys. J., Part II **338**: 104.
9. LOWE, S. A., W. W. ROBERTS, J. YANG, G. BERTIN & C. C. LIN. 1994. Astrophys. J. **427**: 184. (See secs. 7.2, 7.3.)
10. BERTIN, G. 1991. *In* Dynamics of Galaxies and their Molecular Cloud Distributions, F. Combes and F. Casoli, Eds.: 93. Kluwer. Dordrecht, The Netherlands.
11. LIN, C. C., S. A. LOWE. 1990. Ann. N.Y. Acad. Sci. **596**: 80.
12. BERTIN, G. & C. C. LIN. 1995. Spiral Structure in Galaxies: A Density Wave Theory. MIT Press. Cambridge, Mass.
13. BARNES, J. & L. HERNQUIST. 1992. Ann. Rev. Astron. Astrophys. **30**: 705–742 (see esp. p. 712).
14. ROMEO, A. B. 1994. Astron. Astrophys. **286**: 799–806.
15. THOMASSON, M. & K. J. DONNER. 1993. Astron. Astrophys. **272**: 153 (esp. fig. 3 on p. 157).
16. WESTPFAHL, D. J. 1995. Astrophy. J. In press.
17. TOOMRE, A. 1981. *In* The Structure and Evolution of Normal Galaxies, S. M. Fall and D. Lynden-Bell, Eds.: 111. Cambridge Univ. Press. Cambridge, England.
18. TULLY, R. B. 1974. Astrophys. J. Suppl. **27**: 415, 437, 449.
19. MATHEWSON, D. S., P. C. VAN DER KRUIT & W. N. BROUW. 1972. Astron. Astophys. **17**: 468.
20. ROBERTS, W. W. & C. YUAN. 1970. Astrophys. J. **161**: 877.
21. HERNQUIST, L. 1990. Dynamics of Interacting Galaxies, R. Wielen, Ed.: 108–117. Springer-Verlag. Heidelberg.
22. SALO, H. & G. BYRD. 1994. *In* Mass-Transfer Induced Activity in Galaxies, Isaac Shlosman, Ed.: 412. Cambridge Univ. Press. Cambridge, England.
23. RIX, H. W. & M. J. RIEKI. 1993. Astrophys. J., **418**: 123.
24. DRAZIN, P. G. 1992. Nonlinear Systems. Cambridge Univ. Press. Cambridge, England.
25. WOLI, K. C. 1991. Chandrasekhar, S., A Biography: 301. Univ. of Chicago Press. Chicago.
26. BALL, R. 1992. Astrophys. J. **395**: 418–443.
27. KENT, S. M. 1987. Astron. J. **93**: 816, 1062.
28. DONNER, K. J. & M. THOMASSON. 1994. Astron. Astrophys. **290**: 785–795.
29. LINDBLAD, P. O. 1960. Stockholm Obs. Ann. **21**(4).
30. LIN, C. C. 1970. IAU Symp. Proc. No. 38, pp. 377–390.
31. CHANDRASEKHAR, S. 1962. Ellipsoidal Figures of Equilibrium. Yale Univ. Press, New Haven, Conn.
32. DRAZIN, P. & W. REID. 1981. Hydrodynamic Stability, Chap. 7. Cambridge Univ. Press. Cambridge, England.
33. CHANDRASEKHAR, S. 1943. Principles of Stellar Dynamics. New York Academy of Science, New York (reprinted by Dover, New York).
34. VOGLIS, N. & N. HIOTELIS. 1989. Astron. Astrophys. **218**: 1.
35. VOGLIS, N., N. HIOTELIS & P. HOFLOCH. 1991. Astron. Astrophys. **249**: 5.
36. SHU, F. H. 1984. *In* Planetary Rings, A. Brahic and R. Greenberg, Eds.: 513–561. Univ. of Arizona Press, Tucson.
37. SHU, F. H., C. YUAN & J. J. LISSAUER. 1985. Astrophys. J. **291**: 356–376.
38. SALO, H. & LAURIKAINEN. 1995. N-Body Model for M51, Part I. Submitted for publication in Astrophys. J.
39. SALO, H. & LAURIKAINEN. 1993. Astrophys. J. **410**: 586.

# Invariant Spectra of Dynamical Systems[a]

G. CONTOPOULOS,[b,c] N. VOGLIS,[c] C. EFTHYMIOPOULOS,[c]
and E. GROUSOUZAKOU[c]

[b]*Astronomy Department*
*University of Florida*
*Gainesville, FL 32611*

[c]*Astronomy Department*
*University of Athens*
*Panepistimiopolis GR 157 84-Athens*
*Greece*

## INTRODUCTION

In recent years the problem of finite-time Lyapunov numbers has attracted much interest.[1-19] One calculates the value

$$\chi = \chi(t) = \frac{1}{t} \ln \frac{d(t)}{d(0)}, \quad (1)$$

where $d(t)$ is the deviation from a given orbit at time $t$, when the initial deviation is $d(0)$. The usual (maximal) Lyapunov characteristic number (LCN) is

$$\text{LCN} = \lim_{t \to \infty} \chi. \quad (2)$$

The deviations $d(t)$ are assumed to be infinitesimal, and are calculated numerically by solving the variational equations together with the equations of motion.[14]

Usually $t$ is taken to be large, but not very large. In astronomical problems an appropriate time is the Hubble time, $t_H$, which, in the case of stars in galaxies, is about 100 periods. Thus it is of interest to find what are the effective Lauapunov numbers over one Hubble time.[5,11]

However, it is also of interest to find the behavior of the Lyapunov numbers over much shorter times. In fact, the details of the spectrum of values of $\chi(t)$ are washed out as the time $t$ increases.[7] Thus we have adopted the extreme case of $t$ equal to one period, namely one iteration in the case of a map.[10,12] In the case of continuous time we have considered one intersection of a Poincaré surface of section,[13] but even more details can be found if we consider very small time intervals $\Delta t$.[15]

In the following sections we consider (a) two-dimensional (2-D) maps, both conservative and dissipative; (b) 2-D Hamiltonian systems, and (c) four-dimensional (4-D) maps, which correspond to Hamiltonian systems of 3 degrees of freedom.

---

[a]The research was supported in part by the EEC Human Capital and Mobility Program ERB 4050 PL930312. One of the authors (C. E.) was also supported in part by the Greek Foundation of State scholarships.

## SPECTRA OF TWO-DIMENSIONAL MAPS

We consider, first, conservative (area-preserving) 2-D maps of the form

$$x' = f(x, y, K), \quad y' = g(x, y, K) \quad (\text{mod } 1) \tag{3}$$

where $K$ is a measure of nonlinearity. Such maps do not have any escape to infinity. Typical examples are the standard map

$$x' = x + y', \quad y' = y + \frac{K}{2\pi} \sin 2\pi x \quad (\text{mod } 1) \tag{4}$$

and the Hénon map[18]

$$x' = 1 - Kx^2 - y, \quad y' = bx \quad (\text{mod } 1) \tag{5}$$

with $b = 1$ (conservative case).

Then we take the tangent map of (3)

$$dx' = \left(\frac{\partial f}{\partial x} + \frac{\partial f}{\partial y} y_x\right) dx, \tag{6a}$$

$$dy' = \left(\frac{\partial g}{\partial x} + \frac{\partial g}{\partial y} y_x\right) dx, \tag{6b}$$

where the initial slope is

$$y_x = dy/dx, \tag{7}$$

and define the deviations

$$d(0) = (dx^2 + dy^2)^{1/2}, \quad d(1) = (dx'^2 + dy'^2)^{1/2}. \tag{8}$$

Then the one-period Lyapunov number is

$$a = \ln \frac{d(1)}{d(0)}, \tag{9}$$

and we call it a "stretching number."[12]

The distribution of the stretching numbers $a$ is given by the fractional number $dN/N$ of values of $a$ in the interval $(a, a + da)$ after a large number of periods $N$, divided by $da$. The values of

$$S(a) = \frac{dN}{N \, da} \tag{10}$$

constitute the spectrum of the stretching numbers. It is evident that the usual Lyapunov number is the first moment of the values of $a$, namely

$$\text{LCN} = \int_{-\infty}^{\infty} S(a) a \, da. \tag{11}$$

Froeschlé et al.[10] also calculated the second, third, and fourth moments of $S(a)$ for the standard map. But our emphasis[12] has been on the invariance of the spectrum $S(a)$ for sufficiently large $N$.

We found the following results.

1. The spectrum is invariant along the same orbit. That is, if we take the spectrum from $N = 10^6$ iterates (Poincaré consequents) along an orbit and superimpose the spectrum of the next $10^6$ consequents, we find a complete agreement. This is true both for stochastic orbits (FIG. 1) and for ordered orbits (FIG. 2). Both spectra are invariant, but they are different. The LCN in the first case is positive, but in the second case is zero.

If the number $N$ of iterations is small, the spectrum is noisy. In FIGURE 3 the dots indicate the noise of the spectrum from $N = 10^5$ points for an adopted width of the bins $\Delta a = 0.001$. If $N$ is smaller, the noise is larger. However, if $N \geq 10^6$, the spectrum is invariant with a good accuracy.

2. The spectrum does not depend on the initial slope $y_x$ of the deviation $d(0)$. Different initial slopes give the same invariant spectrum.

3. The spectrum does not depend on the initial conditions, if they are in the same connected chaotic domain of phase space.

Thus, instead of taking one orbit for $10^6$ periods, we may take several orbits in the chaotic domain, and superimpose their consequents. In FIGURE 4 we have taken

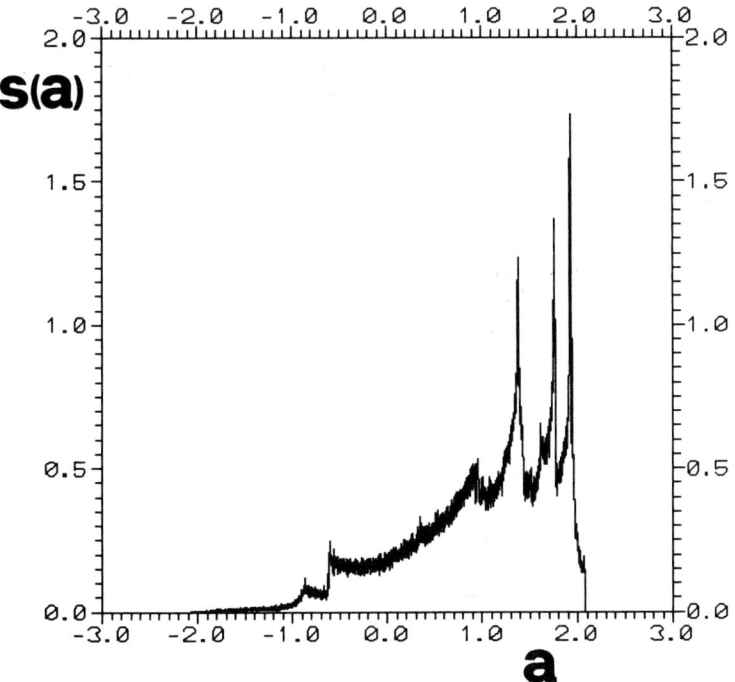

**FIGURE 1.** The invariant spectrum of stretching numbers of a chaotic orbit in the standard map ($K = 5$, $x_0 = 0.1$, $y_0 = 0.5$, $y_{x0} = 0$) calculated for $N = 10^6$ periods. The spectrum of the next $10^6$ periods exactly agrees with that.

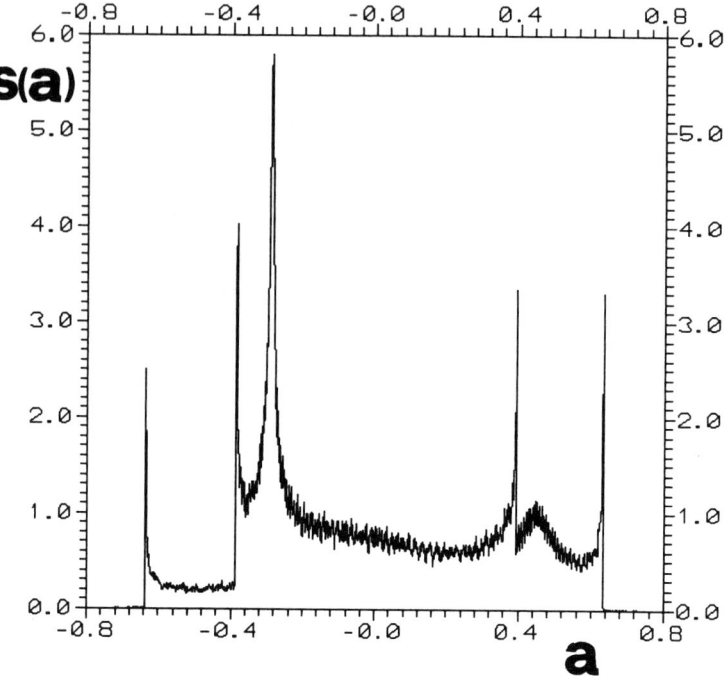

**FIGURE 2.** The same as in FIGURE 1 for a regular orbit ($K = 0.5$, $x_0 = 0.1$, $y_0 = 0.5$, $y_{x0} = 0$).

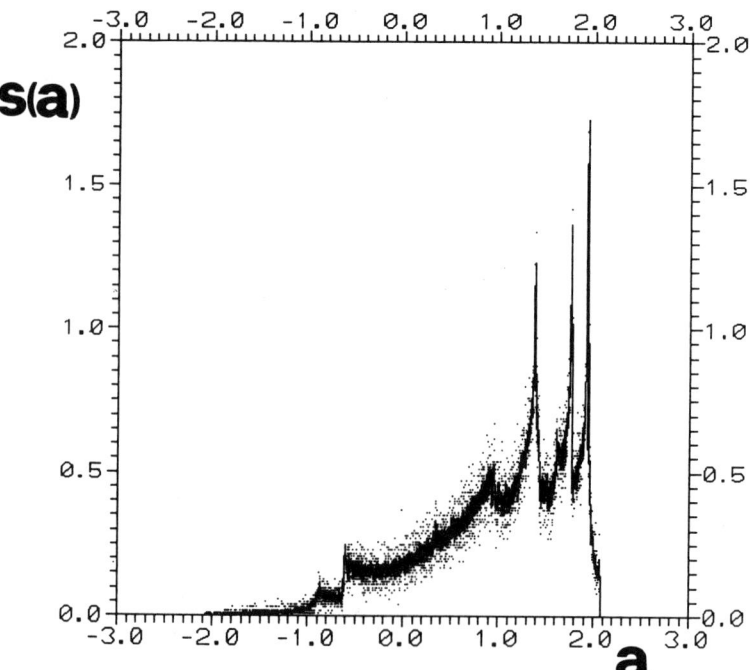

**FIGURE 3.** A comparison of the spectrum of FIGURE 1 with the spectrum derived from $N = 10^5$ periods (*dots*).

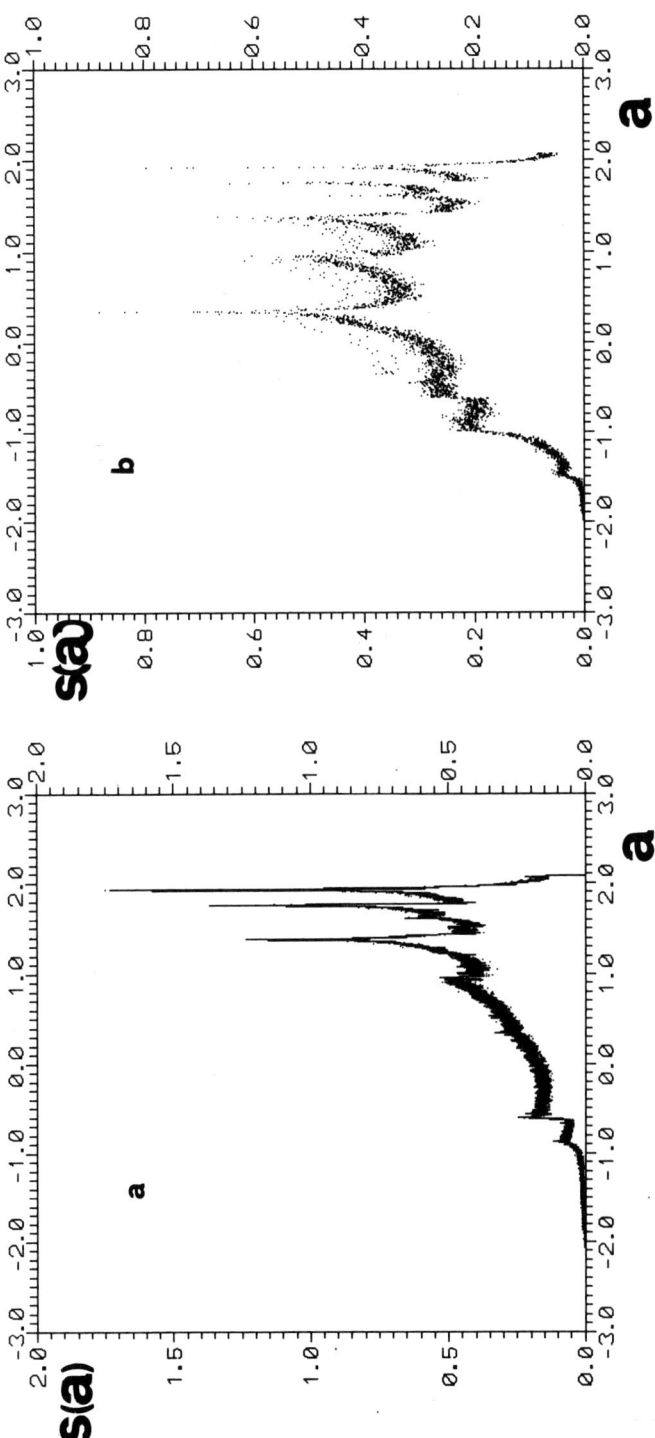

**FIGURE 4.** Composite spectra derived by superimposing 100 × 100 orbits with initial conditions in a box of dimensions 0.1 × 0.1 for 100 periods each. In the first case (**a**) the box is completely in the chaotic region ($x = 0.15 \pm 0.05$, $y = 0.55 \pm 0.05$) and the spectrum coincides with FIGURE 1. In the second case (**b**), however, part of the box is in the ordered region ($x = 0.69 \pm 0.05$, $y = 0.42 \pm 0.05$) and the spectrum is different.

$100 \times 100$ initial conditions equally spaced in a box of size $0.1 \times 0.1$, with initial slope $y_x = 0$, and calculated the orbits for 100 periods. Thus we have $N = 10^6$ points again.

If the box is completely in the chaotic domain, the composite spectrum is the same as the invariant spectrum of one orbit calculated for long times (compare FIG. 4(a) and FIG. 1). If, however, part of the box is in the ordered region, the spectrum is different (FIG. 4(b)).

This result is important because it allows us to find the spectrum, and then derive the Lyapunov characteristic number, from calculations of orbits over short times. However, one should check whether the orbits belong to the same (connected) chaotic region.

This fact was stressed by Kandrup et al.,[11] who found the correct Lyapunov number by sampling orbits in the chaotic domain for short times. On the other hand, short time Lyapunov numbers for ordered orbits are different from zero, in general, and cannot be used in determining the correct Lyapunov number, which is zero. As regards the spectra of ordered orbits, they are the same for initial conditions on the same invariant curve, but they are different for different invariant curves.

Similar results were found in other maps. For example, in the case of the conservative Hénon map there is again an invariant spectrum (FIG. 5). This spectrum is

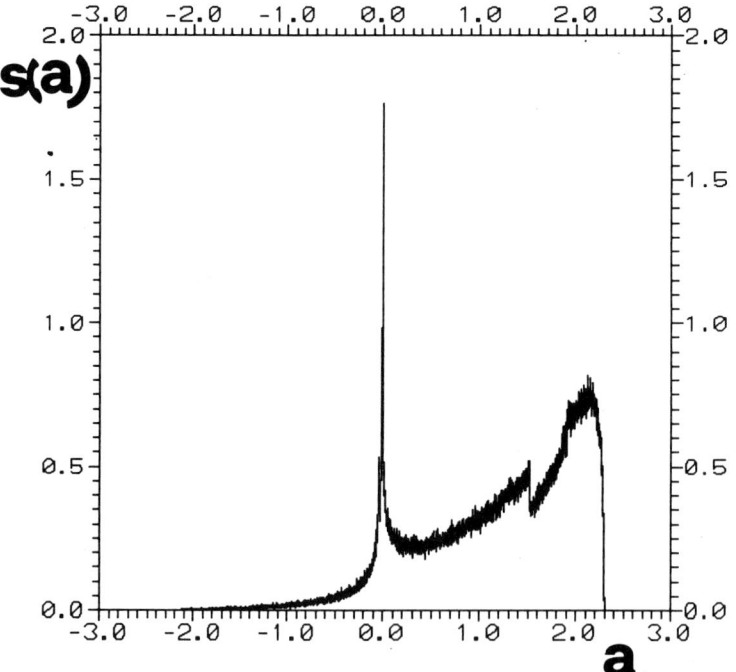

**FIGURE 5.** The invariant spectrum of stretching numbers in the case of the conservative Hénon map ($K = 5$, $b = 1$, $x_0 = 0.1$, $y_0 = 0.5$, $y_{x0} = 0$).

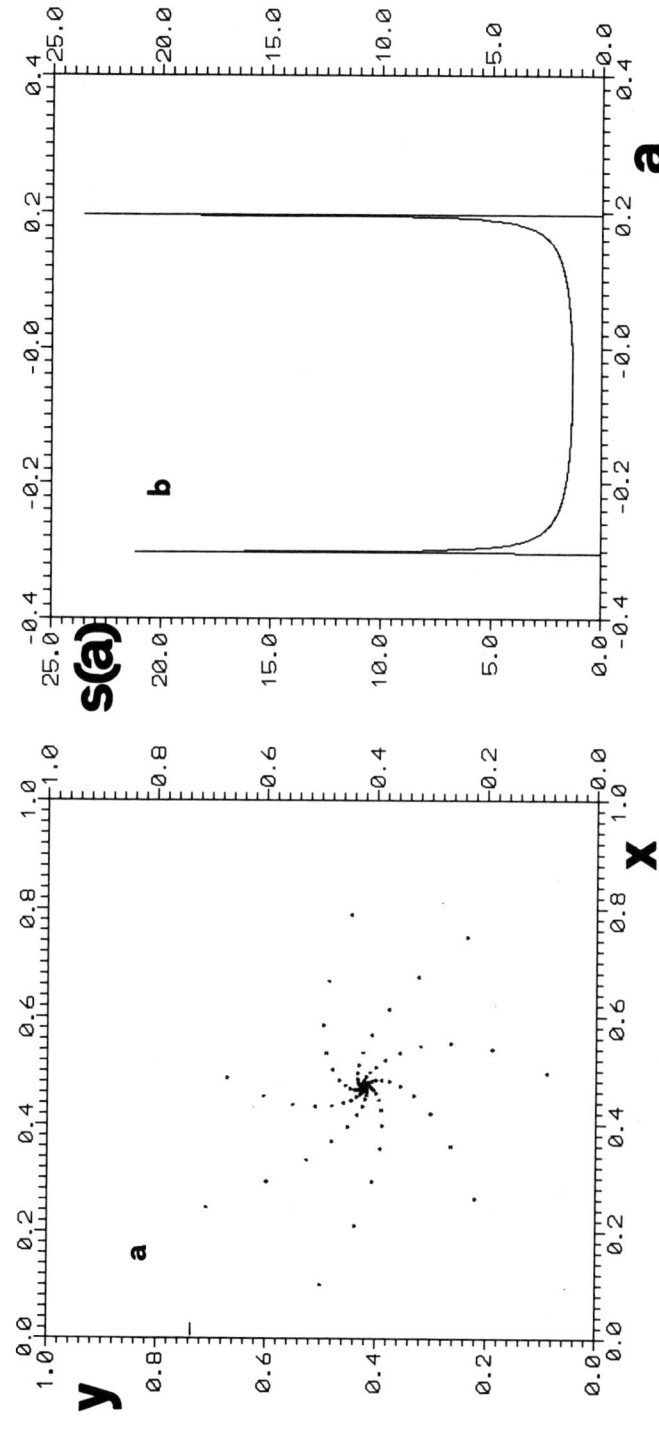

**FIGURE 6.** A dissipative Hénon map ($K = 0.5$, $b = 0.9$) with (**a**) initial conditions ($x_0 = 0.1$, $y_0 = 0.5$, $y_{x0} = 0$), and (**b**) the corresponding invariant spectrum.

rather different from the spectrum of the standard map (FIG. 1). Thus two largely chaotic systems may have different spectra. This is true even if their Lyapunov numbers are equal.

If the constant $b$ of the Hénon map[16] [(5)] is smaller than one, the system is dissipative. Also in this case an invariant spectrum can be defined. In the case of FIGURE 6(a) the successive iterates after about 100 points are concentrated in the large central dot, which is around a stable periodic orbit (point attractor). However, the spectrum of the stretching numbers in the central dot has a well-defined structure. This is shown in FIGURE 6(b), which gives the invariant spectrum for $N = 10^6$ points.

Invariant spectra in dissipative systems also appear when we have a line attractor, or a strange attractor.[12] In all cases the spectra are independent of the initial slope. Furthermore, as long as the initial conditions belong to the same basin of attraction, the spectra are also independent of the initial conditions.

## TWO-DIMENSIONAL HAMILTONIANS

Similar results were found for time-independent Hamiltonian systems. We considered the spectra for mappings defined on a Poincaré surface of section[13] $y = 0$ in the cases of two Hamiltonian systems

$$H = \tfrac{1}{2}(\dot{x}^2 + \dot{y}^2 + Ax^2 + By^2) - \varepsilon xy^2, \qquad (12)$$

and

$$H = \tfrac{1}{2}(\dot{x}^2 + \dot{y}^2 + Ax^2 + By^2) - \varepsilon x^2 y^2. \qquad (13)$$

The first system was studied for $A = 1.6$, $B = 0.9$, $H = 0.00765$, and different values of $\varepsilon$, from $\varepsilon = 4$ to $\varepsilon = 4.5$. For $\varepsilon = 4.5$ the system is highly chaotic, as can be seen from the distribution of $5 \times 10^4$ Poincaré consequents of a particular orbit on a Poincaré surface of section $y = 0$ (FIG. 7). This figure shows a more or less homogeneous distribution of points, except for two main symmetric islands and a number of very small islands. However, the number $N = 5 \times 10^4$ is not sufficient for a completely homogeneous distribution of the consequents.

The spectra of two chaotic orbits, starting at quite different points on the surface of section, are shown in FIGURE 8. The agreement is very good.

We also calculated spectra of ordered orbits inside the islands of FIGURE 7. The spectra are again invariant, but only with respect to initial conditions along the same invariant curve.

Similar results were found for the Hamiltonian (13).[19] In this case the distribution of the points on the Poincaré surface of section is symmetric with respect to the axes $x = 0$ and $\dot{x} = 0$, and it is again homogeneous, except for four islands (FIG. 9). The corresponding spectrum is again invariant (FIG. 10), but quite different from the spectrum of FIGURE 8.

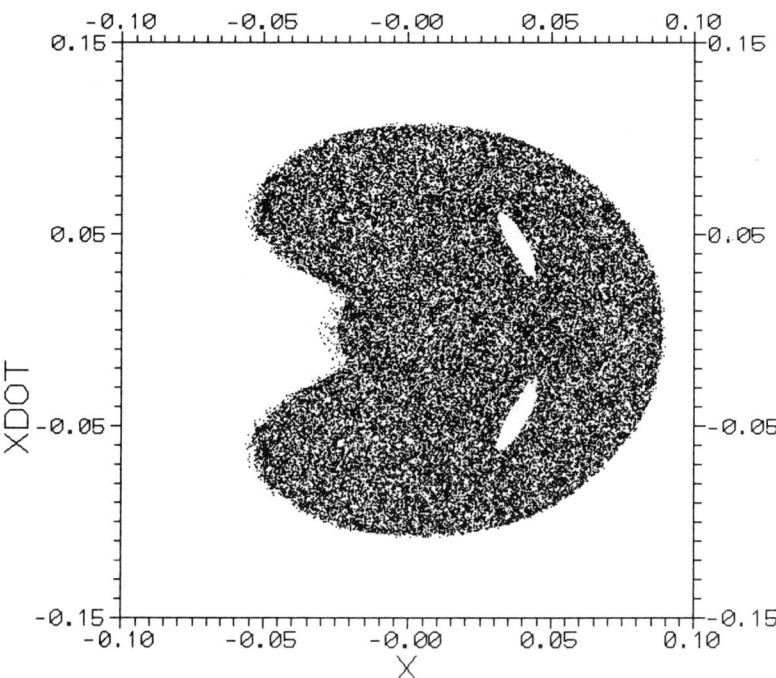

**FIGURE 7.** The distribution of $5 \times 10^4$ consequents on the Poincaré surface of section $y = 0$ ($\dot{y} > 0$) of the Hamiltonian (**12**) with $A = 1.6$, $B = 0.9$, $\varepsilon = 4.5$, $H = 0.00765$, and initial conditions ($x = 0.1$, $y = \dot{x} = 0$, $\dot{y} > 0$).

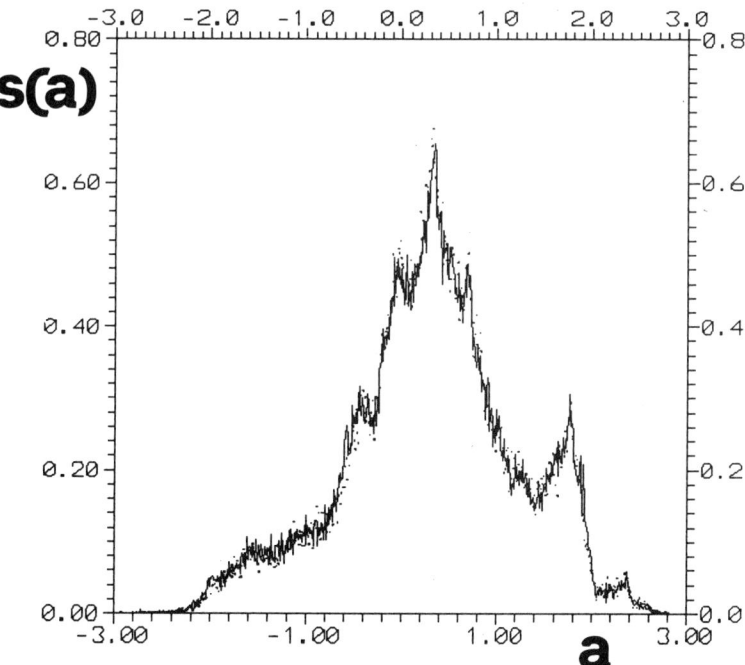

**FIGURE 8.** The spectra of two orbits in the Hamiltonian (**12**) for $N = 10^5$ each. The initial conditions ($x_0 = 0.1$, $y = \dot{x} = 0$, $\dot{y} > 0$; *solid line*) and ($x = 0$, $y = 0$, $\dot{x} = 0.1$, $\dot{y} > 0$; *dots*) belong to the same connected stochastic domain.

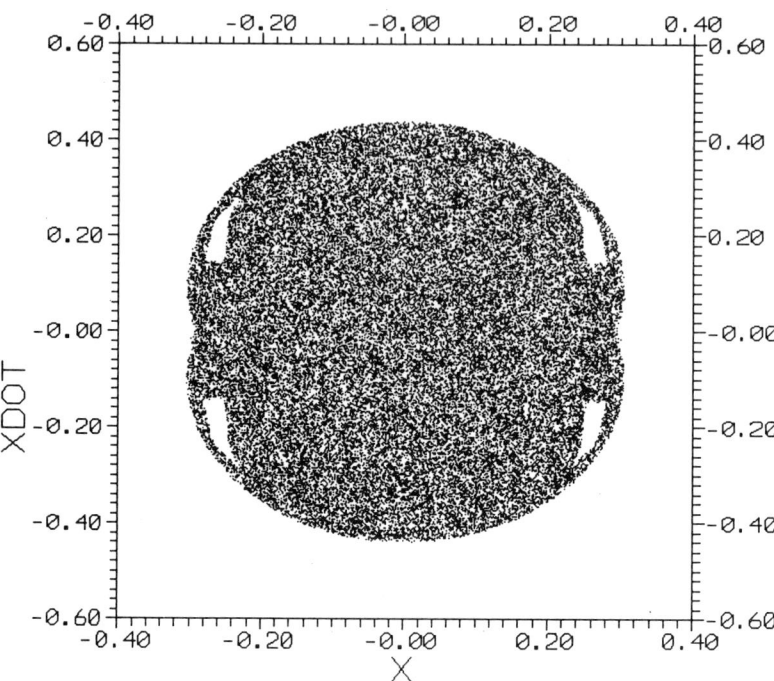

**FIGURE 9.** The distribution of $5 \times 10^4$ consequents on the Poincaré surface of section $y = 0$ ($\dot{y} > 0$) of the Hamiltonian (**13**) with $A = 1.6$, $B = 0.9$, $\varepsilon = 3$, $H = 0.12$, and initial conditions ($x = 0.184$, $y = \dot{x} = 0$, $\dot{y} > 0$).

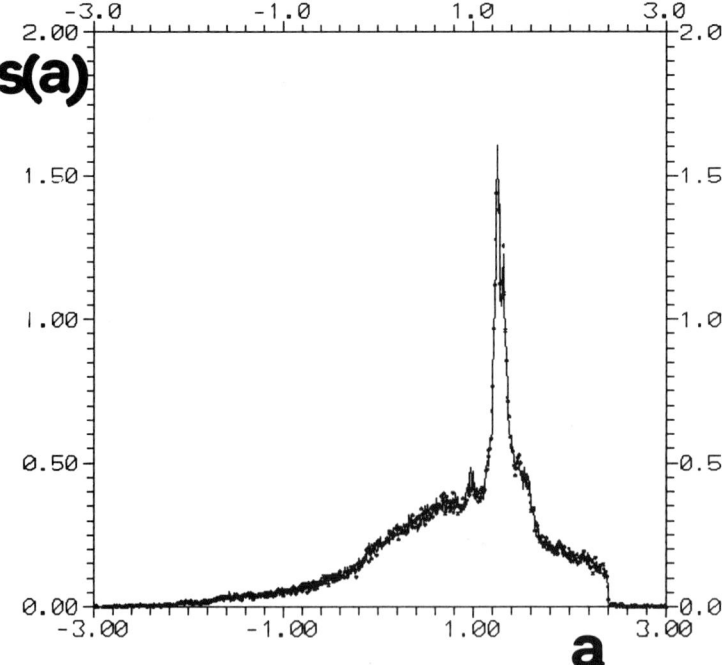

**FIGURE 10.** The spectra of two orbits in the Hamiltonian (**13**) for $N = 10^5$ each. The initial conditions ($x = 0.184$, $y = \dot{x} = 0$, $\dot{y} > 0$) and ($x = 0.0351$, $y = \dot{x} = 0$, $\dot{y} > 0$) belong to the same connected stochastic domain.

## SPLITTING THE SPECTRUM INTO PARTS

One question that comes immediately to mind is, what is the distribution of the points with different values of $a$ on the plane $(x, y)$?

In order to find that, we separated the invariant spectrum (FIG. 1) into nonoverlapping parts and plotted separately the Poincaré consequents corresponding to various intervals of $a$. In FIGURE 11 we show the distribution of $3 \times 10^4$ Poincaré consequents with all possible values of a (FIG. 11(a)), and then the consequents that have $a < 1$ (FIG. 11(b)), $1 < a < 1.5$ (FIG. 11(c)) and $a > 1.5$ (FIG. 11(d)). In order to see better the boundaries of the sets of FIGURE 11(b), (c), (d), we have again calculated the orbit for $N = 2 \times 10^6$ periods and added in FIGURE 11(b), (c) the points with $a = 1 \pm 0.0005$, and in FIGURE 11(b), (c) the points with $a = 1.5 \pm 0.0005$. Thus,

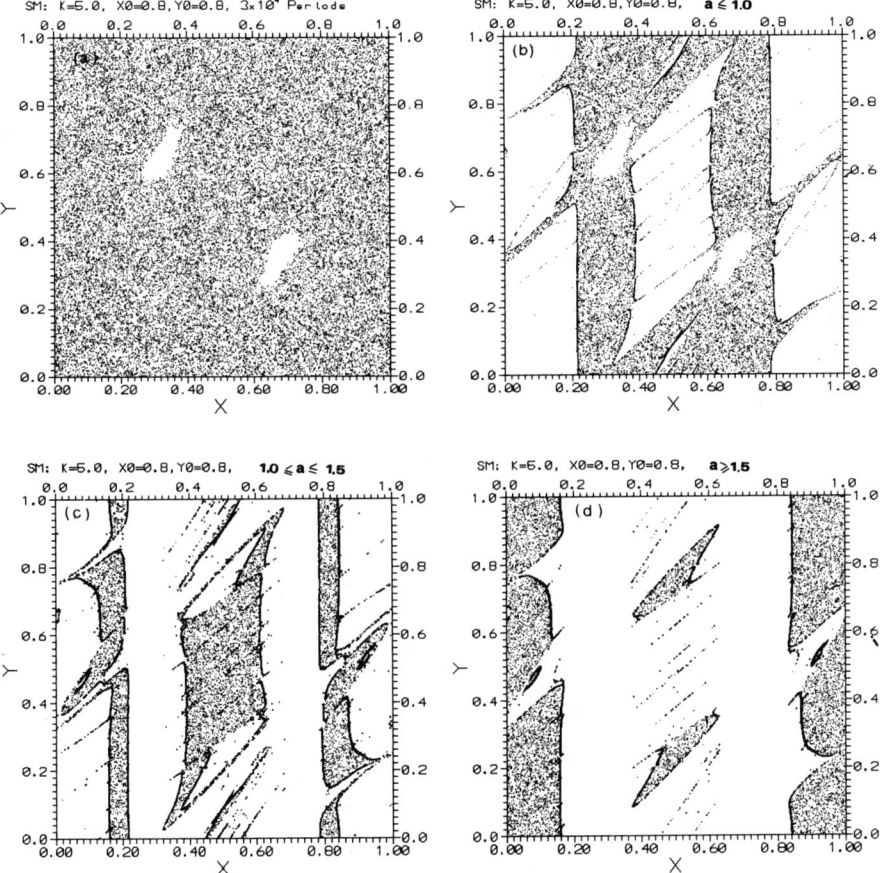

**FIGURE 11.** (a) The (invariant) distribution of $3 \times 10^4$ iterates of the mapping (4) on the plane $(x, y)$ for $K = 5$, and the distributions of the points belonging to different parts of the spectrum of FIGURE 1, namely (b) $a < 1$, (c) $1 < a < 1.5$, and (d) $1.5 < a$.

the boundaries appear darker. The points of FIGURE 11(a) cover the plane $(0, 1) \times (0, 1)$ uniformly except for two empty regions that contain islands of stability. But the sets of points in FIGURE 11(b), (c), (d) are quite complicated. Nevertheless, these sets have certain regularities. They are composed of compact regions, but also contain infinite lines of small thickness, which seem to form Cantor sets. For example, there are some almost vertical boundaries and many lines with roughly the same inclination (0.8). These sets do not overlap. But if we superimpose FIGURE 11(b), (c), (d), omitting the extra points near the boundaries, we find FIGURE 11(a), which gives the overall distribution of $3 \times 10^4$ points on the plane $(x, y)$.

It is remarkable that the distributions of the points in FIGURE 11(b), (c), (d) are invariant. Namely the corresponding figures for two orbits with different initial conditions in the same chaotic domain agree almost exactly with each other. There are only some small discrepancies, especially along the lines that are thinly populated, which are due to the fact that even a number $N = 2 \times 10^6$ is not sufficient to delineate all the details of the boundaries in FIGURE 11(b), (c), (d). At any rate we have enough evidence that the distributions of the points corresponding to specific parts of the spectrum on the plane $(x, y)$ are invariant. Furthermore different parts of the spectrum correspond to different regions in the plane $(x, y)$. The various regions are separated by lines of constant $a$ (e.g., the boundaries of the sets of FIGURE 11(b), (c), (d)). These lines are in general well separated. However, at many points in the $(x, y)$ plane there is a convergence of lines corresponding to various values of $a$. Near these points the value of $a$ as a function of the position varies widely.

Similar results were found for the Hamiltonian systems (12) and (13).[13] The case of the Hamiltonian (12) is remarkable in another way also, because the invariant distributions of the Poincaré consequents corresponding to different intervals (FIG. 12) have no symmetry as in FIGURE 11(b), (c), (d). But although these distributions are quite asymmetric, if we take another initial condition in the same chaotic domain, we find the same distributions.

## AN INVARIANT MEASURE

The existence of cantori with small holes on the plane $(x, y)$ brings about an important question. How long does it take for the successive iterates of an orbit in the stochastic domain to sample uniformly this domain? It is expected that after a sufficiently long time these iterates will fill the stochastic domain smoothly. But will they fill it uniformly?

In order to check this we considered cantori of two kinds:

1. Global cantori that extend from one end of the plane $(x, y)$ to the other.
2. Local cantori that surround islands of stability.

For example, the cantorus of the standard map that is formed when the parameter $K$ in (4) goes beyond $K = 0.97$ (the "golden mean" cantorus) is a global one, extending all the way from $x = 0$ to $x = 1$, while cantori that surround islands of stability and produce the stickiness effect around such islands are local. We remark that the global cantori play an important role when $K$ is somewhat larger than the critical

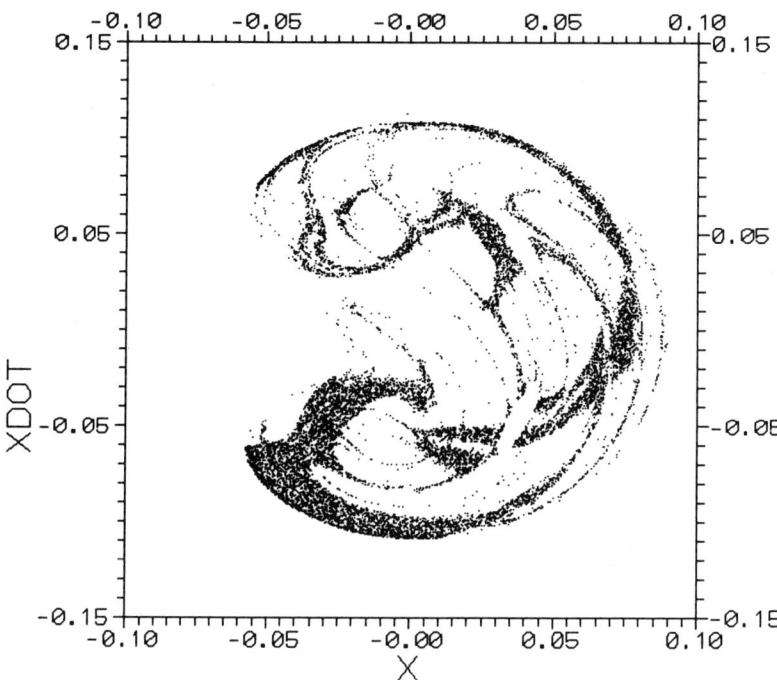

**FIGURE 12.** The (invariant) distribution of the consequents of an orbit in the chaotic domain of the Poincaré surface of section in the Hamiltonian (12) with $a$ between $-1$ and $0$.

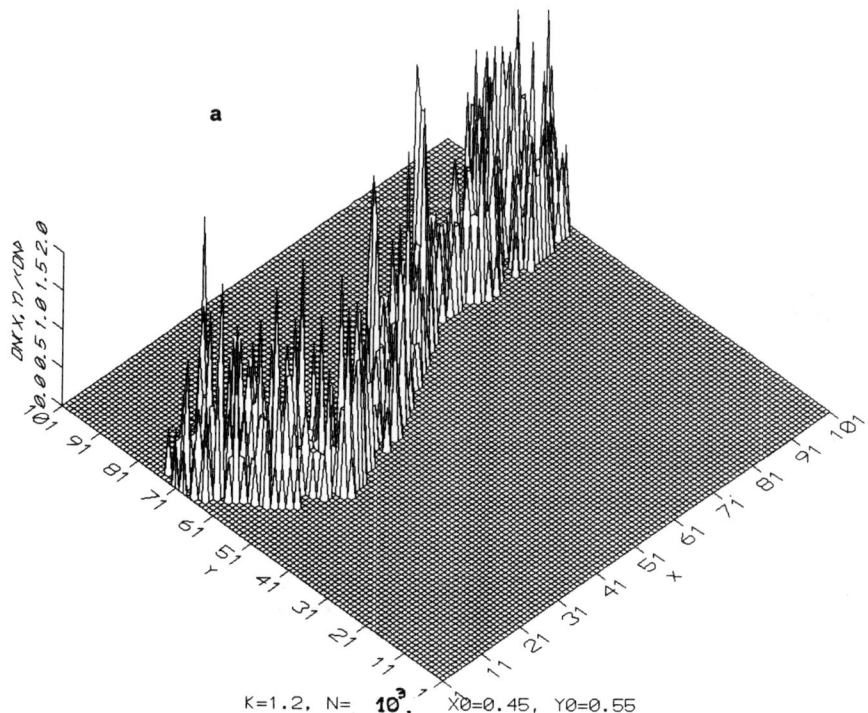

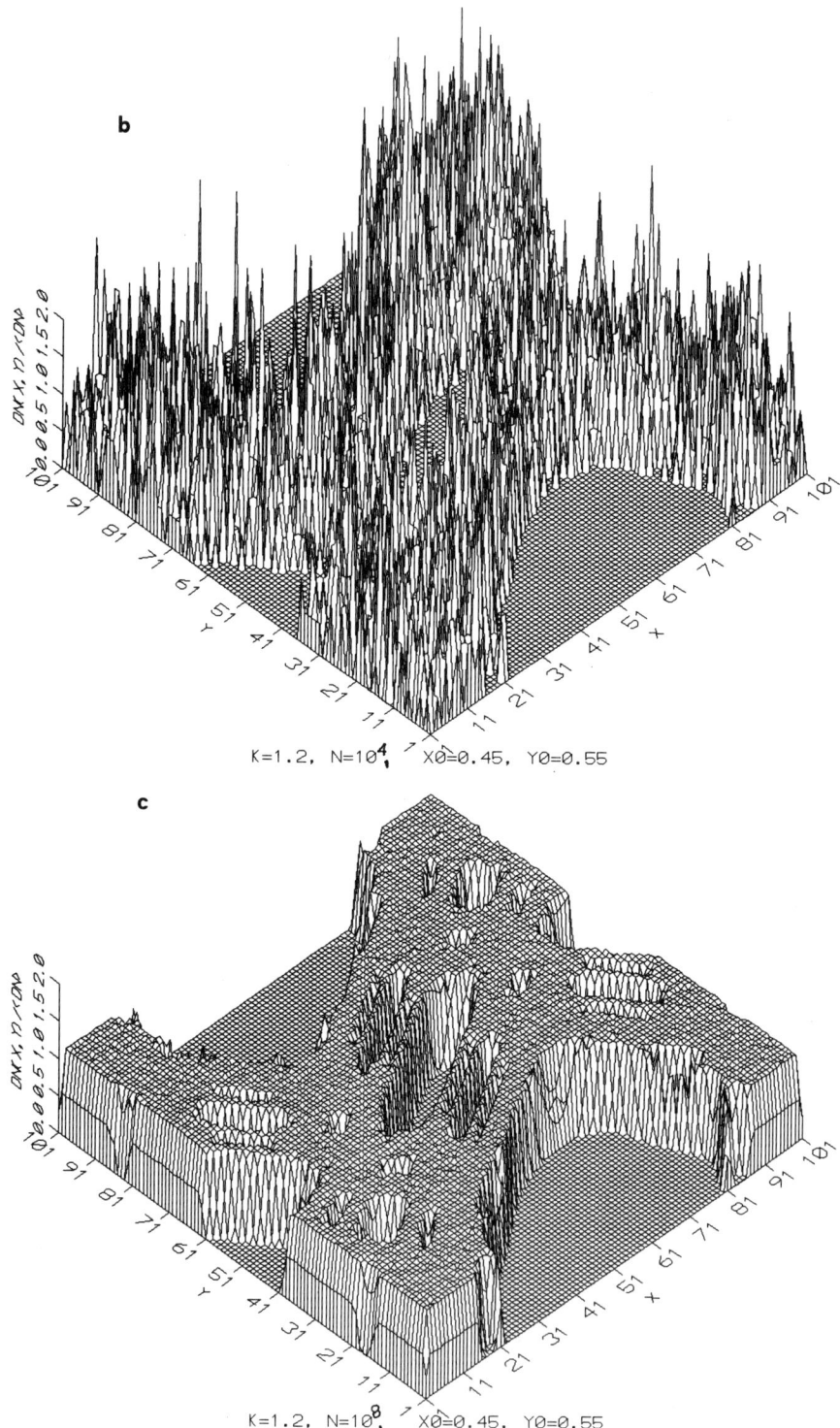

**FIGURE 13.** The densities of the consequents of an orbit starting above the golden mean cantorus in the standard map (4) for $K = 1.2$ and (a) $N = 10^3$, (b) $N = 10^4$, and (c) $N = 10^8$.

value $K = 0.97$, but not much larger, while cantori of the second, local, type exist always, whenever we have islands.

In the case of FIGURE 11(a), the value of $K = 5$ is much larger than the critical value $K = 0.97$, and there is no evidence of difference of density of the consequents above and below the golden mean cantorus. The distribution of the points looks quite uniform, except for the two main islands.

But what happens if $K$ is much smaller? We have studied in detail the case $K = 1.2$, which is larger than $K_{crit} = 0.97$, but not much larger. In this case the golden mean cantorus plays an important role, which is seen in FIGURE 13(a), (b), (c). In these figures we show the densities of the points on the Poincaré surface $(x, y)$ for the same initial condition, but for different values of the number of points $N$.

In FIGURE 13(a) we give the densities of the first $10^3$ consequents of an orbit starting above the golden mean cantorus. The densities are given for all bins of size $(\Delta x, \Delta y) = (0.01, 0.01)$. We see that all the points are above a line that gives approximately the position of the cantorus, but their density varies wildly from one point to the next. The density is zero below the cantorus. This distribution is not invariant, because the distribution of an equal number of consequents from a different initial condition may be very different.

If the number $N$ becomes $N = 10^4$ (FIG. 13(b)), a large number of points is now below the cantorus. Again there are large variations of the density from one point to the next, but we may say that the density of points below the cantorus is now of the same order as above the cantorus.

As $N$ increases, the variations of the density of the consequents in the chaotic region are decreasing. The successive consequents cross the cantorus many times downwards and upwards and bring equal densities.

The final result is shown in FIGURE 13(c) for $N = 10^8$. This number is big enough so that the variations of the density are now insignificant. Thus we have reached an invariant measure with constant density everywhere, except in the islands.

In FIGURE 13(c) we see many islands, where the density is zero. But in the large chaotic domain the density is constant. We may call this measure a "microcanonical measure in the chaotic domain." Strictly speaking this is not a microcanonical distribution because of the existence of the islands. Even in the region where the density is positive and constant there may be very small islands. But if we disregard details smaller than the bin size, we may say that we have achieved a microcanonical distribution "except for the islands."

In FIGURE 13(c) we see also that there is apparently no stickiness around the various islands (no increase of density close to the islands), although such stickiness appears when $N$ is not very large. However, there is probably always some stickiness around the islands, but of very small extent, that would require a much more detailed binning of the plane $(x, y)$ to be distinguished.

As $K$ increases, most of the islands of FIGURE 13(c) disappear. For example in the case of FIGURE 11(a), for $K = 5$, only two islands survive. In this case the chaotic domain is larger than in FIGURE 13(c), and the "microcanonical" invariant measure is reached earlier.

On the other hand, if $K$ becomes smaller than $K = 1.2$, the holes of the cantori become smaller and the transient effects last longer. Such long-time transient distributions of the consequents, due to important cantori (i.e., cantori with small holes) were found by Kandrup and his associates.[19] These transients may be important in particular dynamical systems over long, but not very long, times. In fact, Kandrup and his associates[19] indicate that after a sufficiently long time, these effects disappear and a microcanonical distribution is established.

The theory of the crossing of cantori with small gaps has been developed by Bensimon and Kadanoff[20] and by MacKay et al.[22] However, we found[22] that a modest increase of the perturbation produces a large increase in the gaps of the cantori, and a fast diffusion through them. Therefore, if we fix a confinement time $T$, due to a particular cantorus (or a set of cantori), we have a corresponding critical perturbation $K_{crit}$, such that for $K < K_{crit}$ the confinement is appreciable for a time of order $T$, while for $K > K_{crit}$ the confinement is insignificant.

## SPECTRA OF FOUR-DIMENSIONAL MAPS

The study of four-dimensional symplectic maps is important, because they correspond to conservative Hamiltonian systems of 3 degrees of freedom on a surface of section. Furthermore such systems are expected to exhibit Arnold diffusion. Thus, the various chaotic regions are not separated in general. However, the timescale of Arnold diffusion is extremely long, and one may have long-lived transient phenomena.

Here we study systems of coupled standard maps of the form

$$x'_n = x_n + y_n$$

$$y'_n = y_n + \frac{K}{2\pi} \sin 2\pi x_n - \frac{\beta}{2\pi}$$

$$\times [\sin 2\pi(x_{n+1} - x_n) + \sin 2\pi(x_{n-1} - x_n)] \quad (\text{mod } 1) \quad (14)$$

where $n = 1, 2, 3, \ldots, N$, and is given modulo $N$. This represents a $2N$-dimensional map.[23]

The case $N = 2$ is simpler:

$$x'_1 = x_1 + y'_1, \qquad y'_1 = y_1 + \frac{K}{2\pi} \sin 2\pi x_1 - \frac{\beta}{\pi} \sin 2\pi(x_2 - x_1) \quad (\text{mod } 1) \quad (15)$$

$$x'_2 = x_2 + y'_2, \qquad y'_2 = y_2 + \frac{K}{2\pi} \sin 2\pi x_2 - \frac{\beta}{\pi} \sin 2\pi(x_1 - x_2).$$

This represents a four-dimensional map; $K$ is the nonlinearity parameter and $\beta$ is the coupling parameter.

If $\beta$ is small, the coupling phenomena are small.

In FIGURE 14(a) we see the projections of $N = 10^6$ points on the plane $(x_1, y_1)$ for $\beta = 0.1$ and $K = 3$. The spreading of the points is small in this case. The corresponding spectrum is compared with the spectrum of an ordered orbit (with the same initial conditions) in the uncoupled case ($\beta = 0$) (FIG. 14(b)).

The spectrum of the case $\beta = 0.1$ is similar to the uncoupled spectrum, but the maxima are displaced.

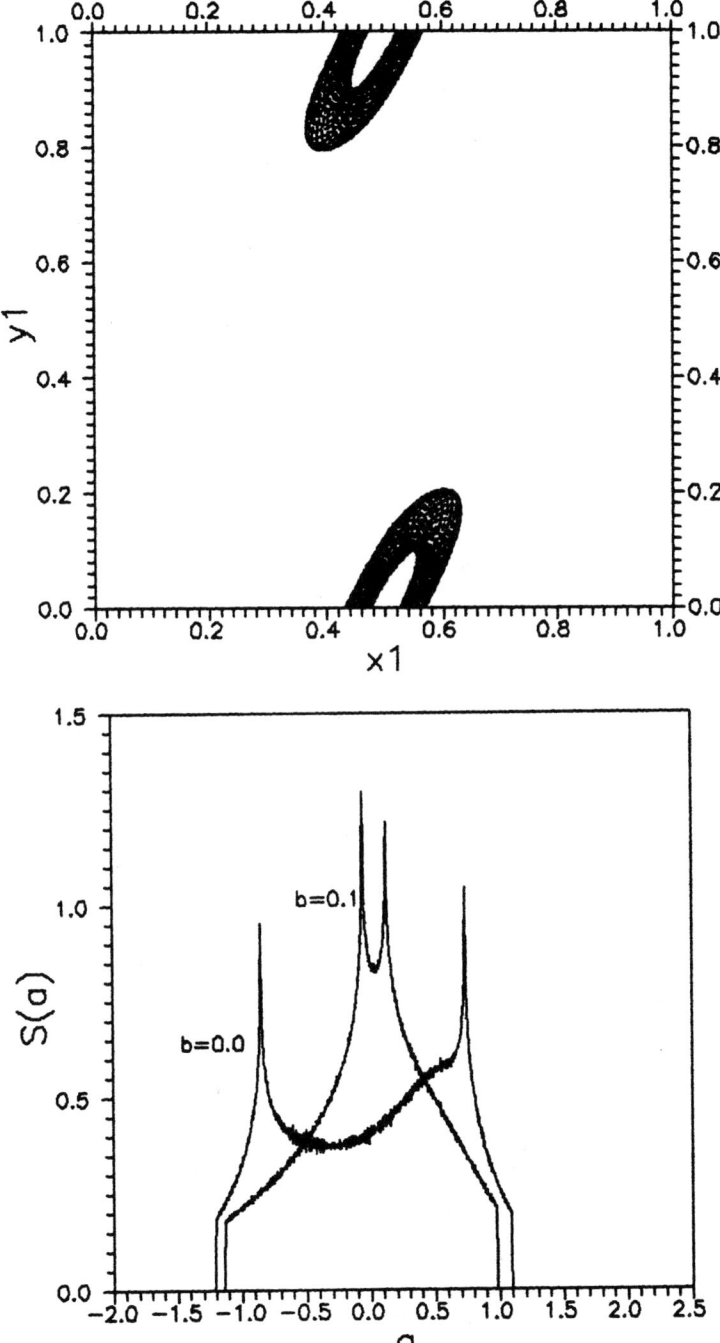

**FIGURE 14.** (a) Projections of the distribution of consequents on the $(x_1, y_1)$ plane for $K = 3$ in the system (15) and $\beta = 0.1$ (initial conditions $x_1 = 0.55$, $y_1 = 0.1$, $x_2 = 0.62$, $y_2 = 0.2$), and (b) the corresponding spectrum, together with the ordered spectrum of the case $\beta = 0$.

As we increase the coupling $\beta$, the spreading of the points becomes larger ($\beta = 0.3$, FIG. 15(a)). The corresponding spectrum is compared to the (invariant) chaotic spectrum of the uncoupled case ($\beta = 0$) (FIG. 15(b)). The spectrum is noisy and much more complicated than the spectrum of the case $\beta = 0.1$, but it has a mean (Lyapunov number) very close to zero.

A dramatic change occurs as $\beta$ changes from $\beta = 0.3$ to $\beta = 0.30513$ (FIG. 16(a)). The points now spread over the whole plane $(x_1, y_1)$, after staying in the neighborhood of the uncoupled invariant curves for a long time. The corresponding spectrum for $N = 10^6$ points is smoother than in the case $\beta = 0.3$, and extends over larger positive values, so that its Lyapunov number is clearly positive (FIG. 16(b)).

Finally, when $\beta = 0.4$ the distribution of the points is uniform over the whole plane $(x_1, y_1)$ and there is no concentration of points anywhere on this plane (FIG. 17(a)). The corresponding spectrum (FIG. 17(b)) is invariant and characterizes the large chaotic region in phase space. Similar spectra appear for larger $\beta$.

The spreading of the points of FIGURE 16(a) after a long time in the whole area $(x_1, y_1)$ is an indication of Arnold diffusion. This phenomenon is similar to the crossing of cantori that we described in previous sections. Thus the spectrum given in FIGURE 16(b) is a transient one. As the number $N$ increases, the spectrum changes and approaches a "final" invariant spectrum similar to that of FIGURE 17(b). Only this "final" spectrum is really invariant.

What happens now if $\beta$ is equal to 0.3 or smaller? Do we have a similar transition to a global chaos as in the case $\beta = 0.30513$? The answer is probably yes, but the time scale is much longer than $10^6$ periods. (For example, when $\beta = 0.30512$, the transition to a global chaos occurs after $2 \times 10^7$ periods). This is due to the fact that Arnold diffusion is extremely slow for small values of the coupling parameter $\beta$.

Similar results, indicating the effect of Arnold diffusion, and an approach to a "final" invariant spectrum were also observed for many degrees of freedom.

## CONCLUSIONS

We have found that *short-time Lyapunov numbers* (or *stretching numbers*) give valuable information about dynamical systems that is not provided by the Lyapunov numbers themselves, or long-time Lyapunov numbers. In particular, we found the following results.

1. The spectra of the stretching numbers are invariant with respect to the initial point along a given orbit and with respect to the initial direction of deviation from this orbit. In the case of chaotic orbits they are also invariant with respect to the initial conditions in the same chaotic domain, while in the case of ordered orbits they are invariant only with respect to the initial conditions along the same invariant curve.
2. We found invariant spectra for conservative maps and for dissipative maps with attractors ranging from point attractors to strange attractors.
3. In the case of Hamiltonian systems we found invariant spectra corresponding to the maps generated on a Poincaré surface of section, with the same properties as just given.

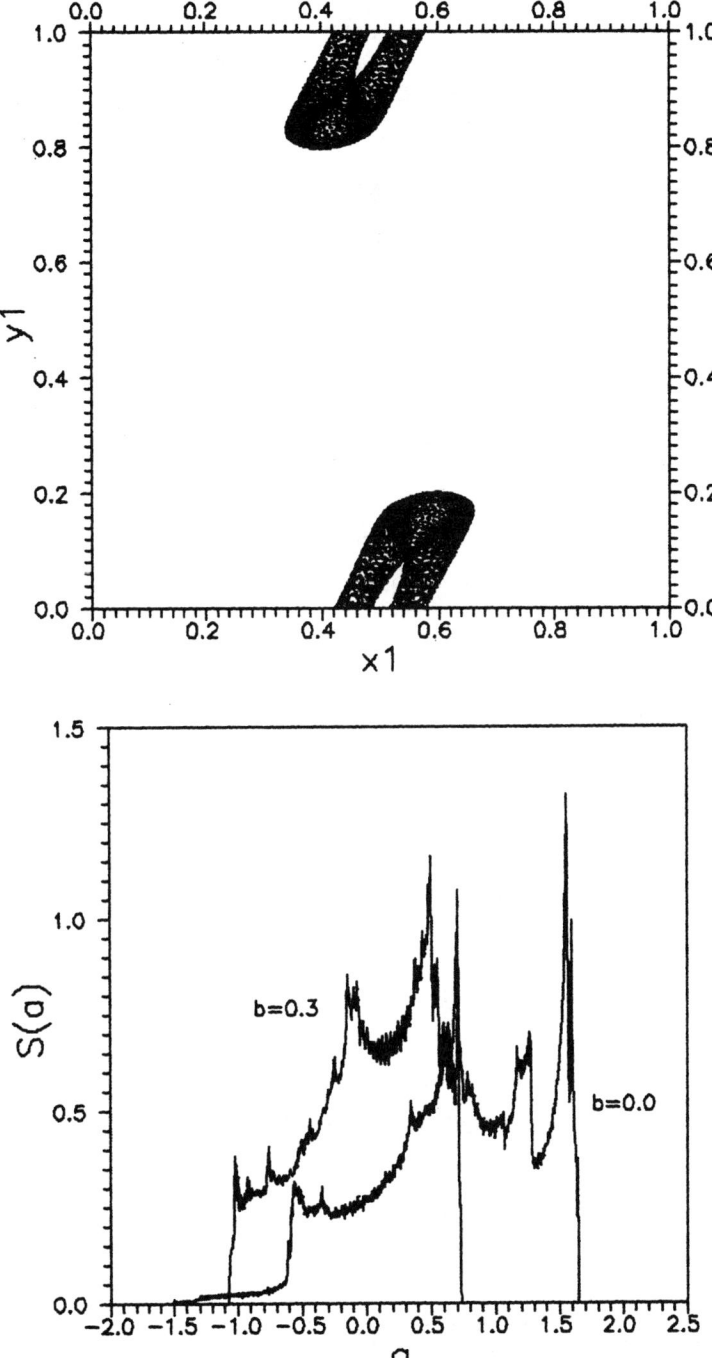

**FIGURE 15.** (a) The same as in FIGURE 14(a) for $\beta = 0.3$. (b) The corresponding spectrum together with the chaotic spectrum of the case $b = 0$.

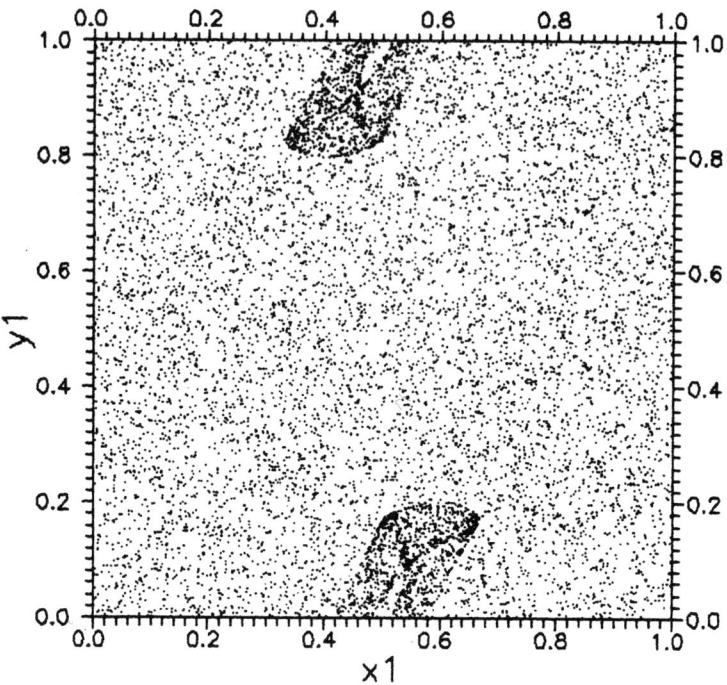
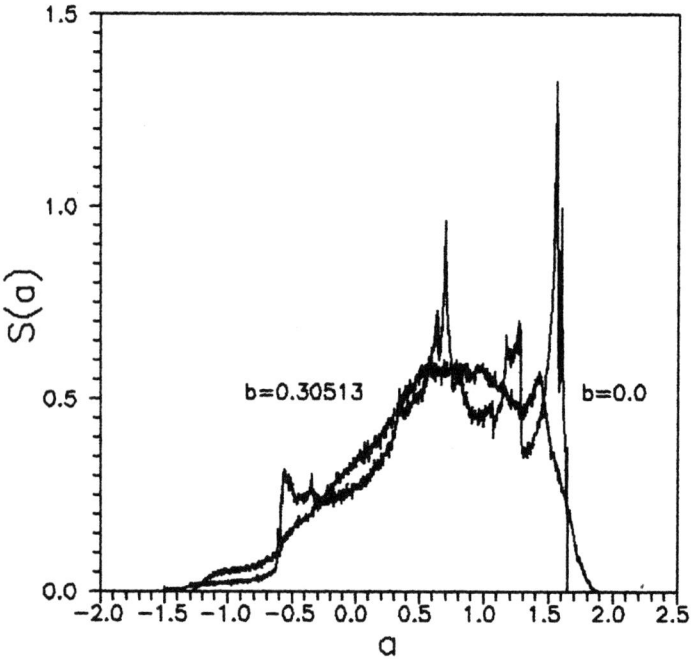

**FIGURE 16.** The same as in FIGURE 15(a), (b) for $\beta = 0.30513$.

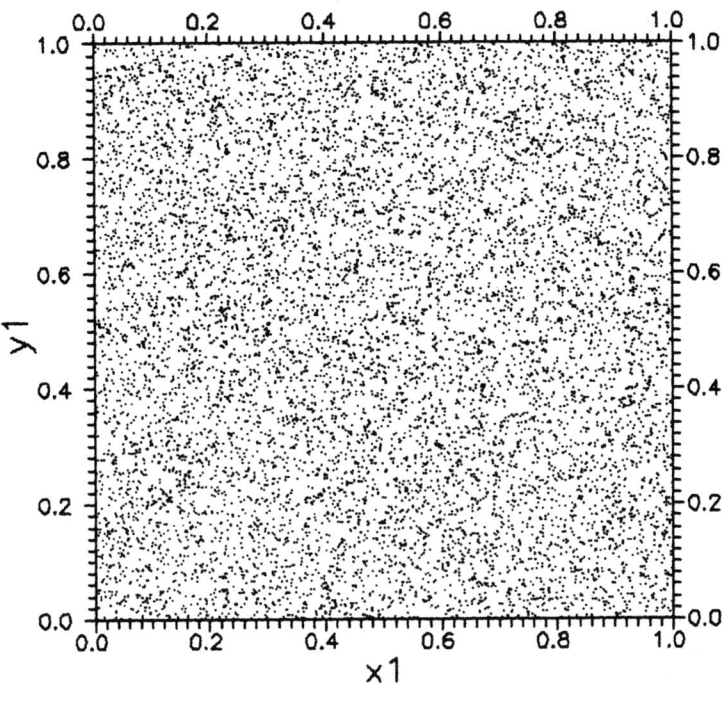

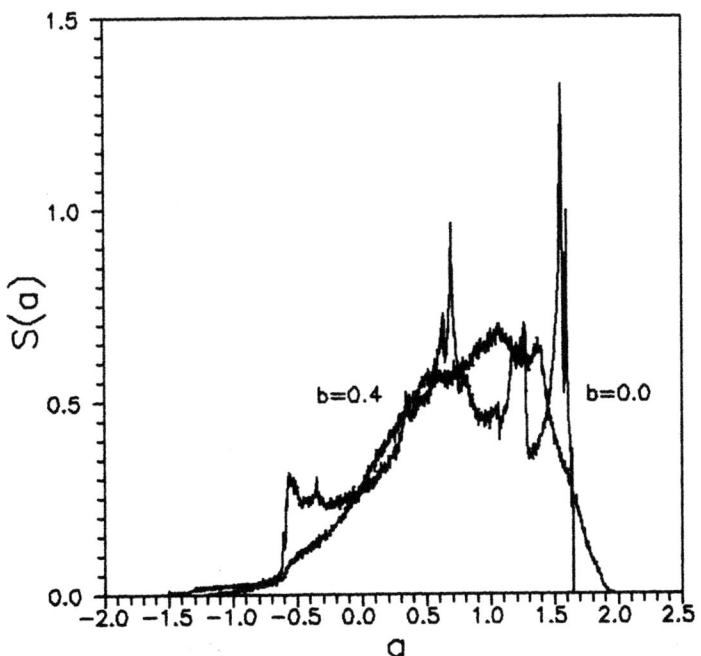

**FIGURE 17.** The same as in FIGURE 15(a), (b) for $\beta = 0.4$.

4. The Poincaré consequents that correspond to particular intervals of values of the stretching number form invariant subsets on the Poincaré surface of section. These subsets do not overlap. They have large compact regions, but also infinite lines of small thickness.
5. The crossing of cantori with small gaps sometimes takes a long time. During that period the distribution of the Poincaré consequents is not homogeneous. But after a sufficiently long time their distribution in the chaotic domain is completely homogeneous. The system becomes microcanonical outside the islands. In particular, the stickiness observed for short times around islands becomes inconspicuous after a long time (although it seems that a very small degree of stickiness persists for arbitrarily long times).
6. Finally, we found invariant spectra in four-dimensional maps, corresponding to Hamiltonian systems of 3 degrees of freedom. In particular, we studied coupled standard maps. If the coupling parameter is small, the chaotic regions are localized for extremely long times. However, for a larger coupling the chaotic regions abruptly increase their size after a finite time. This is an indication of Arnold diffusion. The corresponding spectra change and tend to a "final" invariant form. But this is reached only after a very long time.

When the coupling is sufficiently large, the diffusion is very fast and the approach to the final spectrum is also faster.

Similar results were also found in cases of many degrees of freedom.

## ACKNOWLEDGMENTS

One of the authors (G. C.) acknowledges many useful discussions with Dr. H. Kandrup.

## REFERENCES

1. FUJISAKA, H. 1983. Prog. Theor. Phys. **70**: 1264.
2. GRASSBERGER, P. & I. PROCACCIA. 1984. Physica **D13**: 34.
3. BENZI, R., G. PALADIN, G. PARISI & A. VULPIANI. 1985. J. Phys. A **18**: 2157.
4. GRASSBERGER, P., R. BADII & A. POLITI. 1988. J. Stat. Phys. **51**: 135.
5. UDRY, S. & D. PFENNIGER. 1988. Astron. Astrophys. **198**: 135.
6. CRISANTI, A., G. PALADIN & A. VULPIANI. 1988. J. Stat. Phys. **53**: 583.
7. SEPULVEDA, M. A., R. BADII & E. POLLAK. 1989. Phys. Rev. Lett. **63**: 1226.
8. ABARBANEL, H. D. I., R. BROWN & M. B. KENNEL. 1991. Int. J. Mod. Phys. **B5**: 134.
9. GROBGELD, D., E. POLLAK & J. ZAKRZEWSKI. 1992. Physica **D5**: 368.
10. FROESCHLÉ, C., CH. FROESCHLÉ & E. LOHINGER. 1993. Celest. Mech. Dyn. Astron. **56**: 307.
11. KANDRUP, H. E. & M. E. MAHON. 1994. In Three-Dimensional Dynamical Systems, S. Gottesman, J. Ipser, and H. E. Kandrup, Eds. **751**: 93. N.Y. Academy of Sciences Annals, New York.
12. VOGLIS, N. & G. CONTOPOULOS. 1994. J. Phys. **A27**: 4899.
13. CONTOPOULOS, G., E. GROUSOUZAKOU & N. VOGLIS. 1995. Astron. Astrophys. In press.
14. CONTOPOULOS, G., L. GIORGILLI & A. GALGANI. 1978. Phys. Rev. **A18**: 1183.
15. SMITH, H. & G. CONTOPOULOS. 1995. Spectra and Lyapunov numbers in pulsating systems. This issue.
16. HÉNON, M. 1976. Commun. Math. Phys. **50**: 69.

17. CONTOPOULOS, G. 1971. Astron. J. **76**: 147.
18. SHIRTS, R. B. & W. P. REINHARDT. 1982. J. Chem. Phys. **74**: 5207.
19. MAHON, M. E., R. A. ABERNATHY, B. O. BRADLEY & H. E. KANDRUP. 1995. Mon. Not. R. Astron. Soc. **275**: 443.
20. BENSIMON, D. & I. P. KADANOFF. 1984. Physica **D13**: 83.
21. MACKAY, R. S., J. D. MEISS & I. C. PERCIVAL, 1984. Physica **D13**: 95.
22. CONTOPOULOS, G., H. VARVOGLIS & B. BARBANIS. 1987. Astron. Astrophys. **172**: 55.
23. KANTZ, H. & P. GRASSBERGER. 1988. J. Phys. A. **21**: L127.

# Chaos and Order in Time-Periodic Potentials and the Problem of Structural Stability[a]

HENRY E. KANDRUP,[b,c] ROBERT A. ABERNATHY,
M. ELAINE MAHON,[b] AND BRENDAN O. BRADLEY

*Department of Astronomy*
*University of Florida*
*Gainesville, Florida 32611*

## 1. INTRODUCTION

It would seem obvious that various ideas from nonlinear dynamics can have important and immediate applications to a number of different problems in astrophysics. However, at least within the context of galactic dynamics many of these applications are novel from the viewpoint of the nonlinear dynamicist, in that they entail a somewhat unusual perspective.

As a practical matter, one cannot follow orbits of individual stars within a real galaxy, both because of limited resolution and because the appropriate timescales are long compared with a human lifetime. Moreover, even if one could follow the details of individual orbits, for example, in the context of a numerical simulation, it is not at all obvious that this is the "right" thing to do. In galactic dynamics, the bulk gravitational potential is determined self-consistently by the overall mass distribution, rather than being specified as a fixed entity; but, if one wishes to address this problem of self-consistency, one must eventually focus on the statistical properties of collections of orbits, rather than on individual trajectories.

Another obvious fact is that, although old compared with astronomers, galaxies are, in terms of their natural timescale, relatively young objects. For a galaxy like the Milky Way, a characteristic crossing time $t_{cr} \sim 10^8$ yr, but the entire age of the Universe, $t_H$, is only $\sim 10^{10}$ yr. It follows that, when considering galaxies, one is confronted with objects that are at most $\sim 100 t_{cr}$ in age. This implies, however, that the consideration of an asymptotic $t \to \infty$ limit, implicit in much work in nonlinear dynamics, may not be well motivated for galactic dynamics. Rather, the age of a typical galaxy is sufficiently short that correctly addressing many questions of physical interest may involve a consideration of short time, possibly transient, behavior.

[a]One of the authors (H. E. K.) was supported in part by National Science Foundation Grant PHY92-03333. Another of the authors (M. E. M.) was supported by the University of Florida as a Postdoctoral Reseach Associate. Some of the numerical calculations reported herein were facilitated by computer time provided by IBM through the Northeast Regional Data Center (Florida). Others were facilitated by tiime provided on a KSR by the Parallel Research Laboratory at the University of Florida and on the CM-5 by the Advanced Computing Laboratory at Los Alamos National Laboratory.
[b]Additional address for H. E. Kandrup and M. E. Mahon: Institute of Fundamental Theory, University of Florida, Gainesville, Florida 32611.
[c]Additional address for H. E. Kandrup: Department of Physics, University of Florida, Gainesville, Florida 32611.

For these reasons, the authors have become involved in the formulation of a new approach to the application of nonlinear dynamics to galactic dynamics, a theory of *transient ensemble dynamics*,[1-5] which involves a consideration of the short time behavior of ensembles of orbits, restricting attention to astrophysically relevant timescales $\sim 100 t_{cr}$ and eschewing longer time integrations except to the extent that they provide useful information about shorter time behavior.

That this alternative perspective is of practical importance is readily illustrated by the simple example of nonintegrable trajectories in a two degree of freedom system, characterized by a time-independent Hamiltonian. For such a system, in the $t \to \infty$ limit there are only two classes of orbits, namely regular and stochastic (i.e., chaotic), separated by invariant KAM tori (cf. references 6 and 7). However, on shorter timescales the stochastic orbits oftentimes divide into two relatively distinct subclasses, namely confined and unconfined stochastic orbits, which are separated by cantori,[8] these corresponding to fractured KAM tori associated with the breakdown of integrability.

Because these cantori contain a cantor set of holes,[6-8] it is possible for orbits to pass through them, so that they do not serve as absolute barriers. However, if as is oftentimes the case, the holes are extremely small, the timescale on which orbits wend their way through the maze provided by the cantori (cf. the "turnstile model" of MacKay et al.[9,10]) can be very long, much longer in physical units than the age of the Universe, $t_H$.

Suppose, for example, that one chooses a localized ensemble of initial conditions of given energy $E_0$, corresponding to unconfined stochastic orbits, and then evolves these initial data into the future using the Hamiltonian equations of motion. In this case, one observes[1,2,5] that, for many potentials, the ensemble will exhibit a coarse-grained exponential approach, on a timescale $\ll t_H$, toward a near-invariant distribution, the form of which appears to be approximately time-independent in the sense that any subsequent systematic evolution only proceeds on a timescale $\gg t_H$. In other words, the ensemble approaches a statistical near-equilibrium.

However, this near-invariant distribution need not coincide, even approximately, with the true invariant distribution. Specifically, for energies where substantial measures of both regular and stochastic orbits exist, and where cantori play an important role, one observes typically[4,5] that, on a much longer timescale, the ensemble does indeed exhibit a slow systematic variation as orbits eventually breach the cantori to sample both unconfined and confined phase space regions. Eventually, the ensemble appears to sample a new time-independent distribution corresponding to a true invariant measure, characterized by a microcanonical population of all the stochastic phase space regions. In other words, the ensemble appears to exhibit a slow systematic evolution away from the quasi equilibrium and toward a true equilibrium.

These model calculations, like most work in galactic dynamics, involve various restrictive assumptions which, at one level or another, must fail. An idealized description of a galaxy characterized completely by a time-independent potential necessarily neglects the effects of an external environment, which in general will induce time-dependent perturbations. Moreover, even assuming that the galaxy can be treated as a completely isolated entity, there is no guarantee that a self-consistent evolution described by a collisionless Boltzmann, or Vlasov, equation will yield an efficient pointwise approach toward a time-indepenent equilibrium. Even the idea that the system can be described adequately by a smooth average potential is significant in that it entails the assumption that discreteness effects associated with individual stars and other substructures can be completely ignored.

It would therefore seem reasonable to investigate the problem of structural stability of galactic models by examining the effects of various perturbations that could mimic these realistic complications. Specifically for various sorts of perturbations it is natural, and important, to determine the minimum amplitude of the perturbing influence required to induce significant changes within a time $t_H \sim 100 t_{cr}$.

In this regard, three comments are in order: (1) Even if these influences are negligible for integrable, or near-integrable, potentials, they may prove to be extremely important for strongly nonintegrable potentials admitting global stochasticity. (2) Even very small non-Hamiltonian perturbations could prove important by serving to relax the constraints associated with Liouville's theorem. The crucial point here is that, both Newtonianly[11] and relativistically,[12] the collisionless Boltzmann equation is a (noncanonical) infinite-dimensional Hamiltonian system. (3) One might expect physically that time-dependent perturbations could be important by triggering various sorts of resonant behavior.

Statistical physics, as developed during the twentieth century, suggests a concrete framework in terms of which to model such perturbing influences. The basic idea is to start with a "complete" Hamiltonian description of a large composite entity, including both the system and its surrounding environment, and to trace over the "uninteresting" degrees of freedom so as to obtain a (typically nonlocal) reduced description of the "interesting" degrees of freedom, that is, the system (cf. reference 13). One then finds that, oftentimes, the high-frequency effects associated with the uninteresting degrees of freedom can be modeled as friction and noise, related via a fluctuation–dissipation theorem (cf. references 14 and 15, and references cited therein), whereas the lower frequency effects can be Fourier decomposed into a sum of periodic disturbances.

It is therefore natural, and physically well motivated, to investigate the structural stability of orbits in strongly nonintegrable Hamiltonian systems toward the effects of (1) friction and noise, and (2) periodic driving. The effects of the former have already been described elsewhere.[1,4,5] The aim here is to examine the effects of periodic driving and then to compare these with the effects of friction and noise.

Section 2 begins by introducing various tools that can be used to provide useful characterizations of the short time behavior of stochastic trajectories, both for individual orbits and for ensembles thereof. Section 3 then uses these tools to describe the effects of friction and noise on ensembles of orbits. Section 4 turns to the effects of low-amplitude periodic driving on individual orbits, and Section 5 considers the effects of periodic driving on ensembles of stochastic orbits.

## 2. DISTINGUISHING BETWEEN CHAOS AND ORDER ON SHORT TIMESCALES

For a time-independent Hamiltonian system with two degrees of freedom, in an asymptotic $t \to \infty$ limit there are only two classes of orbits, namely regular and stochastic, and it is completely straightforward to distinguish between these two classes. One way to do this involves computing surfaces of section, and determining whether successive intersections are restricted to curves or whether instead they appear to fill a finite two-dimensional area. Another way involves estimating the Liapounov exponent, $\chi$, which probes the average rate of instability associated with a stochastic orbit. Thus specifically,[16]

$$\chi \equiv \lim_{t \to \infty} \lim_{\delta z \to 0} \frac{1}{t} \log \left( \frac{\delta z(t)}{\delta z(0)} \right), \qquad (1)$$

where $\delta z$ is the Euclidean phase space deviation.

However, if one is interested in the finite time behavior of orbits, the distinction between regular and stochastic is not adequate and, as stressed already, the $t \to \infty$ limit is unphysical. Moreover, for a time-dependent system without a conserved energy, as would arise naturally if the system were perturbed by friction and noise or by a time-dependent periodic driving, surfaces of section are not adequate. Nevertheless, it is still natural to seek operational answers to such questions as: Is the orbit regular or stochastic? And, is the stochastic orbit confined or unconfined?

Direct visual inspection of the form of an orbit provides one simple way of answering these questions that is (by definition) in accord with physical intuition. However, this approach is intrinsically subjective, and hence difficult to quantify, so as, for example, to permit the development of an automatable algorithm that a computer could use to analyze a large ensemble of orbits. Fortunately, though, the authors have found two other, quite different, approaches that when combined together can provide simple, automatable algorithms in terms of which to distinguish among regular orbits, confined stochastic orbits, and unconfined stochastic orbits.

One approach[17] involves a spectral analysis of finite orbit segments. The idea here is, given a discrete data set of (say) coordinates $x$ and $y$ at fixed times $t_j$, to compute discrete Fourier transforms, such as

$$x(\omega_k) = \sum_{j=1}^{N} x(t_j) \exp(-i\omega_k t_j) \qquad (2)$$

and the corresponding $y(\omega_k)$, for some set of frequencies $\{\omega_k\}$ with $k \leq j$. The crucial question then is whether the support for $|x(\omega)|$ and $|y(\omega)|$ is concentrated at a few fundamental frequencies and harmonics thereof, or whether instead the power is broader band.

As is well known to galactic dynamicists,[18,19] regular orbits have spectra given as the sum of near-delta distributions, determined by a few fundamental frequencies. By contrast, unconfined stochastic orbits have broader band power, although much of the power will typically be concentrated near a few principal frequencies. More precisely, spectra for unconfined stochastic orbits seem typically to have at least three different components, namely a broad-band continuum, a set of fuzzed-out peaks at frequencies appropriate for one type of regular orbit, and a different set of fuzzed-out peaks appropriate for another type of regular orbit.

The fact that a stochastic orbit contains substantial power peaking at frequencies appropriate for several different regular orbits is hardly surprising, given the standard interpretation of chaos as arising via resonance overlap. However, this fact *is* useful in that the relative power in the different components provides a concrete, quantifiable diagnostic in terms of which to determine, for example, whether an orbit is "almost a box orbit," "almost a loop orbit," or "a significant admixture of several different orbital types."

This is illustrated in FIGURE 1, which exhibits the configuration space $(x, y)$ trajectory and the associated powers, $|x(\omega)|$ and $|y(\omega)|$, computed using (2) for three different orbits, the details of which are described below. The top orbit clearly looks "boxy," the bottom "loopy," and the middle "more confused," and these differences are reflected by the form of the power spectra. In particular, for the near-box $|y(\omega)|$ has a single steep maximum, for the near-loop two steep maxima, and for the confused case three less steep maxima at or near frequencies where the near-loop and near-box had maxima. Each of these orbits was computed from an initial condition

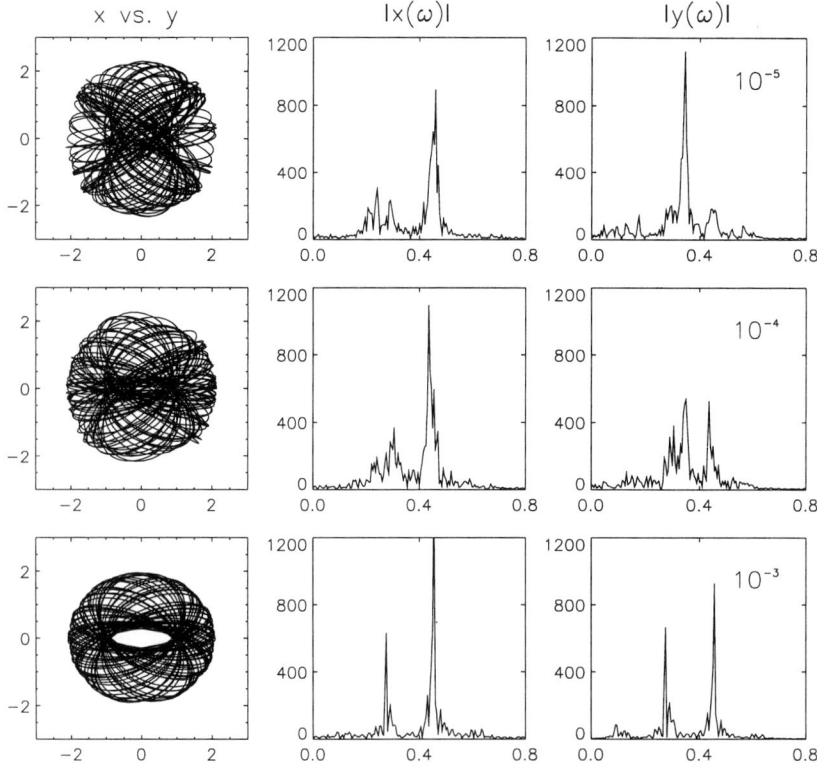

**FIGURE 1.** Configuration space trajectories and power spectra, $|x(\omega)|$ and $|y(\omega)|$, for three different stochastic orbits, computed using (2). All three orbits were generated from the same initial condition, $x = 1.0$, $y = 0.5$, $p_y = 0.0$, and initial energy $E_0 = -0.5$, evolving in the KAMB potential (6), with a core radius $c$ oscillating at frequency $\Omega = 1$. The three orbits correspond, respectively, to perturbations of amplitude $\alpha = 10^{-5}$ **(top)**, $10^{-4}$ **(middle)**, and $10^{-3}$ **(bottom)**.

evoled in a modified anisotropic Kepler (KAMB) potential of the form

$$V(x, y) = -\frac{1}{(c^2 + x^2 + y^2)^{1/2}} - \frac{m}{(c^2 + x^2 + a^2 y^2)^{1/2}}, \quad (3)$$

with $m = 0.3$, $a = \sqrt{0.1} \approx 0.316$, and $c(t)$ oscillating in time about a value $c_0 = 1.0$.

These Fourier spectra serve as effective tools to distinguish unconfined stochastic orbits from other orbits, but it is difficult to distinguish between regular orbits and

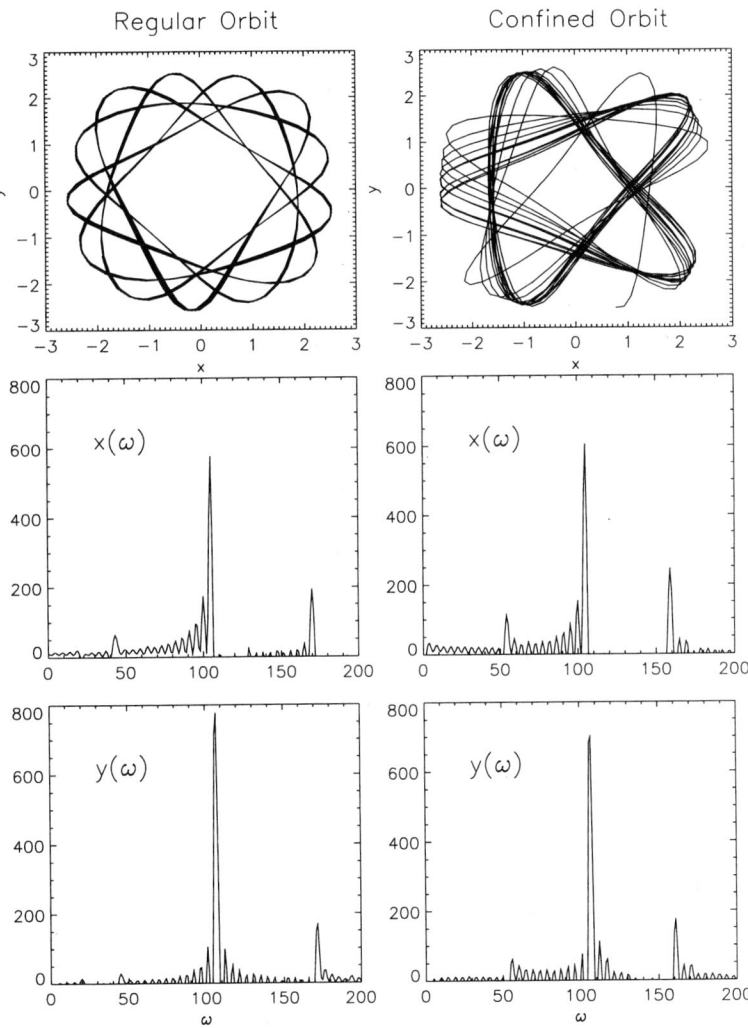

**FIGURE 2.** Configuration space trajectories and power spectra, $|x(\omega)|$ and $|y(\omega)|$, for two different orbits in the $D$-4 potential with $E = 6.0$, one regular and one confined stochastic.

confined stochastic orbits given a discrete set of data points sampling a time interval $t \leq 100t_{cr}$. This is, for example, illustrated in FIGURE 2, which shows the power spectra for two different orbits, one regular and one confined stochastic, computed in the so-called $D$-4 potential,[20] for one particular set of parameter values, namely, $V(x, y) = -(x^2 + y^2) + \frac{1}{4}(x^2 + y^2)^2 - \frac{1}{4}x^2y^2$.

A complementary technique that can be used in distinguishing between order and chaos is the computation of short time Liapounov exponents,[21,22] defined for a time interval $\Delta t$ via a finite time analogue of the usual prescription, that is,

$$\chi(\Delta t) = \lim_{\delta z \to 0} \frac{1}{\Delta t} \log \left( \frac{\delta z(\Delta t)}{\delta z(0)} \right). \tag{4}$$

For regular orbits, $\chi(\Delta t) \to 0$ smoothly as $\Delta t$ increases, a fact well known to nonlinear dynamicists.[16] By contrast, for stochastic orbits, the running $\chi(\Delta t)$ typically shows considerably more irregularity. This is illustrated in FIGURE 3, which exhibits $\chi(\Delta t)$ for four different orbits—one regular, one confined, and two unconfined—in the KAMB potential with $m = 0.3$, $a = \sqrt{0.1} \approx 0.316$, and $c = 1.0$.

By computing these short time Liapounov exponents, it is typically quite simple to distinguish between regular and stochastic orbits, even on relatively short timescales, because of the fact that, for the former, $\chi(\Delta t)$ exhibits little if any structure. However, at the level of individual orbits the distinction between confined and unconfined stochastic is less completely clearcut. Nevertheless, one *can* still distinguish between confined and unconfined stochastic orbits at the level of orbit ensembles by computing a *distribution of short time Liapounov exponents*.[1,3,5]

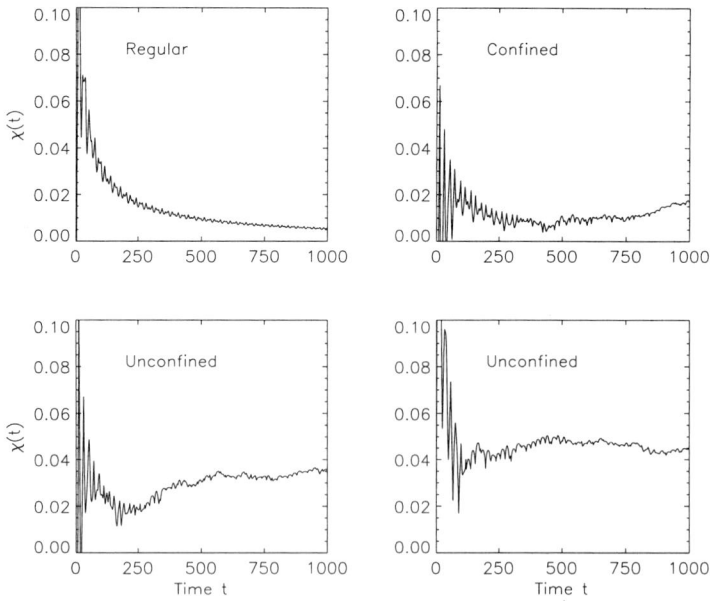

**FIGURE 3.** The short time Liapounov exponent $\chi(\Delta t)$ for four different orbits in the unperturbed KAMB potential (6).

If, for example, one samples a collection of initial conditions corresponding on short times to unconfined stochastic orbits of given energy $E_0$, and integrates in a time-independent potential for a time $\Delta t \sim t_H$, he or she will typically obtain a singly peaked distribution $N[\chi(\Delta t)]$ that is approximately Gaussian, with a dispersion $\sigma \propto (\Delta t)^{-p}$, for $p \leq 1/2$. However, if one considers a set of initial conditions corresponding to substantial numbers of both confined and unconfined orbits, one can instead observe a *bimodal distribution* $N[\chi(\Delta t)]$, which is seemingly well approximated as the sum of two near-Gaussian distributions.

This is, for example, illustrated by the curves in FIGURE 4, again generated for the $D$-4 potential. Here the dashed curves reflect normalized distributions generated by sampling the near-invariant distribution associated with an ensemble of unconfined orbits, whereas the solid curves reflect the corresponding distributions appropriate for a sampling of (an approximation to) the true invariant measure. At the lower energies, cantori are relatively unimportant and the two distributions are quite similar. However, for higher energies cantori become more important and substantial differences begin to emerge. The key point here is that both confined and unconfined stochastic orbits are exponentially unstable, but that the confined orbits are less unstable overall and hence characterized by a distribution of short time Liapounov exponents peaked at much smaller values.

The three diagnostics just described—configuration space, Fourier spectra, and short time Liapounov exponents—are good for at least three reasons: (1) they do not require time-independent evolution equations or an asymptotic $t \to \infty$ calculation; (2) spectral analysis and short time Liapounov exponents are complementary in that they are better suited for different tasks (spectral techniques to single out unconfined stochastic orbits, short time Liapounov exponents to identify regular orbits), but all three diagnostics typically agree with one another and, in appropriate

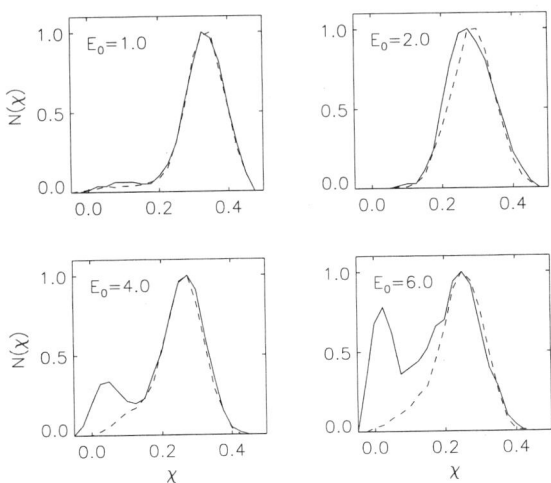

**FIGURE 4.** Distributions of short time Liapounov exponents computed in the $D$-4 potential at varying energies $E$ for a sampling of the unconfined deterministic near-invariant measure (*dashed curves*) and the true deterministic invariant measure (*solid curves*).

limits, with the usual techniques; (3) most importantly, they all coincide with physical intuition. Their applications to an analysis of the effects of friction and noise and of periodic driving is the focus of the remainder of this paper.

## 3. THE EFFECTS OF FRICTION AND NOISE

The structural stability of Hamiltonian trajectories toward the effects of friction and noise was studied by effecting Langevian simulations,[1,23-26] in which the Hamiltonian evolution equations are perturbed by allowing for a dynamical friction, $-\eta \mathbf{p}$, and noise characterized by a "temperature," or mean squared velocity, $\Theta$. This entailed solving equations of the form (cf. references 27 and 28)

$$\frac{d\mathbf{r}}{dt} = \mathbf{p} \quad \text{and} \quad \frac{d\mathbf{p}}{dt} = -\nabla V - \eta \mathbf{p} + \mathbf{F}, \tag{5}$$

where $V(\mathbf{r})$ is a time-independent two-dimensional potential, and the last two terms reflect the effects of the friction and noise. Here $\mathbf{p}$ denotes momentum, $\eta$ is the coefficient of dynamical friction, and $\mathbf{F}$ is a stochastic force (i.e., noise). When performing the integrations, $\mathbf{F}$ is treated as a random variable, the idea being to perform multiple noisy realizations of the same initial data and then to extract statistical properties. Because the force is a random variable, it is characterized completely by its moments. The integrations were effected using an algorithm[29] that samples a near-Gaussian distribution with first and second moments

$$\langle F_i(t) \rangle = 0 \quad \text{and} \quad \langle F_i(t_1) F_j(t_2) \rangle = 2\Theta \eta \delta_{ij} \delta_D(t_1 - t_2). \tag{6}$$

In the first instance,[1,23-25] these integrations were effected assuming a constant coefficient of dynamical friction $\eta$ that, via the fluctuation–dissipation theorem, requires a simultaneous consideration of additive white noise.[27,28] Other alternatives can, should, and are being investigated.[26]

At this stage, an important point needs to be stressed: One knows full well that friction and noise will eventually become important on the natural timescale $t_R \sim \eta^{-1}$, as a generic initial ensemble evolves toward a thermal, that is, canonical, distribution. This is *not* the question of interest here. Rather, the question is whether friction and noise can have substantial effects already on much shorter timescales $\ll t_R$.

To date, large numbers of experiments have been performed for three different model systems, namely the D-4 system,[20] the sixth-order truncation of the three-particle Toda lattice,[30] and the KAMB potential (6). The most significant conclusion is that, when expressed in units of the natural timescale $t_{cr}$, the qualitative and semi-quantitative results derived for the three different potentials prove quite similar. This would suggest that the conclusions are relatively robust, depending only on such gross topological features as the existence of cantori.

When viewed in terms of their effects on the energy, which is a collisionless invariant, the friction and noise serve to induce a classic diffusion process, whereby unperturbed and perturbed orbits only diverge significantly on the natural timescale

$t_R \sim \eta^{-1}$. Thus, for example, the rms change in energy satisfies

$$\delta E_{\rm rms}^2 \equiv \langle |E_{\rm per} - E_0|^2 \rangle = A(E_0)E_0 \Theta \eta t, \tag{7}$$

with $A(E_0)$ a slowly varying function of the unperturbed energy, $E_0$, of order unity.

However, when viewed in the full configuration or momentum space, the effects are more complicated, and depend on orbit class. Unperturbed and perturbed *regular* orbits typically diverge as a power law, $t^p$, in time, albeit with a larger exponent $p$ than is observed for the integrable case of a harmonic oscillator potential, and one only gets macroscopic deviations on a timescale $\sim \eta^{-1}$. By contrast, unperturbed and perturbed *stochastic* orbits diverge exponentially at a rate that is set by the Liapounov exponent $\chi$. The fact that $\chi^{-1} \sim t_{\rm cr}$ thus implies that, in this case, macroscopic deviations can arise already within a time $\ll t_R$. For both regular and stochastic initial conditions, the rms quantities $\delta r_{\rm rms}$ and $\delta p_{\rm rms} \propto (\Theta \eta)^{1/2}$, consistent with theoretical predictions.[23,31]

Naively one might anticipate that, on short timescales, this pointwise instability of stochastic orbits will have no appreciable effects on the statistical properties of ensembles of orbits. Specifically, one might anticipate that the only effects of the friction and noise are to continually deflect a trajectory from one deterministic orbit

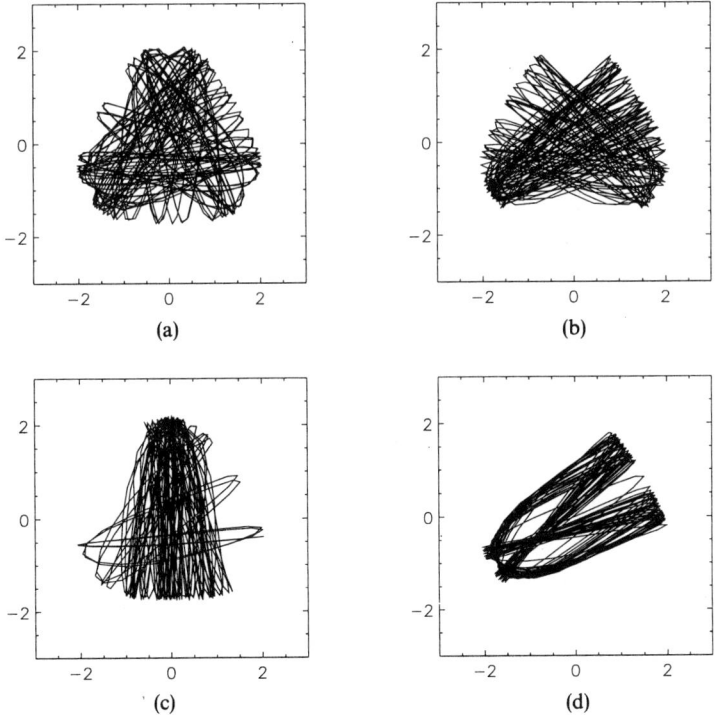

**FIGURE 5.** (a) A Hamiltonian trajectory and (b)–(d) three noisy perturbed trajectories, each computed in the truncated Toda potential from the same initial condition with $E = 20$ for a total time $t = 100$. The perturbed orbits were integrated with $\Theta = 20$ and $\eta = 10^{-6}$.

to another with essentially the same energy and the same statistical properties. This, however, is false. Even very weak friction and noise can also alter the *statistical properties of ensembles of stochastic orbits* on relatively short timescales, for example, by dramatically accelerating the rate of diffusion through cantori.[1,32]

Thus, in particular, if the deterministic near-invariant measure be evolved into the future, allowing for even very weak friction and noise, one observes oftentimes a rapid ($t \ll 100 t_{cr}$) evolution toward a noisy near-invariant measure that is quite different in form from the deterministic near-invariant measure and much closer to the true deterministic invariant measure. In other words, it would appear that even very low-amplitude friction and noise can dramatically accelerate the rate of approach toward a statistical equilibrium.

The obvious point in all this is that friction and noise can induce changes in orbit class, both from unconfined to confined and vice versa. This is, for example, illustrated in FIGURES 5 and 6, which for two different initial conditions, show both the unperturbed Hamiltonian trajectory and three perturbed trajectories. The efficacy of this process depends of course on the relative sizes of the phase space regions containing confined an unconfined orbits. However, when both regions have substantial measure, even very weak friction and noise can induce significant numbers of transitions. For all three potentials, one finds that, for perturbations

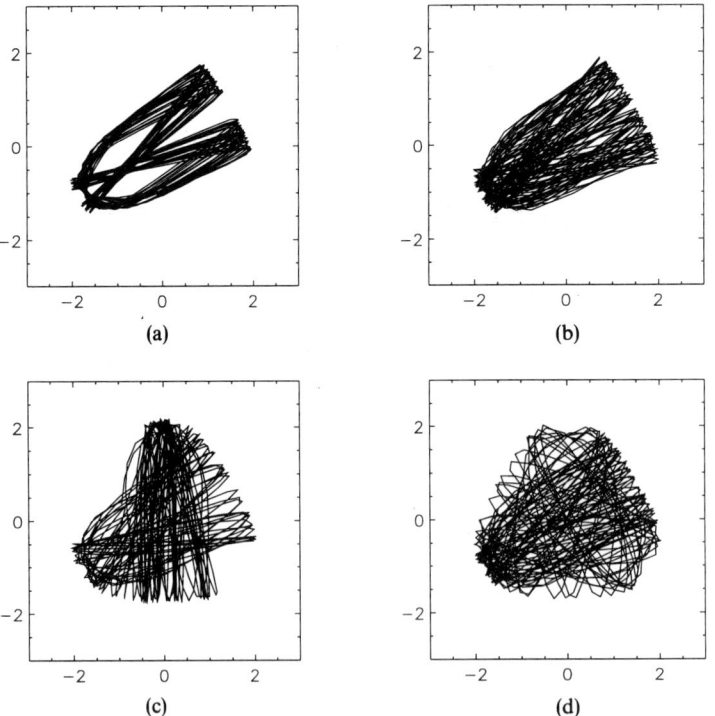

**FIGURE 6.** Same as FIGURE 5, but for a different initial condition, again with $E = 20$.

corresponding to a relaxation time as long as $t_R \sim 10^{12} t_{cr}$, friction and noise only have minimal effects within a time $t \sim 100 t_{cr}$. However, at $t_R \sim 10^9 t_{cr}$ transitions begin to occur, and, for $t_R \sim 10^6 t_{cr}$ such transitions are typically quite common overall, occurring in some energy ranges for as many as $\sim 50\%$ of the orbits within $t \sim 100 t_{cr}$. For $t_R$ as short as $\sim 10^3 t_{cr}$, transitions are so common that the distinction between filling and confined becomes essentially meaningless.

What this implies is that even very weak friction and noise, with $t_R \sim 10^6 - 10^9 t_{cr}$, can significantly alter the statistical properties of ensembles of orbits on timescales $t < t_H \sim 100 t_{cr}$. This, however, is an interesting number since, for "collisionality" within galaxies (i.e., close encounters associated with discreteness effects), $t_R \sim 10^6 - 10^9 t_{cr}$ (cf. reference 33). Indeed, one is led thereby to suppose[1,25] that realistic quasi equilibria must manifest a *detailed balance* for transitions between different orbit classes, as induced by discreteness effects modeled as friction and noise.

Suppose that one has constructed a galactic model that invokes confined stochastic orbits to help support some structure. A good example might be a bar within a spiral galaxy, where, near corotation or some other resonance, few if any regular orbits exist. In this case, if the subsequent evolution entails a net outward flow of orbits through the cantori, crucial support may be lost so that the structure will dissipate. Alternatively, if there is a net trapping of orbits inside the cantori, the structure may be reinforced and hence become more pronounced.

The moral of the story is that even very weak friction and noise can change both the *pointwise* and the *statistical* properties of orbits on relatively short timescales, much shorter in particular than the timescale on which numerical errors can accumulate. This fact also has important potential implications for the problem of "shadowing" numerical orbits.[34,35] Astronomers[36] and other scientists are rightly concerned about the degree to which a numerical simulation of a chaotic system can provide a reasonable characterization of the true solutions to that model system, that is, the extent to which the numerical solutions can shadow the true dynamics. However, there is also another important question of shadowing that needs to be addressed:[37,38] To what extent can the real world, which contains all sorts of blemishes and irregularities that the theorist might like to ignore, shadow either the equations selected to model one's system or their numerical realization?

## 4. PERIODIC DRIVING OF INDIVIDUAL ORBITS

The effects of periodic driving have been investigated already by subjecting the core radius $c$ and the anisotropy parameter $m$ in the KAMB potential (6) to time-dependent oscillations. Specifically, the constant values of $c$ and $m$ were modulated separately by allowing for oscillations of the form

$$c(t) = c_0(1 + \alpha_c \cos \Omega t) \quad \text{and} \quad m(t) = m_0(1 + \alpha_m \cos \Omega t), \tag{8}$$

with $c_0 = 1.0$ and $m_0 = 0.3$. The amplitudes were allowed to assume a broad range of values, $10^{-12} \leq \alpha_{c,m} \leq 10^{-1}$ and the driving frequency was varied in the interval $0.01 \leq \Omega \leq 100$. For typical energies in this potential, both regular and stochastic orbits have most of their power concentrated at frequencies $\omega \sim 0.05-1.0$, so that the values of $\Omega$'s considered allowed for a broad range of resonant couplings. A systematic examination of other potentials is currently underway.[26]

This research entailed an examination of the $(\alpha_{c,m}, \Omega)$ parameter space, studying both the effects of variable amplitude $\alpha_{c,m}$ at fixed driving frequency $\Omega$ and the effects of variable $\Omega$ for fixed $\alpha_{c,m}$. The aim was to determine the qualitative and semiquantitative effects of the periodic driving, both on the orbital energies and on the overall topology of the orbits. Results appertaining to individual orbits are summarized in this section. The results derived for ensembles of orbits are summarized in the following section.

Consider first the effects of periodic driving on the orbital energies. Here variations in $\alpha_{c,m}$ at fixed driving frequency $\Omega$ lead to no surprises.

If a regular orbit be driven at near resonance with one of its natural frequencies, one observes an oscillatory response, oftentimes manifestly multiperiodic, characterized by an amplitude that, for relatively small $\alpha_{c,m}$, scales almost exactly linearly in $\alpha_{c,m}$, and by a period that is nearly independent of $\alpha_{c,m}$. However, as $\alpha_{c,m}$ increases to values as large as $\sim 10^{-3}$, this simple scaling with $\alpha_{c,m}$ typically fails, and one finds that both the amplitude and the characteristic periodicity of the response assume somewhat different forms. When $\alpha_{c,m}$ is raised to sufficiently high values, the orbit exhibits large-scale variations in the total energy, triggering changes to different orbit classes, including unconfined stochasticity.

If instead the regular orbit be driven at a frequency $\Omega$ that is far from resonance with any of the natural frequencies $\omega$, a simple, multiperiodic oscillatory response is not observed. However, it remains true that the amplitude of the response is roughly proportional to $\alpha_{c,m}$. An analogous behavior is observed when a typical unconfined stochastic orbit is subjected to identical pulsations.

Varying $\Omega$ at fixed amplitude $\alpha_{c,m}$ leads to more interesting, and nontrivial, results. Suppose, in particular, that $\alpha_c$ or $\alpha_m$ is nonzero, with a magnitude of order $10^{-3}$ or less, and that the pulsation frequency $\Omega$ is permitted to vary. For a regular orbit, a computation of the maximum energy excursion $\delta E_{\max}(t_H)$ arising within a time $\sim t_H$ (or any other comparable diagnostic of short time changes in energy), will then show clear unambiguous evidence of resonant coupling. Specifically, there is a direct one-to-one correspondence between (1) the frequencies $\Omega$ that trigger especially large responses and (2) harmonics of the unperturbed orbital frequencies $\omega$.

As illustrated in FIGURE 7(a), the observed patterns tend to be particularly simple for the case of loop orbits, but, as illustrated in FIGURE 7(b), one also observes a relatively simple pattern for boxes. In both these cases, the locations of the successive local maxima of $\delta E_{\max}$ with increasing $\Omega$ are separated by intervals $\delta \Omega$ that correlate directly with multiples of the natural frequencies $\omega_i$ associated with the unperturbed orbit.

The response exhibited by a driven stochastic orbit is substantially more complicated, but nevertheless still associated with resonant coupling. As shown in FIGURE 7(c) and (d), if, for an unconfined stochastic orbit, one computes $\delta E_{\max}$ as a function of the driving frequency $\Omega$, he or she will observe a complicated pattern comprised of a large number of relatively narrow bands interspersed in a broad-band continuum. This continuum can be associated tentatively with the broad-band continuum of natural frequencies contributing to the unperturbed power spectra for $|x(\omega)|$ and $|y(\omega)|$. The complicated pattern of multiple maxima can then be interpreted as

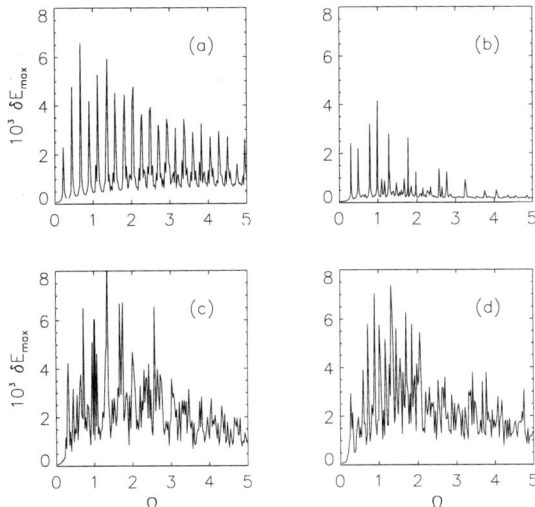

**FIGURE 7.** The maximum energy excursion $\delta E(t_H)$ as a function of driving frequency $\Omega$, computed in a perturbed KAMB potential for (**a**) a regular loop orbit, (**b**) a regular box orbit, and (**c, d**) two unconfined stochastic orbits. In each case, $E_0 = -0.3$ and the core radius $c$ was pulsed with amplitude $\alpha_c = 10^{-3}$.

reflecting complicated couplings with all the different frequencies $\omega_i$ at or near which the unperturbed power spectra $|x(\omega)|$ and $|y(\omega)|$ have particularly significant support.

An examination of FIGURE 7(c) and (d) would also suggest that a plot of $\delta E_{max}(\Omega)$ might exhibit significant structure even on very small scales. A more detailed investigation, one example of which is illustrated in FIGURE 8, indicates that this is in fact true. This figure summarizes orbits computed for the same initial condition as FIGURE 7(d), now pulsing $m$, rather than $c$, with amplitude $\alpha_m = 10^{-3}$. The four different panels were generated by (a) focusing on frequency intervals of width $\Delta\Omega = 5.0$, $0.5$, $0.05$, and $0.005$, (b) computing $\delta E_{max}$ for an equal number of equidistant points within each interval, and then (c) plotting the resulting $\delta E_{max}$. Visual inspection of the output indicates that the amount of structure does not decrease significantly on short scales, this indicating that $\delta E_{max}$ is a very sensitive function of the driving frequency $\Omega$, if not necessarily fractal.[32]

Overall, regular orbits appear robust in form up to frequencies $\alpha_{c,m} \sim 10^{-3}$, a fact confirmed by three different diagnostics, namely (1) a visual inspection of the orbits, (2) an examination of the form of the spectra $x(\omega)$ and $y(\omega)$, and (3) a computation of the short time Liapounov exponent $\chi(\Delta t)$ for time intervals $\Delta t \leq t_H$. At higher amplitudes, one begins to observe more substantive changes: the orbits fuzz out as if being deformed into a confined stochastic orbit, the spectra $|x(\omega)|$ and $|y(\omega)|$ begin to show broader band power, and (3) $\chi(\Delta t)$ begins to shows structure, even if the amplitude of $\chi$ remains small. At sufficiently high amplitudes $\alpha_{c,m}$, one sees transitions to completely different orbit classes, including transitions to unconfined stochasticity.

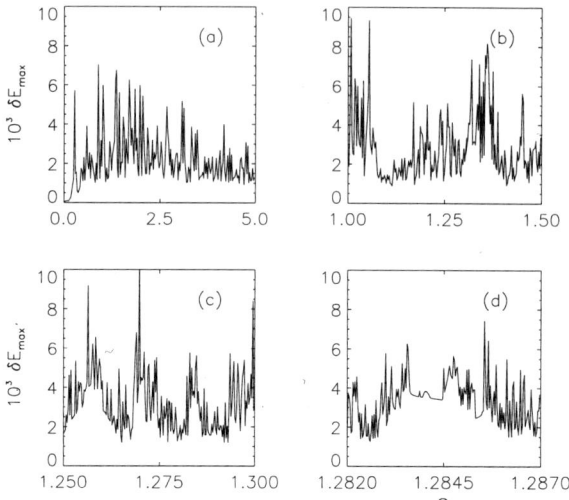

**FIGURE 8.** Small-scale structure of the maximum energy excursion $\delta E(t_H)$, generated from the same initial conditions as FIGURE 7(d), with the anisotropy parameter pulsed with amplitude $\alpha_m = 10^{-3}$. **(a)–(d)** The four panels exhibit, respectively, frequency intervals of size $\Delta \Omega = 5.0, 0.5, 0.05,$ and $0.005$.

Unconfined stochastic orbits are substantially less robust, in the sense that even very small amplitude driving can cause significant qualitative changes. Thus, for example, even relatively low-amplitude driving can transform an orbit from being a near-box either to a near-loop or to a more completely irregular orbit. This is illustrated by the orbits in FIGURE 1, each of which was generated in a pulsed KAMB for the same initial condition, with $E_0 = -0.5$, and with the same driving frequency, $\Omega = 1.0$. The only difference is in the amplitude $\alpha_c$, which for the three successive orbits, takes values $\alpha_c = 10^{-5}, 10^{-4},$ and $10^{-3}$. As is evident from FIGURE 1, these changes in orbit shape correlate with the power spectra $|x(\omega)|$ and $|y(\omega)|$. One also finds that, for these and other orbits, low-amplitude perturbations will typically alter the value of $\chi(t_H)$ by factors of 50% or more.

However, despite this sensitive dependence on the driving amplitude, it is only for values of $\alpha_{c,m}$ as large as $10^{-3}$–$10^{-2}$ that one sees any indications of unconfined stochastic orbits behaving in a fashion that is manifestly near regular. The natural physical interpretation therefore seems to be that the time dependence *per se* is not all that important at relatively low amplitudes. Rather, what matters simply is the fact that the unstable stochastic orbits are being perturbed *somehow*, and that the perturbations are growing exponentially.

## 5. PERIODIC DRIVING OF ENSEMBLES OF ORBITS

The experiments described in this section involved (a) choosing localized ensembles of 400 initial conditions of given unperturbed energy $E_0$, corresponding in the absence of perturbations to stochastic orbits; (b) evolving these initial conditions into the future using the KAMB potential (6), modified by time-dependent perturbations of the form given by (8); and then (c) recording orbital data and $\chi(t)$ at fixed time intervals $\delta t$ for a total time $2t_H$. In visualizing the overall evolution, it is useful to recognize that, for sufficiently weak driving, the energy is almost conserved over time intervals $\sim t_H$, so that one can speak of an evolution restricted to an "almost constant energy hypersurface."

The data were analyzed to answer three different questions about the bulk evolution of the initial ensembles: (1) How does the distribution of energies, $N(\delta E = E - E_0)$, for the orbits within the ensemble evolve over the course of time? (2) Does the time-dependent driving induce significant changes in the distribution of short time Liapounov exponents, $N[\chi(t_H)]$? (3) Does the time-dependent driving change either the rate of approach toward a near-invariant measure, or the form of the invariant measure? In particular, does it trigger substantial diffusion through cantori?

Turn first to the effects on the orbital energy. Suppose that one is considering an ensemble of unconfined stochastic orbits, where the unperturbed power is not dominated initially by a single resonant frequency, and that the driving frequency $\Omega$ is chosen to correspond to a frequency between $\sim 0.05$ and $5.0$. In this case, one finds typically that the effect of the periodic driving is simply to induce a classic diffusion process.

Thus, in particular, the rms change in energy, $\delta E_{\rm rms}(t)$, for individual orbits in the ensemble grows in time at $t^{1/2}$, with an amplitude that is approximately linear in the $\alpha_{c,m}$. In other words,

$$\delta E_{\rm rms}^2 \approx A(E_0, \Omega) E_0^2 \alpha_{c,m}^2 \Omega t, \qquad (9)$$

where $A(E_0, \Omega)$ is a dimensionless function of $E_0$ and $\Omega$. Here the scaling with $t$ and $\alpha_{c,m}$ is very clean, with fractional uncertainty in the exponent $<5\%$. However, one observes substantial variability in the amplitude of the response as $E_0$ and $\Omega$ are changed, so that $A$ can range from values of order unity to values $\ll 1$. In addition, one observes that, at any given instant, the distribution $N[\delta E(t)]$ is approximately Gaussian, with a mean value $\langle \delta E(t) \rangle \approx 0$.

This behavior is exhibited in FIGURE 9. Here the first and second panels display $\delta E_{\rm rms}(t)$ for the same ensemble of orbits, driven with the same frequency $\Omega = 1$ but with variable amplitudes, $\alpha_c = 10^{-4}$ and $10^{-3}$. It is clear that, with the rescaling $\propto \alpha_c^{-1}$, the curves are essentially identical. The lower two panels in this figure exhibit distributions, $N(\delta E)$, generated from the same data at a time $t = 1600$ which, in physical units, corresponds to a time $\sim 0.8 t_H$. The near-Gaussian form should be apparent.

These results rely on the existence of substantial resonant couplings, which in turn implies that $\Omega$ must be of order unity. If instead the driving frequency $\Omega \gg 1$, resonant couplings are far less important and the evolution of $\delta E_{\rm rms}$ is substantially different. Thus, for example, for frequencies $\Omega \geq 10$, one typically observes a small initial jump in $\delta E_{\rm rms}$ roughly proportional to $\alpha_{c,m}$, but thereafter no appreciable systematic changes.

Similarly, if the stochastic ensemble is dominated by a single resonant frequency, as will typically be the case for a collection of confined stochastic orbits, the evolu-

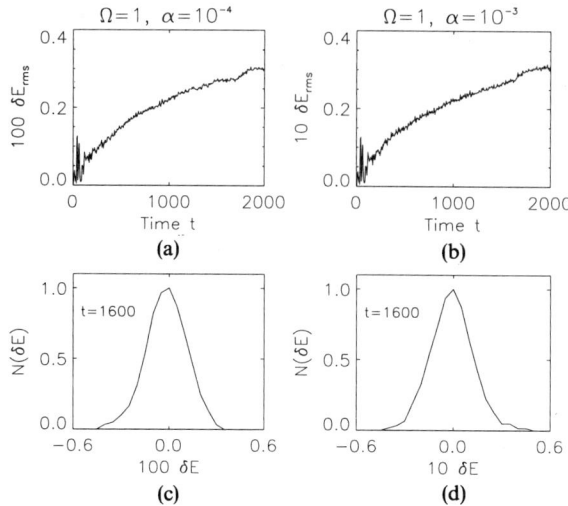

**FIGURE 9.** (a) The rms change in energy $\delta E_{rms}(t)$ for an ensemble of 400 unconfined stochastic orbits with $E = -0.5$, evolved in a perturbed KAMB potential, with the core radius $c$ driven with frequency $\Omega = 1$ and amplitude $\alpha = 10^{-4}$. (b) The identical quantity for the same ensemble evolved with $\Omega = 1$ and $\alpha = 10^{-3}$. (c) The distribution $N(\delta E)$ for the simulation summarized in (a), evaluated at time $t = 1600$, this corresponding to a time $t \sim 0.8 t_H$. (d) The corresponding distribution for the simulation summarized in (b).

tion of $\delta E_{rms}$ is more complicated. In this case, $\delta E_{rms}$ does not grow as $t^{1/2}$ and the distribution $N(\delta E)$ deviates significantly from a Gaussian, being characterized, in particular, by a nonzero mean.

Perhaps the most striking conclusion in all this is that once again, even though periodic driving triggers very complex behavior at the levels of individual orbits, one is led to a simple physical picture for the behavior of stochastic orbit ensembles: Just like friction and noise, a periodic time-dependence triggers a diffusion process in energy space. There is, however, one important difference, namely that, for the case of periodic driving, this response relies crucially on the fact that the orbits are stochastic.

Naively, one might have expected that resonant couplings would be essentially irrelevant when stochastic orbits are subjected to periodic driving, so that the response would be minimal. However, it would seem instead that, overall, resonant couplings are *more important* for stochastic orbits than for regular orbits. If a regular orbit is driven with just the right frequency, it will of course yield a large-amplitude response, but if the driving frequency is substantially displaced from resonance, the response will be relatively small. Stochastic orbits are characterized by broader band power, thus providing them with the ability to resonate with a larger range of frequencies. If, in particular, these stochastic orbits are not dominated by a single resonant frequency, the overall response will be diffusive, triggering a $t^{1/2}$ process that scales linearly in the amplitude of the driving.

Turn now to the effects of periodic driving on the general phase space evolution and the degree of stochasticity. Here one principle conclusion is that pulsing either the core radius $c$ or the anisotropy parameter $m$ has only relatively minor effects on the distribution of short time Liapounov exponents. Thus, in particular, for ensembles of unconfined stochastic orbits one sees no statistically significant changes in $N[\chi(\Delta t)]$ until $\alpha_c$ or $\alpha_m$ is quite large, sufficiently large that substantial changes in energy are observed, so that it no longer makes sense to visualize the evolution as being restricted to an "almost constant energy hypersurface." Moreover, even for $\alpha_c$ and $\alpha_m$ as large as $10^{-2}$, where significant changes in the energy are observed, there is absolutely no evidence that appreciable numbers of unconfined stochastic orbits can become trapped by cantori on timescales as short as $\sim t_H$. For ensembles of confined stochastic orbits, one *can* see changes in the distribution $N[\chi(\Delta t)]$ arising once $\alpha_{c,m}$ is as large as $10^{-3}$, but once again these changes only arise at amplitudes sufficiently large that energy is not approximately conserved.

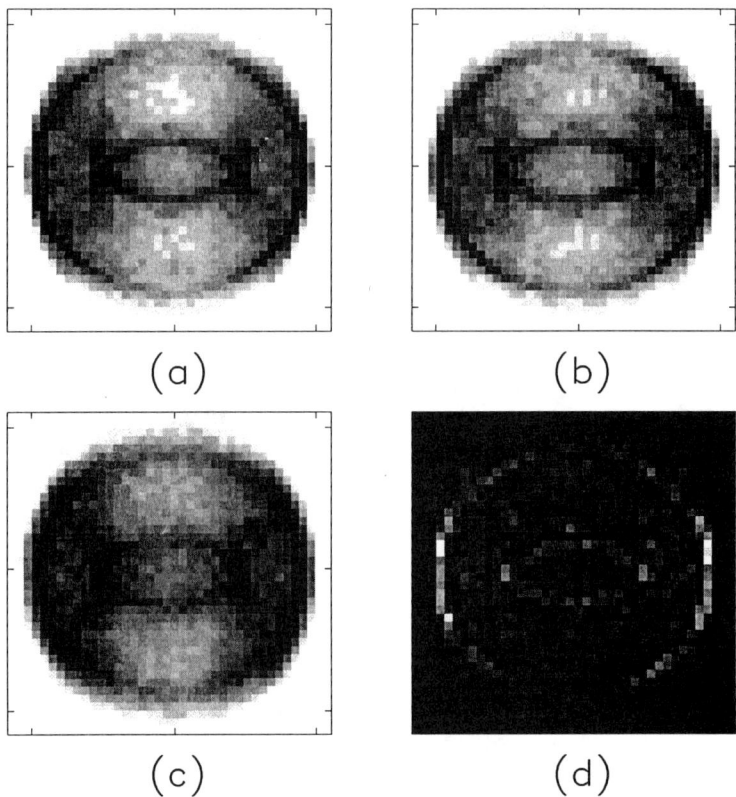

**FIGURE 10.** (a) The deterministic, unpulsed near-invariant measure $f(x, y)$, associated with an initial ensemble of confined stochastic orbits with $E = -0.6$ in the unperturbed KAMB potential. (b) The deterministic, near-invariant measure appropriate for the same ensemble, when the core radius $c$ is pulsed with $\Omega = 1$ and $\alpha = 10^{-3}$. (c) The noisy, near-invariant measure appropriate for the same ensemble, when subjected to friction and noise with $\Theta = 0.6$ and $\eta = 10^{-6}$. (d) The difference between the near-invariant distribution in (a) and (c). The first three plots are normalized identically, with intensity given by a linear scale.

The inference that periodic driving is inefficient in triggering modulational diffusion is further corroborated by an examination of the form of the near-invariant measure associated with the driven ensemble and the rate at which this near-invariant measure is achived. One basic point is that, at least for low-amplitude driving with $\alpha_{c,m} \leq 10^{-2}$, the driven ensembles evolve toward a near-invariant distribution at essentially the same rate $\Lambda(E)$ exhibited by undriven ensembles evolving in a time-independent potential. As for the case of evolution in a time-independent potential,[2,5] one observes a coarse-grained, exponential approach toward a near-invariant distribution on a timescale $\ll t_H$.

The other basic point is that the time-dependent driving does *not* significantly alter the form of the near-invariant distribution. In particular, even for $\alpha_{c,m}$ as large as $10^{-3}$ there is no evidence that unconfined orbits become trapped or that confined orbits become untrapped. There may be some diffusion through cantori triggered by the time-dependent modulation, but this effect, if present, is much weaker overall than the extrinsic diffusion associated with low-amplitude friction and noise.

This fact is illustrated in FIGURE 10, which exhibits projected near-invariant distributions, $f(x, y)$, generated from the same set of 400 initial conditions of initial energy $E_0 = -0.6$ in three different ways. In each case, the data were extracted from a simulation that ran for a total time $t = 2000$. The initial data were created by uniformly sampling a configuration space cell of size $\Delta x = \Delta y = 0.2$, setting $p_y = 0$, and solving for a positive $p_x(x, y, E_0)$. These initial data correspond to an ensemble comprised primarily of confined stochastic orbits, trapped in two "rings," which, however, is contaminated by a number of unconfined orbits. The first panel exhibits the near-invariant measure arising from a deterministic evolution in the time-independent KAMB potential (6).

The second panel exhibits the near-invariant measure arising if instead the initial data are driven with $\alpha_c = 10^{-3}$ and $\Omega = 1.0$. It is clear by visual inspection, and easily corroborated by numerical analysis, that the distributions exhibited in these panels are essentially identical. The third panel exhibits the near-invariant distribution appropriate if, as described in Section 2, the same 400 orbits are evolved in the presence of friction and noise, with $\Theta = 0.6$ nd $\eta = 10^{-6}$. It is easily verified that this final noisy near-invariant measure exhibits systematic, statistically significant differences from the two deterministic near-invariant measures, these corresponding to a net outwards extrinsic diffusion. This is illustrated in the final panel, which exhibits the difference between the unpulsed deterministic near-invariant measure and the noisy near-invariant measure.

## 6. CONCLUSIONS

In summary, the simulations described in these papers lead to at least three potentially significant conclusions:

1. Even very weak friction and noise, corresponding to a characteristic relaxation time $t_R$ as long as $10^6$–$10^9 t_{cr}$ can have significant effects on both the pointwise and the statistical properties of orbits on short timescales $< 100 t_{cr}$, for example, by facilitating extrinsic diffusion through cantori. This would suggest strongly that realistic quasi equilibria, expected to persist for times

$\geq t_H$, must exhibit a detailed balance for transitions between different orbit classes.
2. One cannot necessarily conclude that resonant couplings are irrelevant for stochastic orbits. Indeed, because the Fourier spectra for such orbits are characterized by significant power for a broad range of frequencies, they will in general find it easier to couple to a generic disturbance than will regular orbits, which are characterized by only a relatively few frequencies. If an ensemble of unconfined stochastic orbits is subjected to periodic driving with frequencies between $\sim 0.05$ and $5.0$, it will respond diffusively, with an amplitude of $\delta E_{\rm rms}$ that scales linearly in the perturbing influence $\alpha_{c,m}$.
3. Time-dependent modulation may trigger some diffusion through cantori. However, this effect seems substantially weaker than the extrinsic diffusion triggered by friction and noise, in the sense that the driving amplitude $\alpha_{c,m}$ required to trigger substantial modulational diffusion is substantially larger in natural units than the friction amplitude $\eta$ required to trigger substantial modulational diffusion.

## ACKNOWLEDGMENTS

It is a pleasure to acknowledge useful collaborations and interactions with Salman Habib, Christos V. Siopis, and David E. Willmes, from whom we have learned a great deal.

## REFERENCES

1. KANDRUP, H. E. & M. E. MAHON. 1995. Ann. N. Y. Acad. Sci.
2. KANDRUP, H. E. & M. E. MAHON. 1994. Phys. Rev. E **49**: 3735.
3. KANDRUP, H. E. & M. E. MAHON. 1994. Astron. Astropys. **290**: 762.
4. KANDRUP, H. E. 1995. In Proc. of the 7th Marcel Grossmann Meeting on General Relativity, R. Ruffini, Ed. World Scientific: Singapore. In press.
5. MAHON, M. E., R. A. ABERNATHY, B. O. BRADLEY & H. E. KANDRUP. 1995. Mon. Not. R. Astron. Soc. **275**: 443.
6. LICHTENBERG, A. J. & M. A. LIEBERMAN. 1992. Regular and Chaotic Dynamics. Springer-Verlag. Berlin.
7. OTT, E. 1993. Chaos in Dynamical Systems. Cambridge Univ. Press. Cambridge, England.
8. MATHER, J. N. 1982. Topology **21**: 45.
9. MACKAY, R. S., J. D. MEISS & I. C. PERCIVAL. 1984. Phys. Rev. Lett. **52**: 697.
10. MACKAY, R. S., J. D. MEISS & I. C. PERCIVAL. 1984. Physica D **13**: 55.
11. MORRISON, P. J. 1980. Phys. Lett. A **80**: 383.
12. KANDRUP, H. E. & E. O'NEILL. 1994. Phys. Rev. D **49**: 5115.
13. KUBO, R., M. TODA & N. HASHITSUME. 1991. Statistical Physics II: Nonequilibrium Statistical Mechanics, 2d ed. Springer-Verlag. Berlin.
14. LEBOWITZ, J. L. & E. RUBIN. 1963. Phys. Rev. **131**: 2381.
15. BARONE, P. M. V. B. & A. O. CALDEIRA. 1991. Phys. Rev. A **43**: 57.
16. BENNETIN, G., L. GALGANI & J.-M. STRELCYN. 1976. Phys. Rev. A **14**: 2338.
17. KANDRUP, H. E. & B. O. BRADLEY. 1995. Phys. Rev. E. Submitted.
18. BINNEY, J. J. & D. SPERGEL. 1982. Astrophys. J. **252**: 308.
19. BINNEY, J. J. & D. SPERGEL. 1984. Mon. Not. R. Astron. Soc. **206**: 159.
20. ARMBRUSTER, D., J. GUCKENHEIMER & S. KIM. 1989. Phys. Lett. A **140**: 316.

21. GRASSBERGER, P., R. BADII & A. POLOTI. 1988. J. Stat. Phys. **51**: 135.
22. SEPÚLVEDA, M. A., R. BADII & E. POLLAK. 1989. Phys. Rev. Lett. **63**: 1226.
23. KANDRUP, H. E. & D. E. WILLMES. 1994. Astron. Astrophys. **283**: 59.
24. HABIB, S., H. E. KANDRUP & M.E. MAHON. 1994. Phys. Rev. E. In press.
25. HABIB, S., H. E. KANDRUP & M. E. MAHON. 1994. Astrophys. J. Submitted.
26. KANDRUP, H. E., B. O. BRADLEY & I. POGORELOV. 1995. In preparation.
27. CHANDRASEKHAR, S. 1943. Rev. Mod. Phys. **15**: 1.
28. VAN KAMPEN, N. 1981. Stochastic Processes in Physics and Chemistry. North-Holland. Amsterdam.
29. GRINER, A., W. STRITTMATTER & J. HONERKAMP. 1988. J. Stat. Phys. **51**: 95.
30. TODA, M. 1967. J. Phys. Soc. Japan **22**: 431.
31. PFENNIGER, D. 1986. Astron. Astrophys. **165**: 74.
32. LIEBERMAN, M. A. & A. J. LICHTENBERG. 1972. Phys. Rev. A **5**: 1852.
33. ALLEN, C. W. 1973. Astrophysical Quantities. Athlone Press. London.
34. ANOSOV, D. V. 1967. Proc. Steklov Inst. Math. **90**: 1.
35. BOWEN, R. 1975. J. Differential Equations **18**: 333.
36. QUINLAN, G. D. & S. TREMAINE. 1992. Mon. Not. R. Astron. Soc. **259**: 505.
37. EUBANK, S. & D. FARMER. 1990. *In* Lectures in Complex Systems, E. Jen, Ed. Addison-Wesley: Redwood City, Calif.
38. WILLMES, D. E. 1995. Shadowing and noise in nonhyperbolic systems. This issue.

# Spectra and Lyapunov Numbers in Pulsating Systems[a]

HAYWOOD SMITH, JR.,[b] AND GEORGE CONTOPOULOS[b,c]

[b]*Department of Astronomy*
*University of Florida*
*Gainesville, Florida 32611*

[c]*Department of Astronomy*
*University of Athens*
*Athens, Greece*

## INTRODUCTION

There is experimental evidence[1] that there may be long-lived global oscillations of galaxies. Such oscillations have been seen in the kinetic and potential energies as well as in the density profiles of computed $N$-body systems. The presence of such oscillations is consistent with the virial equations,[2] and at least one detailed self-consistent model[3] has been constructed having such oscillations.

Our particular interest is in the characteristics of orbits in such systems, in which the potential is time-dependent, and especially in manifestations of chaos in some of those orbits. To explore this subject we have chosen the particular case of the so-called "conical" model of Miller and B. F. Smith,[1] which in the unperturbed state has a linear decrease of density with increasing radial distance $R$ and potential of the form

$$V(R) = \frac{GM}{R_L}(2x^2 - x^3 - 2) + \frac{J^2}{2R_L^2 x^2}, \quad R < R_L;$$

$$= -\frac{GM}{R_L}\frac{1}{x} + \frac{J^2}{2R_L^2 x^2}, \quad R \geq R_L. \quad (1)$$

where $x \equiv R/R_L$, $J$ is specific angular momentum, $R_L$ is the limiting radius of the density, $M$ is the mass, and $G$ is the gravitational constant. We have studied orbits under the influence of what those authors termed the "second radial mode," a perturbation whose shape is illustrated in FIGURE 1. The time-dependent term in the potential corresponding to this perturbation has the form

$$\tilde{V}(R, t) = \frac{16}{3} A(1 - \cos \Omega t)\left(\frac{1}{24} - \frac{3}{8}x^2 + \frac{7}{12}x^3 - \frac{1}{4}x^4\right), \quad R < R_L;$$

$$= 0, \quad R \geq R_L. \quad (2)$$

---

[a]One of the authors (G. C.) received support from the European Economic Community Human Capital and Mobility Program under Grant ERB 4050PL930312.

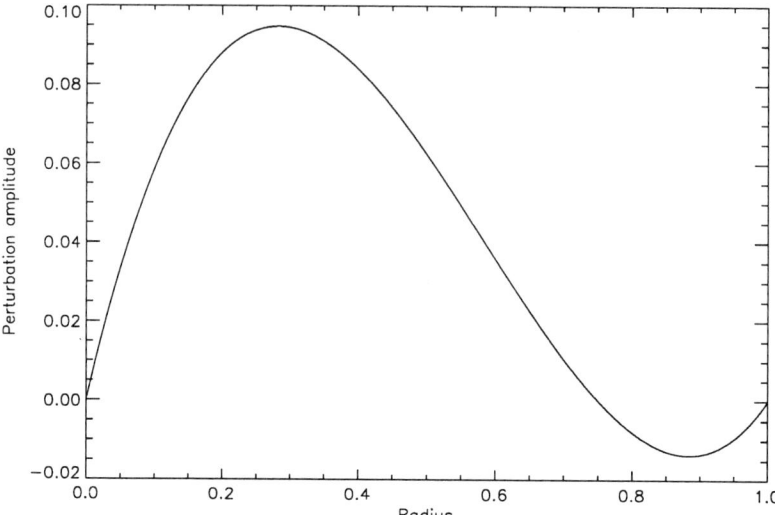

**FIGURE 1.** Magnitude of the perturbing acceleration (in arbitrary units) for Miller and Smith's[1] "second radial mode" as a function of radius (in units of the limiting radius $R_L$).

So far we have concentrated mainly on orbits of low angular momentum; such orbits traverse a substantial portion of the system and in particular pass through the inner regions where the perturbation due to the oscillations is strongest.

## CHARACTER OF THE ORBITS

To identify the character of the orbits we first reduce the problem from three dimensions to one using the vector angular momentum; its direction identifies the orbit plane, and its magnitude $J$ allows us to replace the tangential component of kinetic energy with a centrifugal potential. Then, to allow for the time dependence of the gravitational potential we employ the usual extended phase space, with $t$ and $-E$ as the additional canonical phase-space coordinates. Because the perturbation is strictly periodic we can fold the time dimension onto itself at intervals of the oscillation period $P$. The surface of section used is then the $(R, \dot{R})$ plane sampled at intervals of $P$ ("time slices"), just as in Louis and Gerhard[3] and in Miller and Smith.[1] Integration of the orbits was done with a fourth-order explicit symplectic integrator.[4]

The surface-of-section plots show qualitatively the behavior expected from the somewhat similar model of Louis and Gerhard. For amplitude $A = 0$, where the amplitude $A$ is the ratio of the density amplitude of the oscillation at the center of the system to the unperturbed central density, all orbits are regular as expected, with closed invariant curves. On the other hand, for $A \neq 0$ resonances with the oscillations appear. Along the $R$-axis there are odd $m:1$ resonances that are stable and even ones that are unstable. As $A$ increases, invariant curves at progressively smaller radii begin to break down, starting with the outermost. The surface of

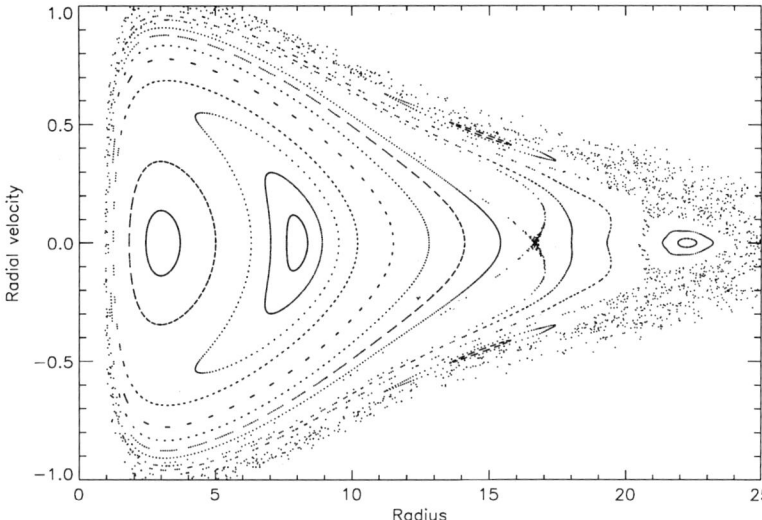

**FIGURE 2.** Poincaré surface of section for $J^2 = 1.0006$, amplitude $A = 0.1$, phase of the perturbation $\varphi = 0.0$. Note in particular the 1:1 resonance at $R = 8$, the 2:1 resonance at $R = 16.7$, the 3:1 resonance at $R = 22.3$, and the closed invariant curve with apocenter distance $R_0 = 18$.

section for $A = 0.1$ sampled at phase angle $\varphi = 0$ is shown in FIGURE 2. We note in particular the chaotic region associated with the 2:1 resonance around $R = 16.7$ and the closed invariant curve extending in from apocenter distance $R_0 = 18$ separating that region from the chaotic regions farther out.

## LYAPUNOV SPECTRA

The importance of calculating finite-time Lyapunov exponents has been stressed by several authors[5-15] in recent years. The spectra of these finite-time Lyapunov exponents are particularly useful.[8-15] The most detailed information is found if we take very small time intervals $\Delta T$, while for longer intervals many of the details are lost.

In the case of mappings the shortest interval is one iteration of the map.[11,14,15] It has been shown[14,15] that if we take a sufficiently large number of iterations the spectrum of the one-period Lyapunov numbers is invariant: (1) with respect to the initial conditions along an orbit; (2) with respect to different initial directions of the perturbation to an orbit; and (3) with respect to the initial conditions of orbits in a given connected chaotic zone.

In the present paper we find short-time Lyapunov spectra for a system depending periodically on the time, using various sampling intervals $\Delta T$. We find that we can derive all the information about such spectra if we calculate the spectrum for very small $\Delta T$, namely the integration timestep $\Delta t$ itself. We start with an orbit having certain initial conditions, then perturb those conditions by a very small amount $\xi_0$ in phase space. After the interval $\Delta T$ we calculate the current phase-space separation between the orbits $\xi$. The *stretching number a* over that interval is then defined as

$$a = \ln (\xi/\xi_0). \tag{3}$$

After renormalizing $\xi$ to the original magnitude $\xi_0$ while preserving its current direction we can carry the process forward over the next interval $\Delta T$, and by continuing this process build up a collection of $a$ values. This accumulation can be done for a single orbit over an extended time interval or for an ensemble of orbits with different initial conditions over a relatively short interval. In the present work we have followed the former procedure. Either way, one then bins the large number of $a$ values thus obtained; the binned distribution $\Delta N(a)$ is then normalized to obtain a spectral density, which we refer to as the Lyapunov spectrum:

$$S(a) = \frac{\Delta N(a)}{N \cdot \Delta a}, \tag{4}$$

where $\Delta a$ is the bin width and $N$ is the size of the sample. The Lyapunov characteristic number is given by the mean of $a$ (with $S(a)$ as a weight function) divided by the sampling interval $\Delta T$.

In what follows we will examine Lyapunov spectra for four types of sample orbits: (1) regular orbits in the time-independent potential; (2) regular orbits in the time-dependent potential, which is an integrable case; (3) orbits from the chaotic region near the 2:1 resonance; and (4) orbits from the chaotic region near the 5:2 resonance, which is separated from the preceding chaotic resonance by the closed invariant curve originating from $R_0 = 18$.

### *Regular Orbit, Time-Independent Potential*

The Lyapunov spectrum for initial conditions ($R_0 = 11.0$, $\dot{R}_0 = 0.0$) is shown in FIGURE 3. The sampling interval $\Delta T$ is 9.5 time units, an arbitrary choice that is one-fourth of the oscillation period; the sample size is $N = 10^5$. The spectrum is almost exactly symmetric around $a = 0$, which is the case for symmetric orbits in an integrable potential when the sample size is sufficiently large. The cusplike "spikes" and broad maxima are also typical. As $\Delta T$ is varied, these features move and appear or disappear, as indicated in FIGURE 4(a)–(d). Furthermore, as we increase $\Delta T$ the width of the spectrum, as measured between the outermost spikes, increases until $\Delta T$ has become quite large and thereafter decreases, which puzzled us at first.

To understand the characteristics of these spectra we need to go in the opposite direction in $\Delta T$, namely to the smallest value feasible, which is the timestep of the integrator, $\Delta t$; for the present calculations this last has been taken to be 0.0125. Although the same features—spikes, broad maxima, symmetry—are present as for the larger values of $\Delta T$, this spectrum is different from those shown previously, as may be seen in FIGURE 5. It turns out that the features in this spectrum can be explained very simply given the orbit dynamics and a certain rather plausible

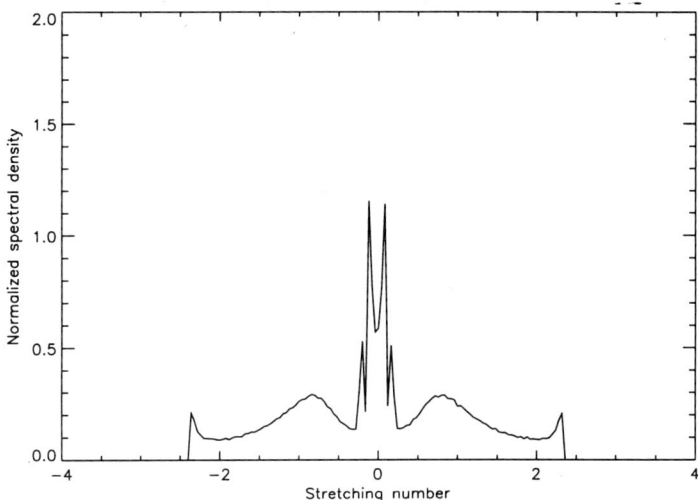

**FIGURE 3.** The Lyapunov spectrum $S(a)$ for the orbit in the time-independent potential ($A = 0.0$) having $R_0 = 11$, $J^2 = 1.0006$. For this spectrum the sampling interval $\Delta T = 9.5$, and the sample size $N = 10^5$.

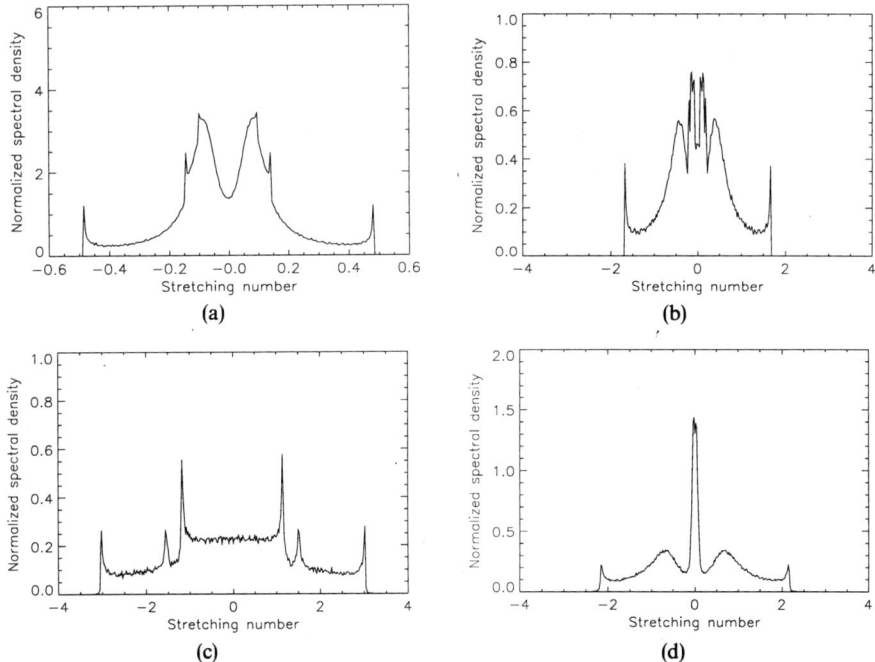

**FIGURE 4.** Lyapunov spectra $S(a)$ for the same orbit as in FIGURE 3 with various values of $\Delta T$, sample size $N = 10^5$. (a) $\Delta T = 1.0$; (b) $\Delta T = 5.0$; (c) $\Delta T = 19.0$; (d) $\Delta T = 38.0$.

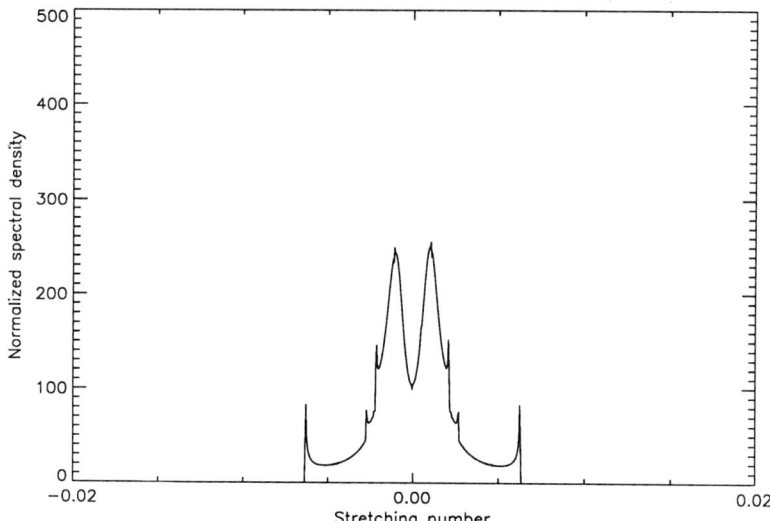

**FIGURE 5.** Same as FIGURE 3 but for $\Delta T = 0.0125$, $N = 10^6$.

assumption concerning the behavior of the perturbation $\xi$, namely, that the direction of the perturbation tends to become aligned with the direction of orbital motion as time passes. This assumption is justified as follows. The end point of the perturbation $\xi$ follows an orbit very close to the unperturbed orbit. Therefore the angle between $\xi$ and the tangent to the unperturbed orbit tends to zero or to 180°. Thus we can treat the perturbation as equivalent over the long term to one directed along the orbit, that is, "down-track." In particular we can consider a perturbation equal to one step of the integrator along the orbit. Then $a$ can be calculated as the logarithm of the ratio of the length of one step in phase space to that of its predecessor, or

$$a = \ln \frac{\Delta s_{i+1}}{\Delta s_i}. \tag{5}$$

But for very small timestep $\Delta t$

$$\Delta s_i = \sqrt{\Delta R^2 + \Delta \dot{R}^2} = \Delta t \cdot \sqrt{\dot{R}^2 + \ddot{R}^2}. \tag{6}$$

We can thus calculate $a$ for each timestep, and the resulting curve $a(t_i)$, shown in FIGURE 6, suffices to completely explain the structure of $S(a)$. The symmetry of $S(a)$ results from the mirror symmetry of $a(t_i)$ around pericenter passage, at approximately $t = 23$. The spikes are due to the turning points of $a(t_i)$, where $\dot{a} = 0$ (local maxima and minima). The broad peaks are attributable to the relatively long stretches of $a(t_i)$ which have $\dot{a} \approx 0$. Comparison of FIGURES 5 and 6 shows that the outermost spikes are located at exactly the extreme $a$ values, and the other spikes coincide with the other maxima and minima. Likewise the broad peaks are at the $a$ values of the long, nearly horizontal stretches of $a(t_i)$.

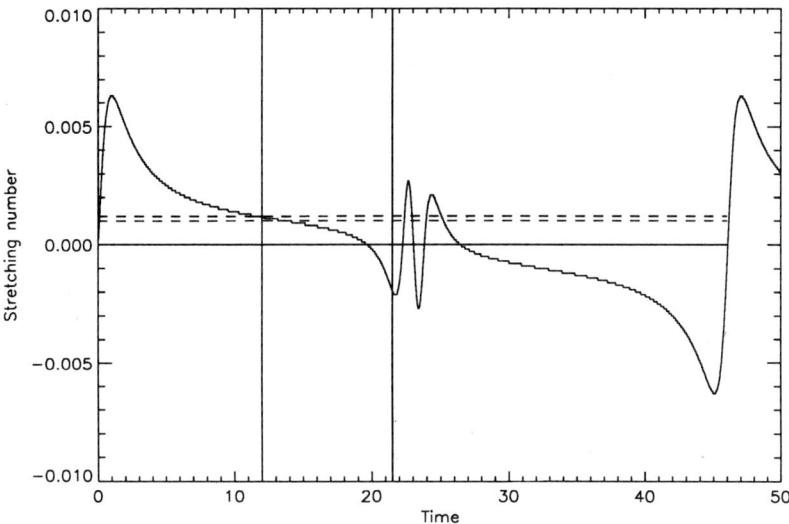

**FIGURE 6.** Stretching number $a$ as a function of time for the same orbit as in FIGURE 3. The *dashed horizontal lines* represent sampling of the curve $a(t_i)$ at a given value of $a$; the two *solid vertical lines* represent a step in the convolution of $a(t_i)$ with a box function of width $\Delta T = 9.5$.

Even more remarkably, the spectra for larger $\Delta T$ can be generated in a very straightforward way from $a(t_i)$ for $\Delta t = 0.0125$. Because $a$ is a logarithm we can add the values over $m$ consecutive timesteps to get the value for $\Delta T = m \cdot \Delta t$:

$$a = \ln \frac{\zeta_m}{\zeta_0} = \ln \frac{\zeta_m}{\zeta_{m-1}} + \ln \frac{\zeta_{m-1}}{\zeta_{m-2}} + \cdots + \ln \frac{\zeta_1}{\zeta_0} \quad (7)$$

or

$$a = \sum_{j=1}^{m} a_j. \quad (8)$$

This means that $a(t_i)$ for larger $\Delta T$ is a convolution with a box function of width $\Delta T$ over $a(t_i)$ for small $\Delta t$. In this way we can explain both the changes in structure and the changes in width. One rather trivial consequence of this property is the fact that, if we take $\Delta T$ equal to precisely the orbital period, the symmetry of the curve guarantees that $a$ will always be zero. We can understand the decrease in width for large values of $\Delta T$ in light of this fact. (Of course this only holds for the time-independent case, for which we can expect $a(t_i)$ to be symmetric because of the strict periodicity of the orbit.)

Incidentally, it would of course be possible to go to shorter timesteps. However, the resulting curve would only be a scaled-down version of this one. Obviously so long as one has a sufficiently short timestep to resolve the orbit very finely, the $a(t_i)$ curve will not reveal anything new. (In fact we have checked this point with a range of values of $\Delta T$ around $\Delta t$.) It is only when $\Delta T$ becomes comparable to the timescale associated with fine structure in $a(t_i)$ that the curve changes its shape; at that point one begins to lose information.

We can explain the behavior of $a(t_i)$ more completely in light of the orbital dynamics, for example, by locating the zeros of $a(t_i)$. We start with the step $\Delta s_i$, which can be rewritten as

$$\Delta s_i = \Delta t \sqrt{\dot{R}_i^2 + \left(-\frac{\partial V}{\partial R}\right)_i^2}. \qquad (9)$$

To simplify notation, let $(\partial V/\partial R) = V'$. The stretching number is then given by

$$a_i = \ln \frac{\Delta s_{i+1}}{\Delta s_i} = \ln\left[1 + \frac{(\dot{\Delta s_i})}{\Delta s_i} \cdot \Delta t\right] \cong \frac{(\dot{\Delta s_i})}{\Delta s_i} \cdot \Delta t. \qquad (10)$$

To evaluate $(\dot{\Delta s_i})$ we take the derivative along the trajectory, that is, the directional derivative. The stretching number is zero when this derivative is zero; this condition leads to the equation

$$\dot{R} V'(V'' - 1) = 0. \qquad (11)$$

There are three possible sets of roots:

1. $\dot{R} = 0$, which occurs at the maximum and minimum of $R$;
2. $V' = 0$ (or $\ddot{R} = 0$), which corresponds to the maximum and minimum of $\dot{R}$; and
3. $V'' = 1$, the root(s) of which (in $R$) may or may not lie between the minimum and maximum of $R$. (Because 1 is dimensionless while $V''$ is not, necessarily the location of the root(s) will depend upon the choice of units for $R$ and $\dot{R}$. This curious fact is traceable to the form of the step $\Delta s$, which combines $\Delta R$ with $\Delta \dot{R}$ without a conversion factor having units of time. By suitable choice of the time unit one can emphasize either $\Delta R$ or $\Delta \dot{R}$; the corresponding spectra will then be different, and in particular the features attributable to this root (these roots) can be made to appear or disappear.)

From the preceding and from the nature of the orbital motion several points about $a(t_i)$ are obvious:

1. There must be at least one extremum between each pair of zeros.
2. The roots for $V' = 0$ must lie between those for $\dot{R} = 0$ in time (i.e., along the orbit).
3. The condition $V'' = 1$ is the same as $\ddot{R}' = -1$, which is a condition on the slope of the acceleration; because it is negative it must be satisfied (if at all) in the inner regions, most probably inside the radius at which centrifugal balance obtains, that is, where $\ddot{R} = 0$. A key question is whether the root for (3) lies within the range $(R_{\min}, R_{\max})$. It turns out that the determining parameter for this question (apart from the choice of units, as was noted earlier) is the square of the angular momentum, $J^2$. As shown in FIGURE 7(a), for an orbit with initial conditions $(R_0 = 16.7, \dot{R} = 0)$ there is a pair of very large, sharp spikes in $a(t_i)$ around the pericenter, while for the same $R_0$ but $J^2 = 3.0006$ (FIGURE 7(b)) those spikes have all but disappeared. At a slightly

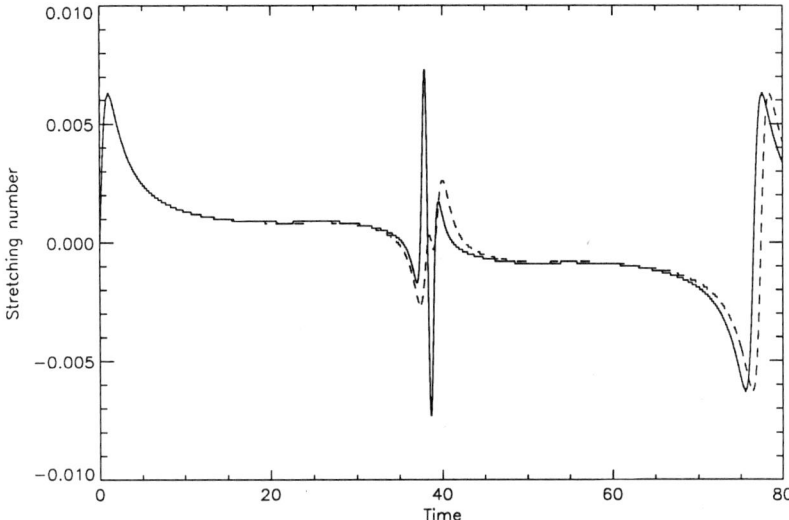

**FIGURE 7.** Stretching number $a$ as a function of time for two orbits in the time-independent potential having $R_0 = 16.7$ but different values of $J^2$; the sampling interval in both cases is $\Delta T = 0.0125$. The *solid curve* is for $J^2 = 1.0006$, and the *dashed curve* is for $J^2 = 3.0006$.

larger value of $J^2$ they do indeed disappear completely. It seems that, as $J^2$ is increased, the orbit does not penetrate as far into the centrifugally dominated region and thus avoids the features associated with the root (3).

### *Regular Orbit, Time-Dependent Potential*

With $A \neq 0$—in this case, $A = 0.1$—and a regular orbit we find that the Lyapunov spectrum retains its symmetric shape (provided once again that a sufficiently large sample is used). The $a(t_i)$ curve is no longer strictly periodic because the orbit expands and contracts under the influence of the perturbation and thus has a varying orbital period. This aperiodicity shows up quite clearly in FIGURE 8, where we have folded $a(t_i)$ over itself with a periodicity equal to the period of the unperturbed orbit of the preceding section. Experimentation with other folding periods produced no greater degree of success. Because the period of the osculating orbit varies over the orbit itself, it is not to be expected that $a(t_i)$ retain the strict symmetry of the time-independent potential. The symmetry in the present instance must result from an averaging over the longer periodicity associated with the oscillation-driven variation in the osculating orbit. The spectrum for this case with $\Delta T = 9.5$ is given in FIGURE 9.

### *Irregular Orbit near 2:1 Resonance, Time-Dependent Potential*

We now consider two orbits from the chaotic region close to the 2:1 resonance at $R_0 = 16.7$, once again for $A = 0.1$. Orbit 1 has initial conditions ($R_0 = 16.6$, $\dot{R}_0 = $

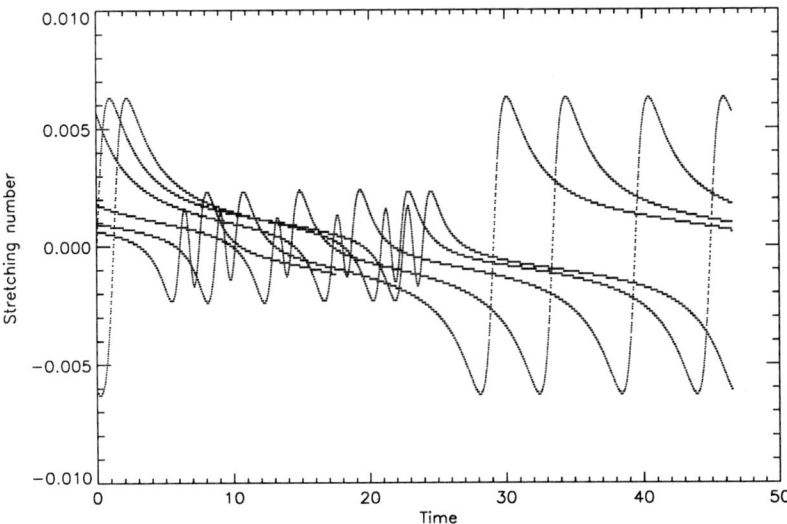

**FIGURE 8.** Stretching number $a$ as a function of time for the orbit having $R_0 = 11.0$, $J^2 = 1.0006$ in the time-dependent potential with oscillation amplitude $A = 0.2$. The $a(t_i)$ curve has been folded back on itself at intervals of $t = 46.5$, approximately the period of the same orbit in the time-independent potential.

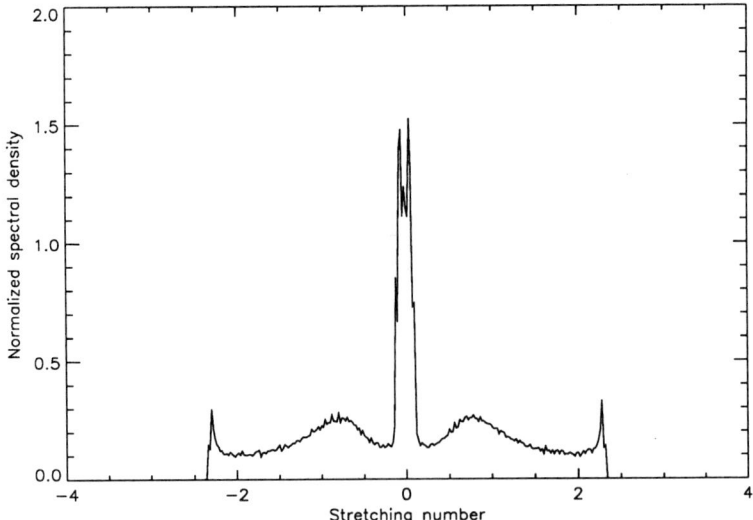

**FIGURE 9.** Lyapunov spectrum for the orbit of FIGURE 8 with sampling interval $\Delta T = 9.5$.

+0.02), and Orbit 2 has ($R_0 = 16.56$, $\dot{R}_0 = +0.05$). These orbits are weakly stochastic, with the most obvious stochastic behavior in a region of the surface of section near the apocenter; compare FIGURE 10(a) and (b). The Lyapunov characteristic numbers are nearly the same ($1.29 \times 10^{-3}$ and $0.98 \times 10^{-3}$, respectively, for

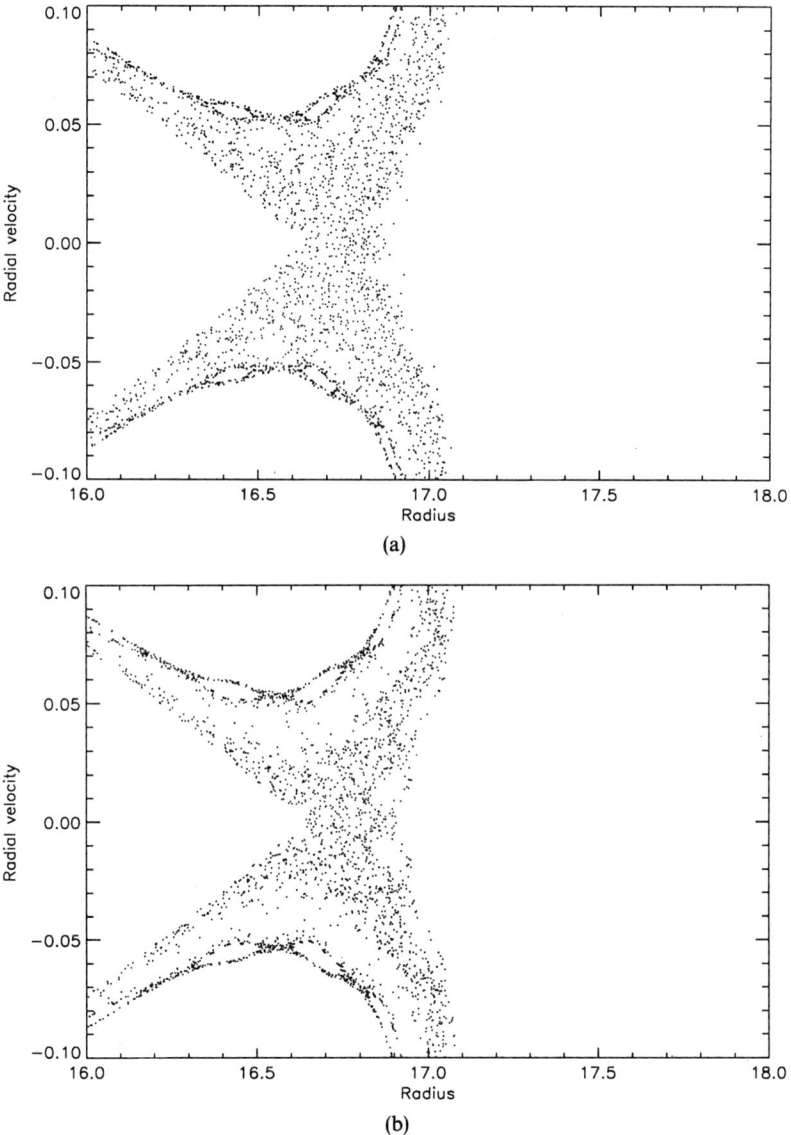

**FIGURE 10.** Surfaces of section for two orbits near the 2:1 resonance at $R = 16.7$ in the time-dependent potential with oscillation amplitude $A = 0.1$. (**a**) Orbit 1, $R_0 = 16.6$, $\dot{R}_0 = +0.02$; (**b**) Orbit 2, $R_0 = 16.56$, $\dot{R}_0 = +0.05$.

$N = 10^6$ and $\Delta T = 9.5$), and the corresponding Lyapunov spectra are virtually identical (FIG. 11). If we look at the respective spectra for $\Delta T = \Delta t$, we see that they are almost identical to each other and that there is an asymmetry in the broad peaks in the center (FIG. 12).

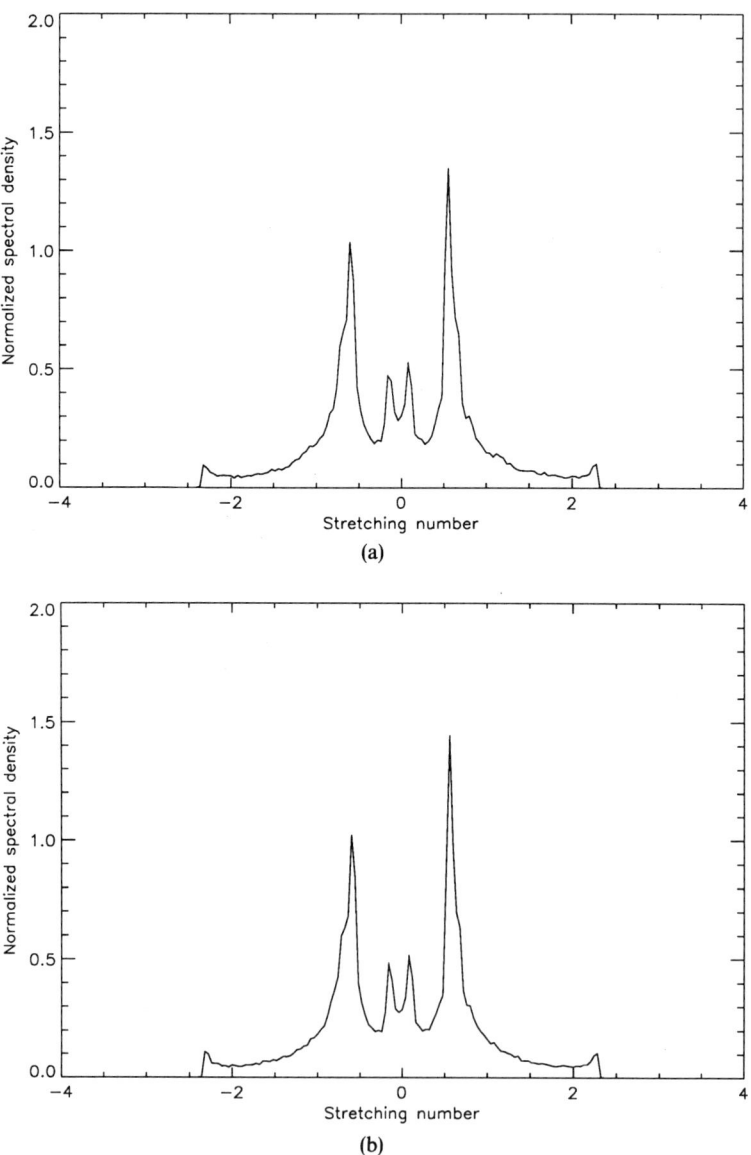

**FIGURE 11.** Lyapunov spectra for the two orbits represented in FIGURE 10(a) and (b), $N = 10^6, \Delta T = 9.5$.

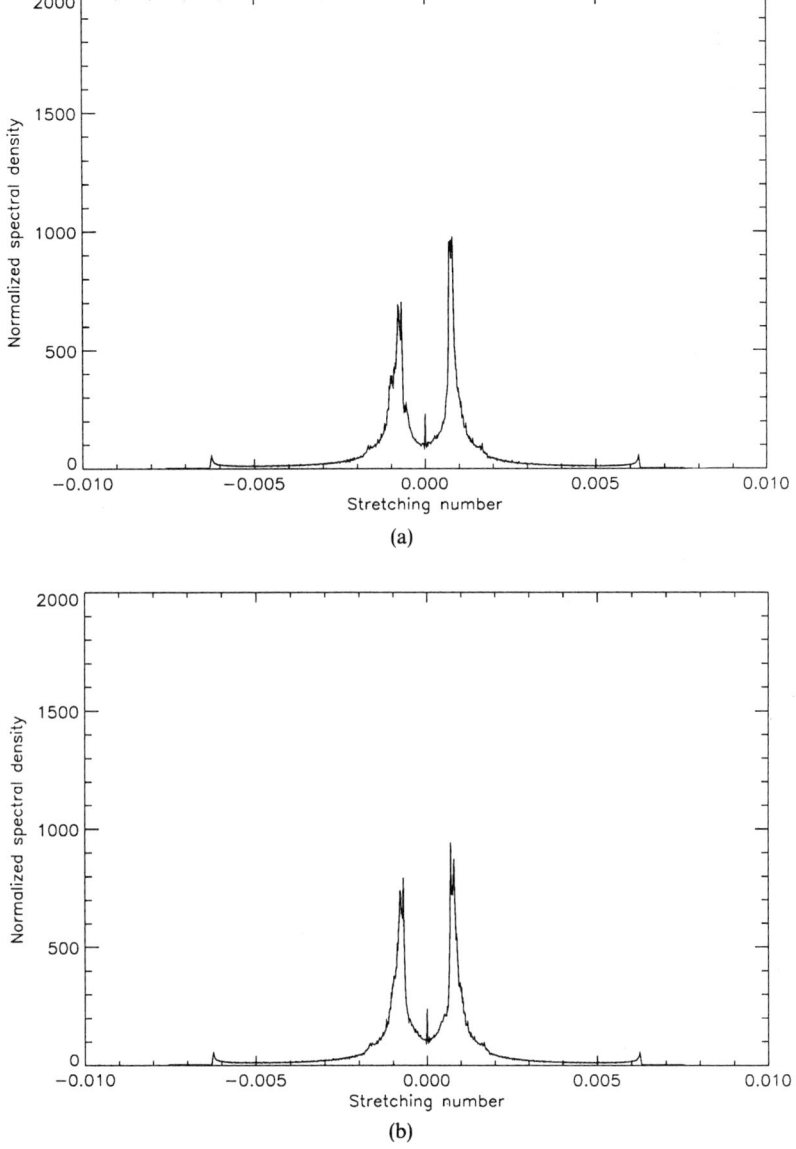

FIGURE 12. Same as in FIGURE 11 but for $\Delta T = 0.0125$.

*Irregular Orbit near the 5:2 Resonance, Time-Dependent Potential*

We now consider an orbit near the 5:2 resonance having initial apocenter distance $R_0 = 19.2$. This orbit is, like those around the 2:1 resonance, weakly chaotic; however, this chaotic region is separated from the previous one by a closed invariant curve for $R_0 = 18$. The Lyapunov spectrum for this orbit for $\Delta T = 9.5$, shown in FIGURE 13, is noticeably different from those for the preceding two orbits. The same is true for the spectrum for $\Delta T = \Delta t$, shown in FIGURE 14. Interestingly, the Lyapunov characteristic number for this orbit, $1.27 \times 10^{-3}$, is nearly the same as that for Orbit 1 from the other chaotic region, despite the fact that the spectra are different.

## SUMMARY

This is the first study of Lyapunov spectra in a time-potential. Our study has thus far yielded several results that in our opinion are likely to be true for all spherically symmetric oscillating systems.

1. The spectra of regular orbits are symmetric around $a = 0$ as a consequence of the approximate symmetry of the orbital motion in the $(R, \dot{R})$ plane and the time-symmetry of the oscillation.
2. The features in those spectra—the spikes and the broad maxima—can be readily understood on the basis of the $a(t_i)$ curve derived from the "downtrack" perturbation.

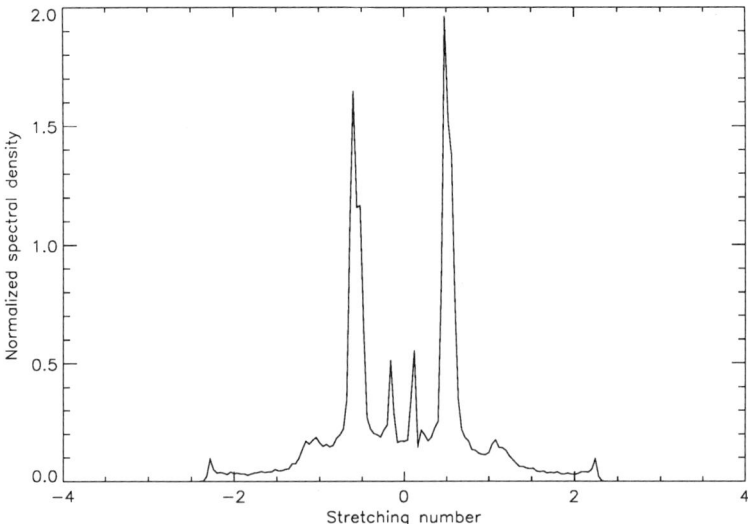

**FIGURE 13.** Lyapunov spectrum for an orbit in the chaotic region near the 5:2 resonance, namely that with $R_0 = 19.2$, $J^2 = 1.0006$. The sampling interval is $\Delta T = 9.5$.

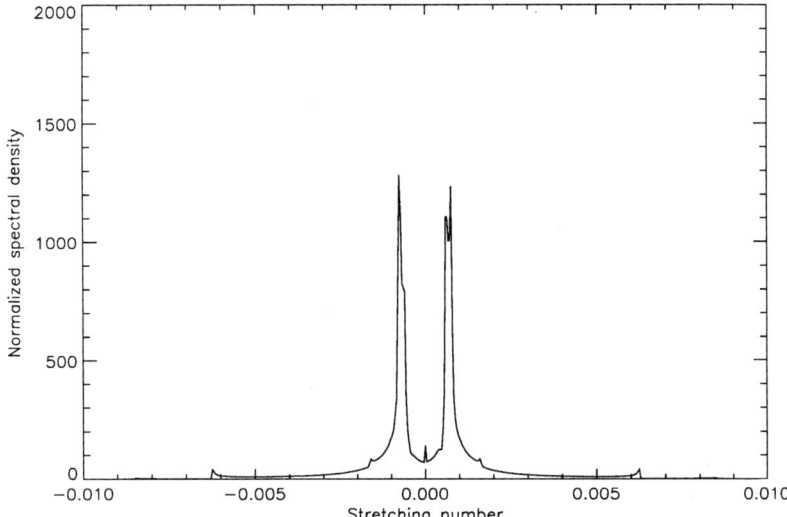

**FIGURE 14.** Same as in FIGURE 13 but for sampling interval $\Delta T = 0.0125$.

3. The features of the latter can generally be understood from the orbital dynamics; one can rather easily predict where the zeros of $a(t_i)$ will occur on the osculating orbit and can roughly anticipate how many will be present.
4. The spectra of irregular orbits are the same for orbits belonging to the same stochastic region but will in general be different for orbits belonging to stochastic regions that are separated by a closed invariant curve. This will be true even if the Lyapunov characteristic numbers should happen to be the same.

Finally, we note that there is a potential complication in the study of these orbits, especially those with relatively large apocenter distances. In several cases that we have encountered, the oscillations led to escape from the system. Indeed, there are reasons to anticipate that *all* such orbits will sooner or later lead to escape. This "opening up" of the phase space would of course render invalid the assumption of compactness on which many results from ergodic theory are based. But there is already a large biliography on such problems, which are called collectively "chaotic scattering."[16-20] We plan to study this problem later. A key question is the escape timescale in relation to the relevant physical timescales, in particular the Hubble time.

## ACKNOWLEDGMENTS

The authors gratefully acknowledge stimulating discussions with H. Kandrup and R. H. Miller on these matters.

## REFERENCES

1. MILLER, R. H. & B. F. SMITH. 1994. Celest. Mech. **59**: 161.
2. CHANDRASEKHAR, S. & D. D. ELBERT. 1972. Mon. Not. R. Astron. Soc. **155**: 435.
3. LOUIS, P. D. & O. E. GERHARD. 1988. Mon. Not. R. Astron. Soc. **233**: 337.
4. CANDY, J. & W. ROZMUS. 1991. J. Comp. Phys. **92**: 230.
5. FUJISAKA, H. 1983. Prog. Theor. Phys. **70**: 1264.
6. BENZI, R., G. PALADIN, G. PARISI & A. VULPIANI. 1985. J. Phys. A **18**: 2157.
7. GRASSBERGER, P., R. BADII & A. POLITI. 1988. J. Stat. Phys. **51**: 135.
8. UDRY, S. & D. PFENNIGER. 1988. Astron. Astrophys. **198**: 135.
9. SEPÚLVEDA, M. A., R. BADII & E. POLLAK. 1989. Phys. Rev. Lett. **63**: 1226.
10. GROBGELD, D., E. POLLAK & J. ZAKRZEWSKI. 1992. Physica D **56**: 368.
11. FROESCHLÉ, G., C. FROESCHLÉ & E. LOHINGER. 1993. Celest. Mech. Dyn. Astron. **56**: 308.
12. KANDRUP, H. E. & M. E. MAHON. 1995. In Three-Dimensional Dynamical Systems, H. E. Kandrup, S. T. Gottesman, and J. R. Ipser, Eds.: 93. N. Y. Acad. Sci. New York.
13. KANDRUP, H. E. & M. E. MAHON. 1994. Astron. Astrophys. **290**: 762.
14. VOGLIS, N. & G. CONTOPOULOS. 1994. J. Phys. A **27**: 4899.
15. CONTOPOULOS, G., N. VOGLIS, C. EFTHYMIOPOULOS & E. GROUSOUZAKOU. 1995. Invariant Spectra of Dynamical Systems. This issue.
16. ECKHARDT, B. & C. JUNG. 1986. J. Phys. A **19**: L829.
17. JUNG, C. & H. J. SCHOLZ. 1987. J. Phys. A **20**: 3607; also 1988. J. Phys. A **21**: 2301.
18. ECKHARDT. B. 1987. J. Phys. A **20**: S971; also 1988. Physica D **33**: 89.
19. HÉNON, M. 1988. Physica D **33**: 132; also 1989. La Recherche **20**: 490.
20. JUNG, C. & S. POTT. 1989. J. Phys. A **22**: 2925.

# Arnold Diffusion and Equipartition in an Oscillator Chain[a]

ALLAN J. LICHTENBERG

*Department of Electrical Engineering and Computer Sciences
and the Electronics Research Laboratory
University of California, Berkeley, California 94720*

## 1. INTRODUCTION

Near-integrable Hamiltonians have the form $H = H_0 + \varepsilon H_1$, with $H_0$ integrable, $H$ not integrable, and the perturbation strength $\varepsilon \ll 1$. The nature of the motion in systems with three or more degrees of freedom is reasonably well understood. Stochastic and integrable (KAM) trajectories are intimately comingled in the $2N$-dimensional phase space. Stochastic layers form near the separatrices associated with resonances of the motion among the degrees of freedom. For a sufficiently strong perturbation, resonance overlap leads to motion across the layers and the presence of strong stochasticity. In the limit of weak perturbation, resonance overlap does not occur, and the stochastic motion is along the resonance layers. This motion is the consequence of a fundamental geometric fact: for $N > 3$, the $(2N - 1)$-dimensional resonance layers are not isolated from each other by the $N$-dimensional KAM surfaces. As a result, all stochastic layers are connected into a single complex network—the Arnold web. The web consists of an intricate system of "freeways, streets, sidewalks, and cracks" that permeates the entire phase space, intersecting or lying infinitesimally close to every point. For any initial condition within the web, the subsequent stochastic motion will eventually intersect every finite region of the phase space, even the predominantly stable regions where the fraction of stochastic initial conditions is small, and even in the limit as the perturbation strength $\varepsilon \to 0$.

The merging of stochastic trajectories into a single web was proved by Arnold for a specific nonlinear Hamiltonian,[1]

$$H = \tfrac{1}{2}I_R^2 + \tfrac{1}{2}I_S^2 + \varepsilon(\cos\theta_R - 1)(1 + \mu \sin\theta_S + \mu \cos t), \tag{1}$$

with $\mu \ll \varepsilon$. Arnold further showed that the rate of diffusion along the resonance layers could be estimated in the form

$$|\dot{I}| \sim f(\varepsilon, \mu) e^{-a/\varepsilon^{1/2}}, \tag{2}$$

where the exponential function is an approximation to a Melnikov–Arnold (MA) integral, for $\varepsilon \ll 1$, with $a \sim 1$. For $\varepsilon \ll 1$, the diffusion is seen to be exceedingly slow, and would be computationally unobservable. Note that the estimate (2) is for $N = 3$. We return to the more general problem in Section 2.

---
[a]The author would like to acknowledge the support of Office of Naval Research Grant N00014-89-J-1097.

Despite its slowness, Arnold diffusion has been postulated as the mechanism for various physical phenomena. For astrophysical phenomena the diffusive timescale can be very long, and the mechanism has been proposed to explain the emptying of resonant planetary rings and asteroid belts.[2,3] For particle dynamics, in which the fundamental periods can be very short, the mechanism has been proposed as an explanation for beam blowup in accelerator storage rings[4] and for particle loss out of electron-cyclotron-resonance-heated (ECRH) magnetic mirrors.[5]

From a practical point of view, there are two major questions concerning Arnold diffusion in a particular system: (1) What is the relative measure of stochastic trajectories in the phase-space region of interest? and (2) For a given initial condition, how fast will the system diffuse along the stochastic Arnold web? For large $N$, in weakly coupled systems, the first question can be answered by a geometrical argument, to indicate that most initial conditions are in the stochastic web.[6] To see this we note that if a fraction $f < 1$ of the phase space in a single dimension lies on KAM curves, and if the coupling is sufficiently weak, then we would expect a fraction $f^N$ of the $N$-dimensional phase space to be regular. This is a consequence of the fact that an initial condition that lies in the stochastic region of any of the $N$ degrees-of-freedom is sufficient for stochasticity. The stochastic fraction $1 - f^N$ thus tends to 1 as $N$ becomes large. Strong coupling only enhances this effect, as observed numerically.[7] In this paper we concentrate our attention on question 2. In Sections 3 and 4 we consider this question for two systems, coupled standard maps, for which the parameters of the Arnold diffusion mechanism can be well controlled, and the Fermi–Pasta–Ulam (FPU) oscillator chain, which historically has been very important in investigating the transition to statistical energy equipartition.

If three resonances can be locally isolated to be of dominant importance, then a method exists for calculating the rate of diffusion along a local resonance layer, known as the three-resonance model.[8] The model has been used to analytically calculate the local diffusion rate,[6,8] and, in particular, for two coupled standard maps (see (15)), with good agreement obtained with numerical results.[9] The three-resonance model predicts a diffusion rate in agreement with (2), with explicit values of the constants. This calculation has been extended to higher dimensions by Wood et al.,[10] and is presented in Section 3.

Fermi et al.[11] reported the first numerical study of a chain of coupled oscillators, with a quartic nonlinearity (see (22)). They observed, for a particular initial energy distribution, that the oscillator chain did not relax, as expected from statistical mechanics, to the equipartition state, but displayed a persistent recurrence to the initial condition. The recurrences were later explained in terms of beating among the system modes. A theoretical prediction of a threshold to equipartition was obtained by Izraelev and Chirikov[12] using an "overlap" criterion for the modes. They predicted a critical energy $E'_c$, of the initial excitation, for widespread stochasticity. For energy in a few low-frequency modes they found $E'_c \propto N$, where $N$ is the number of oscillators (number of modes) in the chain.

Subsequently there have been many studies of the interchange of energy among modes, and of energy thresholds to give approximate equipartition among modes.[13–15] Pettini and Landolfi[14] have studied the dependence of the time to equipartition on the energy of the excitation at relatively high energies, finding a relaxation time $\tau \propto N$ and a threshold for equipartition $E'_c \propto N$. The observed scaling agreed with the mode overlap hypothesis. However, equipartition that takes place on a slower timescale, at lower energy, would not have been observed. Kantz et al.[15] studied the transitions and timescales to equipartition for two different types of initial conditions. For energy in initial low-frequency modes around mode number $\gamma$, and $\gamma \propto N$, they found a transition to equipartition at a critical energy $E_c$, independent of $N$. For $E > E_c$ a time scale $\tau \propto N$ (at constant $E$) was required to attain some constant measure of equipartition.

DeLuca et al.[16] developed a theoretical description that was compared to numerical results. They found a transition at which Arnold diffusion transfers energy to the high-frequency modes, leading to equipartition. The transition was predicted to occur at a critical energy $E = E_c$, independent of $N$, which was confirmed numerically. The beat frequency, $\Omega_B \sim \gamma E/N^2$ with $\gamma$ the mode number initially containing the energy, plays a centrally important role in the analysis. We discuss this analysis in Section 4.

The mKdV equation approximately describes low-frequency modes of the FPU system for strong spring coupling if $N$ is sufficiently large. Driscoll and O'Neil[17] showed that an instability of a low-frequency mode (soliton) of the mKdV equation corresponds to exponential growth in the FPU system. A similar relation was also found to exist between the existence of an instability in the sine–Gordon equation and equipartition in an oscillator chain corresponding to the discretized sine–Gordon equation.[18] We compare the onset of the instability with the FPU dynamics in Section 4.

## 2. ESTIMATES AND SCALING OF ARNOLD DIFFUSION

Although most problems involving diffusion along resonances are concerned with relatively large perturbations, in order that the diffusion be computationally observable, considerable effort has gone into calculations for small $\varepsilon$. In this regime, upper bounds to the diffusion rate can be obtained for large $N$. Generalizing (1) in the form

$$\tfrac{1}{2}I_R + \tfrac{1}{2}I_S^2 + \varepsilon(\cos\theta_S - 1)(1 + \mu F(\theta_R)), \tag{3}$$

where $(I_R, \theta_R)$ is an $N - 1$-dimensional phase space and $F(\theta_R)$ is an analytic function

$$F(\theta_R) = \sum f_k \cos(k \cdot \theta_R), \quad |f_k| \sim e^{-\sigma|k|}, \tag{4}$$

then, provided $\mu \ll \varepsilon$, with $\varepsilon$ sufficiently small, an upper bound on the Arnold diffusion rate can be found. Necessary conditions that the unperturbed Hamiltonian be positive definite and include a separatrix are met in (3), as in Arnold's original example. Nekhoroshev[19] first found such a bound

$$|\dot{I}| \lesssim |\omega| |I| \varepsilon^{1+\delta} \exp(-c/\varepsilon^\delta), \quad \delta = \frac{2}{3N^2 - N + 8}, \quad c \sim 1. \tag{5}$$

This bound, however, severely overestimates the diffusion rate. Chirikov[8] numerically examined a Hamiltonian, similar to (3), finding a transition, with decreasing $\varepsilon$, from $\delta = \frac{1}{2}$ to $\delta = \frac{1}{4}$. As we shall see in the Section 3, $\delta = \frac{1}{2}$ corresponds to the three-resonance model[6,8] of Arnold diffusion, where explicit calculation can be made at large $\varepsilon$; $\delta = \frac{1}{4}$ corresponds to $\delta = 1/2(N-1)$ for the particular case $N = 3$. Chirikov developed a heuristic argument that the latter scaling should hold for any $N$. Using a perturbation scheme developed by Benettin, et al.[20] Giorgilli,[21] and Lochack[22] have shown that the Chirikov scaling is an upper bound, which is, in some sense, optimal (approximates the actual diffusion).

Following Lochack, the $k$th harmonic $f_k$ of the coupling term in (3) contributes a term to the Melnikov–Arnold integral for the change in $H$

$$\delta H \sim \mu e^{-\sigma|k|} |k|^{-N-1} \exp(-c/\varepsilon^{1/2} |k|^{N-2}), \tag{6}$$

where $c$ is a constant of order unity, and we have used the diophantine condition for the spacing or rationally independent resonances

$$|\omega \cdot k| \approx |k|^{-(N-2)}. \tag{7}$$

Maximizing (6) with respect to $|k|$ one finds $|k| \approx \varepsilon^{-1/2(N-1)}$, such that (6) becomes

$$\delta H \sim \mu \varepsilon^{1/2} \exp - (\sigma + c)/\varepsilon^{1/2(N-1)}. \tag{8}$$

This gives the numerical scaling with $\varepsilon$, found by Chirikov for $N = 3$ and small $\varepsilon$. For the diophantine condition to hold, we, in fact, need

$$\varepsilon^{1/2} \ll |k|^{-(N-2)}, \tag{9}$$

a condition that requires a very small $\varepsilon$ as $N$ becomes large. This regime is generally not accessible to numerical computation.

In addition to the restriction on the size of $\varepsilon$, the upper bound estimates are local estimates near a given unperturbed solution. The scaling can radically change near a multiple resonance of the unperturbed system. For example, if there are $m$ unperturbed resonances or an $m$th order resonance, then the diophantine condition becomes

$$|\omega \cdot k| \approx \frac{1}{|k|^{d-1}}, \quad d = N - m. \tag{10}$$

Then $d$ replaces $N$ in (10), greatly reducing the diffusion rate. A cartoon illustration of a portion of a global phase space is given in FIGURE 1, showing a few resonances with regions of intersecting resonances. Where the unperturbed resonances of high order exist the resonance structure thins, leaving slow diffusion along the thin layers.

To investigate global diffusion at larger $\varepsilon$ it is useful to employ a system that has uniform properties in a coarse-grained sense. The standard map, described by the equations

$$I(t + 2\pi) = I(t) + K \sin \theta(t)$$
$$\theta(t + 2\pi) = \theta(t) + I(t + 2\pi), \quad \text{mod } 2\pi, \tag{11}$$

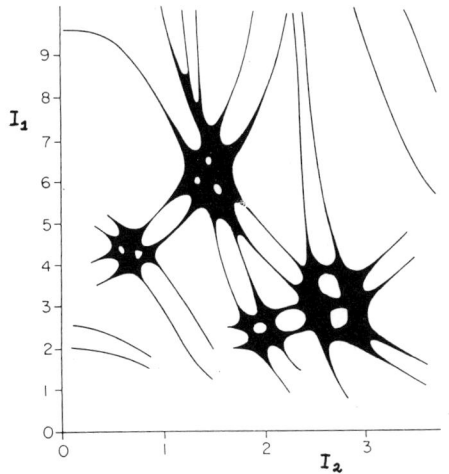

FIGURE 1. Schematic of stochastic resonance bands and their intersections.

where $I$ is the action and $\theta$ is the phase, has the important property that it is $2\pi$ periodic in both angle and action. The mapping is symplectic and has a Hamiltonian form

$$H(\theta, I, t) = \frac{I^2}{2} - K \cos \theta \delta_{2\pi}(t), \qquad (12)$$

where $\delta_{2\pi}(t)$ is a periodic $\delta$-function with period $2\pi$. As can be seen in FIGURE 2, generated by iterating a number of initial conditions with stochasticity parameter $K = 0.8$, the phase space consists of regions of stochasticity (area-filling trajectories) surrounding island chains of rational frequency. The regions of stochasticity are separated by regular motion on phase-spanning KAM curves. The largest region of stochasticity (thick dark region), is the "primary stochastic region," and the thinner

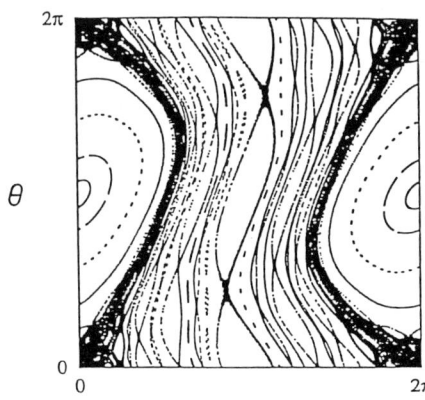

FIGURE 2. The standard map for $K = 0.8$, using a number of initial conditions to illustrate librational and rotational KAM curves and various stochastic layers.

regions around smaller islands are "secondary stochastic regions." The KAM curves consist of two types, librational motion about fixed points (closed curves on the phase plane) and rotational motion (open curves spanning $2\pi$ in the phase $\theta$). For $K \gtrsim 0.9716...$, the final rotational KAM curve is destroyed, such that global diffusion in a coupled set of mappings is across resonance rather than the Arnold diffusion along resources.

Konishi and Kaneko[23] studied global diffusion in a set of coupled standard mappings with nearest neighbor coupling,

$$I_i(t + 2\pi) = I_i(t) + K \sin [\theta_{i+1}(t) - \theta_i(t)] - K \sin [\theta_i(t) - \theta_{i-1}(t)]$$
$$\theta_i(t + 2\pi) = \theta_i(t) + I_i(t + 2\pi) \tag{13}$$

and with global coupling,

$$I_i(t + 2\pi) = I_i(t) + \frac{K}{\sqrt{N-1}} \sum_{j=1}^{N} \sin [\theta_j(t) - \theta_i(t)]$$

$$\theta_i(t + 2\pi) = \theta_i(t) + I_i(t + 2\pi). \tag{14}$$

The nearest neighbor coupling is analogous to the FPU oscillator chain that has been used to study the closely related problem of the approach to energy equipartition in many-degree-of-freedom systems, which we consider in Section 4. For nearest neighbor coupling, Konishi and Kaneko investigated the diffusion for $0.2 \leq K \leq 1$, over a range of $N$ values. They found for $N > 3$ that the diffusion coefficient for the actions $I$ fitted an exponential $D \propto e^{-c/\varepsilon^\delta}$ ($\varepsilon \equiv K$) with $\delta \simeq 0.45$ independent of $N$. This is quite close to $\delta = 0.5$ predicted from a three-resonance model. For global coupling the diffusion coefficient was fitted to a power law $D \propto \varepsilon^\eta$ finding $\eta \approx 5$, which is also independent of $N$.

In a recent calculation Chirikov and Vecheslavov[24] have estimated the rate of global diffusion for $N$ sufficiently large and $\varepsilon$ not too small, in a Hamiltonian of the form used for upper bound estimates, finding $D \propto \varepsilon^\eta$, $\eta \approx 6.5$, independent of $N$. Using the nearest neighbor coupled standard mapping, as given by Konishi and Kaneko, they numerically found diffusion that scaled reasonably closely to their theoretical results over a limited range of $K \equiv \varepsilon$. However, as described earlier, the three-resonance model also fits the data. The mappings used in these calculations do not separate the coupling strength $\mu$ from the nonlinearity $K$, and therefore quantitative calculations cannot be performed. In the next section we present results of coupled mappings in which $\mu \ll K$, as required for the validity of the MA integral calculations.

## 3. CALCULATING ARNOLD DIFFUSION IN COUPLED STANDARD MAPPINGS

Here we follow the work of Wood et al.[9,10] to quantitatively compare calculated and numerical diffusion rates in a system of weakly coupled standard maps. We consider a system of the following form:

$$I_1(t + 2\pi) = I_1(t) + K_1 \sin \theta_1(t) + \mu \sin (\theta_1(t) + \cdots + \theta_m(t))$$
$$\theta_1(t + 2\pi) = \theta_1(t) + I_1(t + 2\pi), \quad \mod 2\pi$$
$$\vdots \quad (15)$$
$$I_n(t + 2\pi) = I_n(t) + K_n \sin \theta_n(t) + \mu \sin (\theta_n(t) + \cdots + \theta_{m-1}(t))$$
$$\theta_n(t + 2\pi) = \theta_n(t) + I_n(t + 2\pi), \quad \mod 2\pi,$$

where a total of $n$ maps are coupled together in groups of $m$, with $2 \le m \le n$; each map is coupled to itself and the next $m - 1$ maps in cyclical order. Note that this system has $N = n + 1$ degrees of freedom. We leave the structure of the individual maps nearly unchanged by making the coupling strength $\mu$ small, and control the number of interacting resonances through the number of coupling phases. The nonlinearity parameters $K_i$, $1 \le i \le n$, can also be varied independently of the coupling. The mapping equations (15) do not have a complete symplectic form, unless $m = n$. However, they are volume preserving and are also reversible, that is, the mapping transformation can be expressed as a product of two involutions. Volume preservation ensures that there are no sources or sinks, and reversibility ensures that at least some KAM surfaces are preserved under the coupling perturbation. The physical nature of the diffusion, arising from a three-resonance model,[9,10] indicates that the most important aspect of the KAM structure is that associated with the uncoupled maps, which are symplectic.

For two coupled maps the action space can be exhibited by plotting (in action) the crossings of a surface of section in the phases. Taking the surface at $\theta_1 = \theta_2 = \pi$, for parameters $K_1 = K_2 = 0.8$ and $\mu = 0.01$, in FIGURE 3 the positions of $10^3$ particles are shown after $2 \times 10^6$ mapping iterations for each particle. The widths of the primary stochastic regions are shown in gray, and positions of a few other secondary stochastic regions are noted by lines, but their widths are not resolved. Although not completely obvious from the figure, the positions of the particles are all in the connected web of stochasticity, some being in the thinner secondary layers (not shown). The region labeled A is a primary librational region in both maps, and the two regions labeled B are in a librational region of the period 2 island chain in one map and in a primary librational region in the other map.

To determine local diffusion the values of the controlling parameters must be closely specified. To do this, the driven map is coupled to the driving maps, which are themselves uncoupled and therefore confined to their respective primary stochastic regions, and $\mu$ is sufficiently small that the diffusion is determined at a given value of $I_1$. The global symplectic structure is not relevant for this calculation. The reduced mapping is

$$I_1(t + 2\pi) = I_1(t) + K_1 \sin \theta_1(t) + \mu \sin (\theta_1(t) + \cdots + \theta_m(t))$$
$$\theta_1(t + 2\pi) = \theta_1(t) + I_1(t + 2\pi), \quad \mod 2\pi \quad (16)$$

for one map, and

$$I_j(t + 2\pi) = I_j(t) + K_j \sin \theta_j(t)$$
$$\theta_j(t + 2\pi) = \theta_j(t) + I_j(t + 2\pi) \quad (17)$$

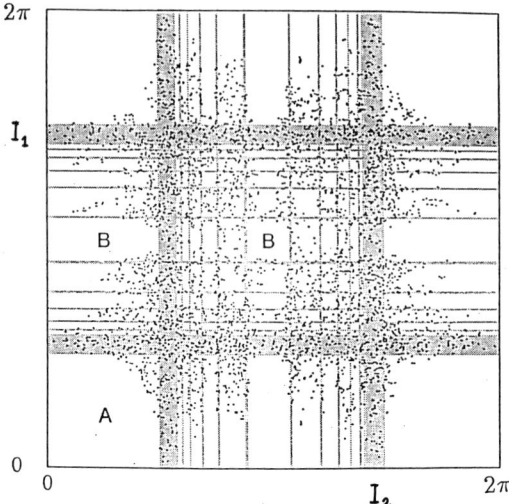

**FIGURE 3.** Arnold diffusion in the $2\pi$ periodic $(I_1, I_2)$ phase plane for a surface of section at $\theta_1 = \theta_2 = \pi$, with $K_1 = K_2 = 0.8$ and $\mu = 0.01$. One thousand particles were started in the primary stochastic region of both maps, and run for two million iterations.

for all other maps, with $j$ going from 2 to $m$. To understand the effect of increasing the number of coupling phases, we first consider the effect of increasing the number of driving maps from 1 to 2. The doubly driven map effectively receives kicks at twice the rate of the singly driven map. Since the two driving maps are not correlated, the mean square local diffusion rate doubles. Occasionally by chance, the two driving maps are approximately in phase, and deliver a single large kick to the driven map. These two effects are added to give the total increase in the diffusion rate. To increase the number of driving maps we sum the probabilities of the various kinds of interactions when one or more of the driving terms coincide to give one large kick. We take the case of $p$ driving phases in primary stochastic layers and $d = r + s$ rotation phases and secondary resonance stochastic phases. The local diffusion coefficient along any of the $d$ actions $\mathbf{I} = (I_1, \ldots, I_d)$ can be written in the form of a modified binomial expansion

$$D_{pd}(\mathbf{I}) = \frac{\mu^2}{2K_jT} \sum_{i=1}^{p} \binom{p}{i-1} \left(\frac{\Delta t}{T}\right)^{p-i} \left(1 - \frac{\Delta t}{T}\right)^{i-1} A_{\mathrm{MA}}^2(p+1-i)$$

$$\times \frac{1}{\pi} \int_{-\pi/2}^{\pi/2} \cos^{2(p-i)}\theta' \, d\theta', \tag{18}$$

where $T \approx 2\pi^2/K_j$, and $Q_0 = \omega_d/K_j^{1/2}$, with $\omega_d = I_1 + I_2 + \cdots + I_d$, mod $2\pi$ for $0 < \omega_d < \pi$ and $\omega_d = 2\pi - [(I_1 + I_2 + \cdots + I_d) \bmod 2\pi]$ otherwise. The peak value of the MA integral for $k = p + 1 - i$ stochastic drives is $A_{\mathrm{MA}}(k)$. This coincides with the usual set of MA integrals as defined in references 2 and 4, $A_{\mathrm{MA}}(k) = A_{2k}$, where

$A_{2k}$ is determined by the recursion relation for MA integrals

$$A_l = \frac{2Q_0}{l-1} A_{l-1} - A_{l-2}, \qquad (19)$$

where $A_1 = 2\pi e^{\pi Q_0/2}/\sinh \pi Q_0$ and $A_2 = 2Q_0 A_1$. For large $Q_0$ this can be expressed in the closed form

$$A_l = \frac{(2Q_0)^{l-2}}{(l-1)!} A_2. \qquad (20)$$

The binomial coefficients in (18) give the number of coincidences of the phases, that lead to the coincidence enhancement of the MA integrals. The factor $(\Delta t/T)^{p-i}$ is the probability of $p - i$ enhancement peaks of width $\Delta t$ occurring simultaneously within the interval $T$, with $\Delta t/T \simeq K$. The last ($i = p$) term in (14) is due to the incoherent summing of the kicks of the $p$ driving phases. The integral accounts for the sharpening of the peak by additional coincidences and can be evaluated as

$$\frac{1}{\pi} \int_{-\pi/2}^{\pi/2} \cos^{2k}\theta \, d\theta = \frac{(2k-1)!!}{2^k k!}. \qquad (21)$$

Using these relations we plot in FIGURE 4 the local spread in action for $K_j = 0.2$ and $m - 1$ between 1 and 5, and compare with corresponding numerical values, obtaining excellent agreement. The dashed curve gives the upper-bound scaling, from (8), with the coefficients matched to the exact result at $m = 2$. As expected, for this relatively large value of the perturbation parameter $\varepsilon = K = 0.2$, the upper-bound result lies well above the numerical and calculated values.

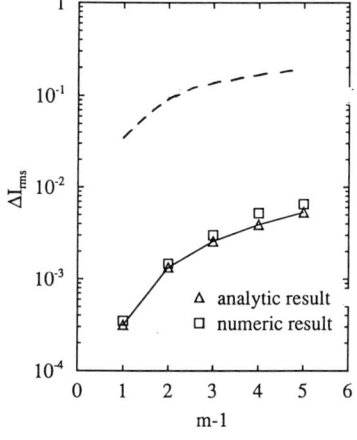

**FIGURE 4.** Local Arnold diffusion distance versus number of driving phases for $K_j = 0.2$, comparing numerical and analytical result. The *dashed curve* is a heuristic upper bound.

We now examine the global diffusion in action across the higher dimensional phase space over long times. The $n$ maps are coupled together in groups of $m$. As a special case, we choose the same value of $K$ for every map. The distinction between driven and driving maps is lost, because a single map may be both driven and driving at different times in the diffusion process. The diffusion rate also varies greatly with the region of phase space that a driven particle occupies. The primary stochastic region can drive Arnold diffusion across rotational orbits and across the stochastic regions associated with the secondary resonances. Libration orbits are filled with diffusing particles, but do not contribute to global diffusion because the motion averages to the fixed point of a librational region. The global diffusion is determined from the probabilities of trajectories being in the various accessible regions of phase space, and from the local rate of diffusion in each region. The numerical values of the global diffusion are given in FIGURE 5, both for the symplectic case ($m = N$) and the nonsymplectic case ($m < N$). The difference probably derives from the additional symmetry in the symplectic case. Computed analytic values of $\Delta I_{rms}$ for two and for three phases in the coupling are also shown in the figure. These values lie somewhat below the numerically determined values for the symplectic maps, which is reasonable since the ergodicity assumption, which slows the diffusion, is not fully realized numerically.

The results for both local and global diffusion both show $m$-dependence indicating that the $N$-independent results of Konishi and Kaneko, and of Chirikov and Vecheslavov have not been obtained. The diffusion may approach an asymptote at large $N$, where the power-law scaling with $\varepsilon$ might apply. This would require consideration of smaller values of $K$, since for $K = 0.2$ the driving phases in the coupling term act collectively like a random phase (diffusion independent of $K$) at large $N$.

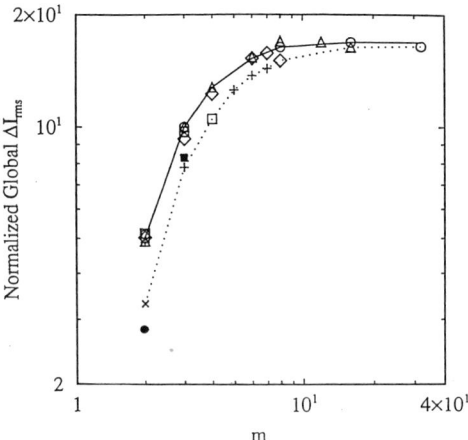

**FIGURE 5.** Numerically determined global Arnold diffusion distance for various values of $m$ and $N$, normalized by dividing by $N^{1/2}$. The *solid line* is drawn through the points with $m < N$, while the *dashed line* is for $m = N$. Calculated values are shown as *solid figures*. ($K = 0.8$, $\mu = 0.01$.)

## 4. ARNOLD DIFFUSION AND EQUIPARTITION IN THE FPU SYSTEM

The quartic FPU Hamiltonian, representing a linear chain of equal masses coupled by nonlinear springs with a quartic nonlinearity,[6,10] is given by

$$H = \sum_{i=1}^{N} \tfrac{1}{2}p_i^2 + \tfrac{1}{2}(q_{i+1} - q_i)^2 + \tfrac{1}{4}\beta(q_{i+1} - q_i)^4. \tag{22}$$

We consider the case of strong springs ($\beta > 0$) and fixed boundaries $q_0 = q_{N+1} = 0$. In the linear case ($\beta = 0$) the chain of oscillators may be put in the form of $N$ independent normal modes, and is therefore integrable and nonergodic. The normal modes (sinusoids) are related to the mass positions ($q$'s) by

$$q_i = \left(\frac{2}{1+N}\right)^{1/2} \sum_{\alpha=1}^{N} \sin\left(\frac{\pi i \alpha}{N+1}\right) Q_\alpha \qquad i = 1, \ldots, N \tag{23a}$$

$$\Omega_\alpha = 2 \sin\left(\frac{\pi \alpha}{2(N+1)}\right), \tag{23b}$$

where $\Omega_\alpha$ is the frequency of the $\alpha$th linear mode and $Q_\alpha$ its amplitude, with $\alpha = 1, \ldots, N$. We refer to those coordinates as the "linear modes" of the system, even though they are not separable modes if we include the quartic term. The FPU Hamiltonian in the linear mode coordinates is calculated to be

$$H = \sum_{i=1}^{N} \tfrac{1}{2}(P_i^2 + \Omega_i^2 Q_i^2)$$

$$+ \frac{\beta}{(8N+8)} \sum_{r,s,m,n} G(r, s, m, n) Q_r Q_s Q_m Q_n, \tag{24}$$

where $r, s, m, n$ are integers running from 1 to $N$ and $G$ is defined by

$$G(r, s, n, m,) \equiv \Omega_r \Omega_s \Omega_n \Omega_m \sum_P B(r \pm s \pm n \pm m). \tag{25}$$

$B$ is given by

$$B(\alpha) = \begin{cases} 1 & \text{if } \alpha = 0 \\ -1 & \text{if } \alpha = \pm 2(N+1) \\ 0 & \text{otherwise} \end{cases} \tag{26}$$

and $\sum_P$ in (25) represents the sum over all eight permutations of the signs in front of $s$, $n$, and $m$. We see then that there is a selection rule for mode coupling via the quartic term.

To numerically study the FPU system, initial conditions have been used for which all the energy is concentrated in some few modes around some low-frequency mode $\gamma$, or in which a packet of modes has been used of width $\Delta\gamma$ in which both $\gamma$ and $\Delta\gamma$ are proportional to $N$. In numerical experiments the time average of the linear energies $E_i$, $i = 1, \ldots, N$, is usually calculated. The information entropy,

$$S = -\sum_{i=1}^{N} e_i \ln e_i, \tag{27}$$

where the $e_i = E_i/\sum_1^N E_i$ are the normalized average energies, is then used to determine the effective number of modes by

$$n_{ef} \equiv \exp S. \qquad (28)$$

DeLuca et al.[16] developed theoretical descriptions for the energy spreading among modes, valid in various energy ranges, which were compared to numerical results. Both the transition, with energy, to observe equipartition and the timescale required to attain a constant measure of equipartition were studied. The main idea is that resonant interaction of a few low-frequency modes in which most of the energy resides can lead to local superperiods (low-frequency beat oscillations) that are stochastic. The transition to stochastic local interaction occurs at

$$R \equiv (N+1)\frac{6\beta}{\pi^2} E \simeq 1, \qquad (29)$$

where $R$ is the ratio of nonlinear to linear energy in a transformed reduced Hamiltonian.[16] Since $R \propto N + 1$, the energy at which this transition occurs becomes vanishingly small as $N \to \infty$. The stochasticity at $R \simeq 1$ corresponds to the appearance of an elliptic and hyperbolic pair of fixed points in a low-frequency resonance, giving rise to a local libration frequency $\Omega_b \simeq 6.2\gamma R/N^3 \simeq 3.7\beta\gamma E_\gamma/N^2$ around the elliptic fixed point. The interaction between $\Omega_b$ and a second low-frequency resonance produces a stochastic separatrix layer connecting the hyperbolic points, which becomes large at $R \simeq 4$. The criterion that $R$ be sufficiently large to give widespread stochasticity within a resonance could be referred to as an "overlap criterion." The system is, however, intrinsically degenerate, such that the stochasticity arises from a complicated set of interactions among the modes (see [6, sec. 2.4]). The overlap of these four modes does not lead to equipartition on computationally observable timescales. Numerically it is found that increasing $R$ results in an increase in the effective number of low-frequency modes that are stochastically interacting, giving approximately $n_{ef} \simeq 0.5R + 1$. This is readily understood in terms of local four-mode interactions in which an increasing number of neighboring mode resonances can form as their associated energies exceed $R \sim 1$.

In addition to this stochastic diffusion among the low-frequency modes we expect the stochasticity to drive Arnold diffusion along stochastic layers to other modes, with the particular channels and rates to be determined by the coupling coefficients. Since the driving frequency for diffusion is associated with $\Omega_b$, a fundamental timescale for numerical observations is

$$\tau_b = \Omega_b^{-1} \propto N^2/\gamma\beta E_\gamma. \qquad (30)$$

Using resonant normal form perturbation theory to isolate the most important coupling to the high-frequency modes, it is found that, above a critical value $E > E_c$, the ratio of the stochastic drive frequency $\Omega_b$ to the lowest frequency high mode number resonance frequency (resonance between two high-frequency modes) $\Delta\Omega_{12} \simeq \gamma\pi^2/N^2$ becomes of order unity. The Arnold diffusion in the three-resonance model depends exponentially on the frequency ratio as $\exp(-\pi\Delta\Omega_{12}/2\Omega_b)$. Hence for $\Omega_b \sim \Delta\Omega_{12}$ we would expect to obtain strong diffusion of energy to high-frequency modes, and equipartition on computationally observable timescales. Numerically the transition to observable diffusion occurs at a value of $E_c \approx 3$, for $\beta = 0.1$, corresponding to $\Omega_b/\Delta\Omega_{12} \approx 0.3$. Since the coupling involves two high-frequency modes that initially have little energy, the resulting increase in energy is at first exponential, but later follows the usual diffusive scaling. A typical mode spectrum for $E > E_c$ is given in FIGURE 6, showing the approach to equipartition. Energy first couples strongly into a few high-frequency modes by Arnold diffusion, and then spreads to neighboring modes. The energy increase in the two modes constituting the resonances $\Delta\Omega_{12}$ are shown in FIGURE 7. The solid line has slope 1, characteristic of normal diffusion. For values of $E \gg E_c$ (but $E/N < 1$ such that diffusion across resonance is not involved) we would expect the exponential factor to be relatively unimportant, and the timescale to equipartition to scale with $\tau_b$ as in (30). There is also a size-dependent timescale found to be proportional to $N^{1/2}$. In FIGURE 8 we plot $n_{ef}(t)$ versus normalized time $\propto \tau_b N^{1/2}$ over a range of

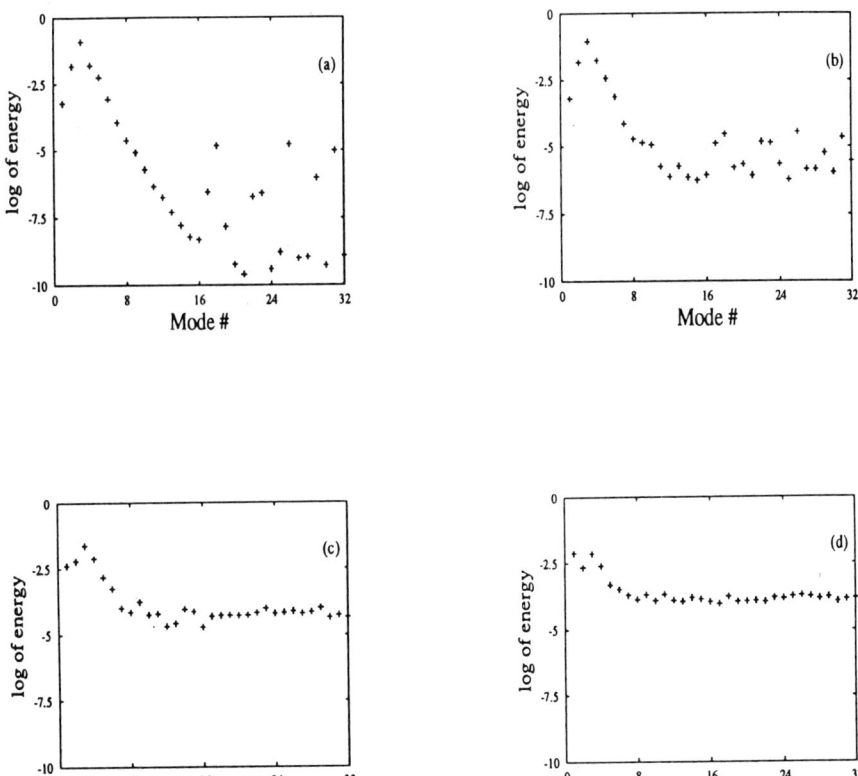

**FIGURE 6.** Natural log of average mode energies for $R = 8$ and $N = 32$ ($E > E_c$), at four normalized times $\tau = \gamma\pi^2 t/3(N + 1)$, showing approach to equipartition (a) $\tau = 2000$, $n_{ef} = 6.3$; (b) $\tau = 10,000$, $n_{ef} = 8.7$; (c) $\tau = 40,000$, $n_{ef} = 19.0$; (d) $\tau = 78,000$, $n_{ef} = 25.5$.

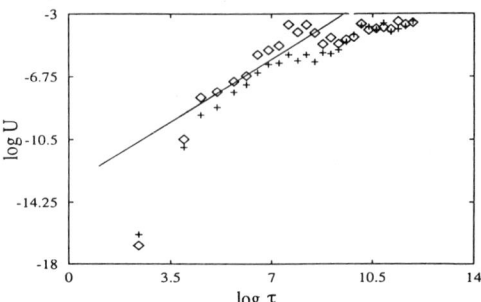

**FIGURE 7.** Natural logarithm of the energy of mode 29 (*diamonds*) and mode 31 (*plusses*) versus natural log of $\tau = \gamma\pi^2 t/3(N+1)$.

$10 < E < 1000$ and $16 < N < 1024$, with all of the data falling within the expected error. This reinforces our picture of the fundamental importance of Arnold diffusion in the equipartition process.

A Taylor expansion of the FPU equation of motion of a given mass point, leads to the mKdV partial differential equation that approximately describes low-frequency modes of the FPU system for $\beta > 0$ and $N$ sufficiently large. Driscoll and O'Neil found an onset, with increasing energy, of an instability in a soliton solution of the mKdV equation that corresponds to exponential growth of a low-frequency mode of the FPU system. Comparing the rescaled parameters of the mKdV equation, as found by Driscoll and O'Neil, to the parameters governing the interaction among the low-frequency FPU modes, the rescaling gives the relationship $R \simeq (8/\pi^2)(\gamma/2)^2 q^2 K^2(q^2)$, where $q$ is the argument of the cnoidal function that describes the soliton, $\gamma - 1$ the number of nodes of the cnoidal function, and $K(q^2)$ is the complete elliptic integral. The instability appears for $q^2 \simeq 0.25$ ($K = 1.7$), for $\gamma = 2$, obtaining $R \simeq 0.6$, which is approximately the same value as that which produced a separatrix

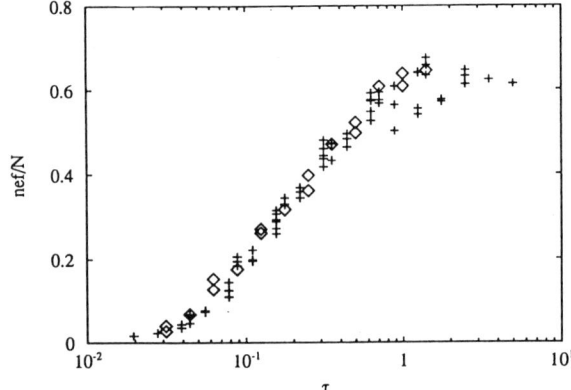

**FIGURE 8.** $n_{ef}/N$ versus $\tau = Et/N^{2.5}$ for initial conditions with $16 \leq N \leq 1024$, $10 \leq E \leq 1000$, and with $E/N < 1$.

layer in the local resonance interaction. However, this instability does not lead to transfer of energy to the high-frequency modes and to equipartition, on computationally observable timescales.

## 5. CONCLUSION

In conclusion, we have examined Arnold diffusion for two systems: (1) weakly coupled standard mappings, for which the nonlinearity and coupling parameters can be separately controlled and explicit calculation of the diffusion rate can be found; and (2) the FPU coupled oscillator chain for which Arnold diffusion occurs indirectly through the interaction of resonances. In both cases, the results have been used to shed light on the various general calculations of the upper-bound scaling of Arnold diffusion. The results indicate that in the parameter ranges that are computationally accessible the nonlinearity parameter that governs the diffusion rate is much larger than that allowed in the upper-bound calculations. In both cases a three-resonance model can be used to understand the diffusion. In the case of weakly coupled standard maps, a quantitative calculation can be made that agrees well with numerical results. The scaling of diffusion that satisfies a power law in $\varepsilon$ and is independent of $N$, which has been estimated for sufficiently large $\varepsilon$ and $N$, was not reached in the weakly coupled mapping system and therefore remains to be investigated.

In the FPU coupled oscillation chain the timescale to attain some fraction of equipartition scaled as $\tau \propto N^{5/2}/\beta E$, which scales strongly with $N$ and inversely with the perturbation strength $\beta E$. This scaling corresponds, approximately, to the stochastic drive of three-resonance Arnold diffusion. Stochasticity among low-frequency resonances was seen to coincide with an instability in the mKdV differential equation that approximates the low-frequency modes of the oscillator chain. This occurs at energies that are generally much lower than the critical energy $E_c$ determining the Arnold diffusion to the high-frequency modes. The energy at which the low-frequency interaction first occurs decreases with $N$, while the Arnold diffusion becomes computationally observable at a critical energy that is independent of $N$.

## ACKNOWLEDGMENT

J. DeLuca, M. Lieberman, S. Ruffo, and B. Wood contributed to work reported here.

## REFERENCES

1. ARNOLD, V. I. 1964. Russ. Math. Surv. **18**: 85.
2. BERRY, M. V. 1978. *In* Topics in Nonlinear Dynamics, Vol. 46, S. Jorna, Ed.: 16. American Institute of Physics. New York.
3. CHIRIKOV, B. V. 1971. Research Concerning the Theory of Nonlinear Resonance and Stochasticity, CERN translation 71-40, Geneva.

4. GERASIMOV, A., F. M. ISRAILEV, J. L. TENNYSON & A. B. TEMNYKH. 1986. The Dynamics of the Beam-Beam Interaction, Inst. Nucl. Phys.: 86–100 Novosibirsk.
5. HOWARD, J. E., A. J. LICHTENBERG, M. A. LIEBERMAN & R. H. COHEN. 1986. Physica **20D**, 259; see also. 1986. Phys. Fluids **29**: 1061.
6. LICHTENBERG, A. J. & M. A. LIEBERMAN. 1992. Regular and Chaotic Dynamics, 2nd ed. Springer-Verlag, New York.
7. FALCIONI, M., U. M. B. MARCONI & A. VULPIANI. 1991. Phys. Rev. **A44**: 2263.
8. CHIRIKOV, B. V. 1979. Phys. Rep. **52**: 263.
9. WOOD, B. P., A. J. LICHTENBERG & M. A. LIEBERMAN. 1990. Phys. Rev. A. **42**: 5885.
10. WOOD, B. P., A. J. LICHTENBERG & M. A. LIEBERMAN. 1994. Physica D **71**: 132.
11. FERMI, E., J. PASTA & S. ULAM. 1965. *In* Collected Papers of E. Fermi. E. Segre, Ed. Univ. of Chicago. Chicago.
12. IZRAILEV, F. M. & B. V. CHIRIKOV. 1966. Sov. Phys.-Dokl. **11**: 30.
13. LIVI, R., M. PETTINI, S. RUFFO & A. VULPIANI. 1987. J. Stat. Phys. **48**: 539.
14. PETTINI, M. & M. LANDOLFI. 1990. Phys. Rev. A **41**: 768.
15. KANTZ, H., R. LIVI & S. RUFFO. 1994. J. Stat. Phys. In print.
16. DELUCA, J., A. LICHTENBERG & M. LIEBERMAN. 1994. Chaos. In print.
17. DRISCOLL, C. F. & T. M. O'NEIL. 1976. Phys. Rev. Lett. **37**: 69; also. 1978. Rocky Mt. J. Math. **8**: 211.
18. FOREST, M. G., C. G. GOEDDE & A. SINHA. 1992. Phys. Rev. Lett. **68**: 2722.
19. NEKHOROSHEV, N. N. 1977. Russ. Math. Surv. **32**: 6.
20. BENETTIN, G., L. GALGANI & A. GIORGILLI. 1987. Commun. Math. Phys. **113**: 87.
21. GIORGILLI, A. 1988. Ann. Inst. H. Poincare **48**: 423.
22. LOCHAK, P. 1990. Phys. Lett. A **143**: 39.
23. KONISHI, T. & K. KANEKO. 1990. J. Phys. A: Math. Gen. L715.
24. CHIRIKOV, B. V. & V. V. VECHESLAVOV. 1992. Theory of the Fast Arnold Diffusion in Many-Frequency Systems, Inst. Nucl. Phys., Novosibirsk. Preprint.

# Universal Properties of Escape[a]

CHRISTOS V. SIOPIS,[b,d] HENRY E. KANDRUP,[b,e]
G. CONTOPOULOS,[b,d] and RUDOLF DVORAK[c]

[b]*Department of Astronomy*
*University of Florida*
*Gainesville, Florida 32611*

[c]*Institut für Astronomie*
*Universität Wien*
*Türkenschanzstrasse 17, A-1180 Wien, Austria*

## INTRODUCTION: MOTIVATION

This paper is the continuation of work by Contopoulos,[1] Contopoulos and Kaufmann,[2] Contopoulos et al.,[3] and Siopis et al.,[4] on the properties of escapes of orbits from Hamiltonian systems described by two-dimensional time-independent potentials. The first system to be studied[3] was characterized by the Hamiltonian

$$H_1 \equiv \tfrac{1}{2}(\dot{x}^2 + \dot{y}^2 + x^2 + y^2) - \varepsilon x^2 y^2 = h_1. \tag{1}$$

This Hamiltonian was chosen because it may perhaps represent the central part of a deformed galaxy. In fact, it describes two harmonic oscillators, coupled via the quartic term $\varepsilon x^2 y^2$, where $\varepsilon$ is a parameter that determines the strength of the coupling. For a fixed value of $\varepsilon$, escape from the system is possible only when the total energy $h_1$ is higher than a certain critical escape value. Alternatively, for a fixed value of $h_1$, escape is possible only when $\varepsilon$ exceeds a critical value $\varepsilon_{\text{esc}}$. In this work, the energy of the system was held fixed, but the coupling parameter $\varepsilon$ was allowed to vary. The form of the potential is symmetric with respect to the origin and, for $\varepsilon > \varepsilon_{\text{esc}}$, the equipotential surfaces allow for four channels of escape (FIG. 1(a)).

The fact that escape is possible energetically gives no information about when an orbit will escape. However, the time of escape can be important for a number of different problems, both in astronomy and other fields. For instance, if a star has enough energy to escape, but it takes more than a Hubble time for this to happen, for most practical purposes it should be considered as being bound to the galaxy. Similar situations are encountered in plasma physics, molecular dynamics, heavy ion physics, and so forth.[5–8]

To better characterize the escape properties of the system, several numerical experiments were performed. Each experiment involved solving the equations of motion for an initially localized ensemble of orbits with $\varepsilon > \varepsilon_{\text{esc}}$. The value of $\varepsilon$ was

---

[a]Three of the authors (C. V. S.) (G. C.) and (R. D.) were supported in part by the European Community Human Capital and Mobility Program ERB4050 PL930312. The fourth author (H. E. K.) was supported in part by the National Science Foundation Grant PHY92-03333.

[d]Also Astronomy Department, University of Athens, Panepistimiopolis, 157 83 Athens, Greece.

[e]Also Department of Physics and Institute for Fundamental Theory, University of Florida, Gainesville, Florida 32611.

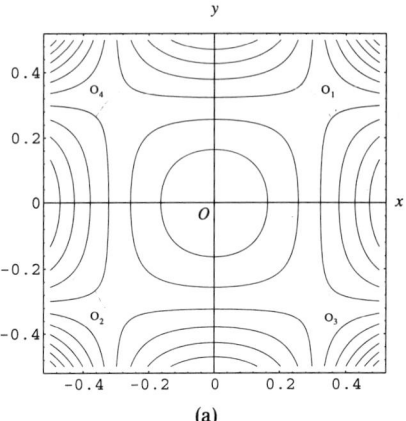
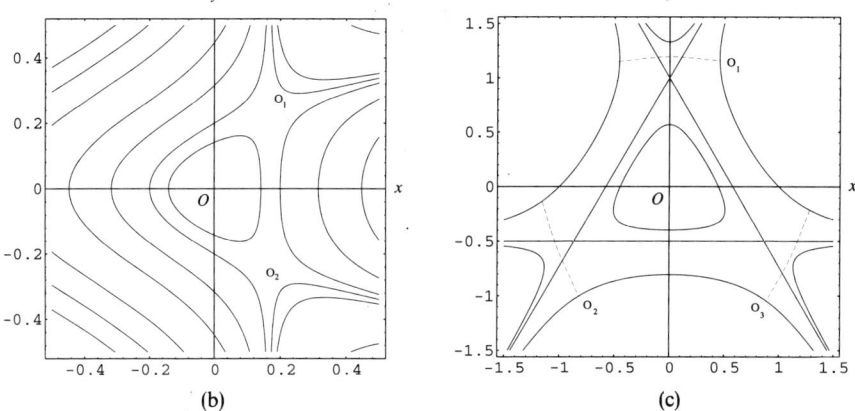

**FIGURE 1.** Equipotential surfaces of the potential that corresponds to (a) $H_1$ for $\varepsilon = 5.26$ and different values of $h_1$; (b) $H_2$ for $\varepsilon = 3.0$ and different values of $h_2$; (c) $H_3$ for $\varepsilon = 1.3$ and different values of $h_3$. $O_1, O_2, O_3$, and $O_4$ represent Lyapunov unstable periodic orbits.

fixed during any given realization, but was different for different realizations. Each orbit was followed for a maximum time or until it escaped from the system, whichever happened first. If the orbit was observed to escape, the time of escape was recorded (a particle has escaped if it crosses outwards a Lyapunov orbit at one opening of the curves of zero velocity[9]). The sampling of the space of initial conditions, for any given ensemble and value of $\varepsilon$, was chosen to yield a wide range of escape times.

Statistical analysis of the escape times for a large number of orbits showed that there is a significant qualitative change of the behavior of the system, for values of $\varepsilon$ beyond a critical value $\varepsilon_2 > \varepsilon_{\text{esc}}$. More precisely, for $\varepsilon \in [\varepsilon_{\text{esc}}, \varepsilon_2]$, the probability per unit time, $p(\varepsilon, t)$, for an orbit to escape from the system at time $t$, is a monotonically decreasing function of time, apparently tending to zero at large times. However, for $\varepsilon > \varepsilon_2$, the value of $p(\varepsilon, t)$ seems to asymptote toward a constant, nonzero value, $p_\infty(\varepsilon)$. This rather abrupt change of behavior happens within a narrow interval around $\varepsilon_2$ and is reminiscent of a phase transition.

In addition, a simple scaling behavior was observed for three different quantities, namely:

1. $p_\infty(\varepsilon)$, the asymptotic probability of escape;
2. $T(\varepsilon)$, the time that is required for $p(\varepsilon, t)$ to converge toward $p_\infty(\varepsilon)$, for a subset $S$ of fixed phase space volume, from which the initial conditions were sampled; and
3. $T(r)$, the time that is required for $p(\varepsilon, t)$ to converge toward $p_\infty(\varepsilon)$, as a function of the characteristic size $r$ of $S$, for a fixed value of $\varepsilon$.

Specifically, it was found that all three quantities have a power-law dependence on the order parameter $\varepsilon - \varepsilon_2$, viz.,

$$p_\infty(\varepsilon) \propto (\varepsilon - \varepsilon_2)^\alpha, \qquad (2)$$

with $\alpha = 0.49 \pm 0.05$ and $\varepsilon_2 = 4.90 \pm 0.01$;

$$T(\varepsilon) \propto (\varepsilon - \varepsilon_2)^{-\beta}, \qquad (3)$$

with $\beta = 0.39^{+0.14}_{-0.06}$, and

$$T(r) \propto r^{-\delta}, \qquad (4)$$

with $\delta = 0.08 \pm 0.03$. It is interesting to note that, within statistical uncertainties, $\alpha - \beta - \delta = 0$.[3]

Once these results were established, an obvious question was whether this scaling behavior is due to the symmetries of this particular potential, or whether it is generic. In order to probe this question, a similar study[4] was performed for a second Hamiltonian:

$$H_2 \equiv \tfrac{1}{2}(\dot{x}^2 + \dot{y}^2 + x^2 + y^2) - \varepsilon x y^2 = h_2. \qquad (5)$$

This Hamiltonian differs from $H_1$ in that the coupling term is cubic instead of quartic. The equipotential surfaces now allow for two channels of escape (FIG. 1(b)). The results of the numerical experiments suggest that the same power-law scaling for $p_\infty(\varepsilon)$ and $T(\varepsilon)$ persists. Moreover, the scaling exponents were found to be $\alpha = 0.55 \pm 0.10$ and $\beta = 0.50^{+0.15}_{-0.10}$. A comparison with the values of $\alpha$ and $\beta$ that were found for $H_1$ indicates that, given the uncertainties, the exponents could be identical.

In order to better establish the conclusions obtained in the two previous cases, the same analysis was performed for yet another potential, now allowing for three channels of escape. This entails a variation of the Hénon–Heiles Hamiltonian[10] (FIG. 1(c)):

$$H_3 \equiv \tfrac{1}{2}(\dot{x}^2 + \dot{y}^2 + x^2 + y^2 - \tfrac{2}{3}y^3) + \varepsilon x^2 y = h_3. \qquad (6)$$

The difference from the original Hénon–Heiles form is the coefficient $\varepsilon$ in front of the coupling term $x^2 y$, that allows for variations of the strength of the coupling.

The object of this paper is to report on the findings from a study of the escape properties of this modified Hénon–Heiles potential. The results are essentially the same as those obtained for the cubic and quartic cases, that is, (1) the same power-law scaling persists, and (2) the numerical values of the exponents for this case ($\alpha = 0.45 \pm 0.05$ and $\beta = 0.37^{+0.10}_{-0.07}$) agree with the other two cases to within statistical uncertainties. This result strengthens the possibility that $\alpha$ and $\beta$ could in fact be identical in all three potentials. This would be a strong indication that $\alpha$ and $\beta$ are universal constants.

One additional issue was also addressed in this work, namely the constancy of the asymptotic escape probability at much later times. In the previous calculations it was assumed that, for a given $\varepsilon > \varepsilon_2$, $p(t)$ eventually reaches a constant asymptotic value $p_\infty(\varepsilon)$. However, if the numerical experiments are performed for longer times, one finds that the escape probability is not actually constant, but slowly decays with time. This may be related to the topological structure of phase space and, more specifically, to the presence of "islands" of stability, inside and near which orbits can be trapped, at least temporarily, even for perturbations $\varepsilon > \varepsilon_2 > \varepsilon_{esc}$. Various attempts to fit the decay curve, suggest that $p_\infty$ in fact falls off as a power-law function of time, that is, that, for large enough times, $p(t) \propto t^{-\mu}$, where $\mu = 0.31 \pm 0.05$, seemingly independent of $\varepsilon$.

## THE NUMERICAL EXPERIMENTS

### A Description of the Experiments

As explained in the preceding discussion, the main objective of the present work is to check whether the escape properties of the Hénon–Heiles potential, described by $H_3$, agree with the ones found for $H_1$ and $H_2$. To that end, the same calculations performed for the two previous cases were repeated for $H_3$. The total energy of the system was chosen to be $h_3 = 1/6$ [cf. (6)], which means that the escape perturbation is $\varepsilon_{esc} = 1$. When $\varepsilon < \varepsilon_{esc}$, no particles can escape. For $\varepsilon > \varepsilon_{esc}$, three channels of escape open (FIG. 1(c)), each bridged by a Lyapunov unstable periodic orbit. If an orbit crosses a Lyapunov orbit moving outwards, it will escape to infinity.[9] This was the criterion used to decide whether a particle had escaped from the system.

The initial conditions were chosen from the $(x - \dot{x})$ plane, with $y = 0$, and $\dot{y} > 0$ uniquely determined from (6). FIGURE 2 shows the $(x - \dot{x})$ plane for $\varepsilon = 1.3$, with the shading used to encode the time at which a particle with given initial conditions ($x$, $\dot{x}$) will escape from the system. It is obvious that, for some regions, the escape time is a smooth function of the initial conditions while, for other regions, small changes of the initial conditions lead to very different escape times. Only the latter regions were considered in this work. For each set of experiments, ten square cells, with side $\Delta x = \Delta \dot{x} = 0.05$, were chosen in regions that exhibit extreme sensitivity to initial conditions. Each cell was sampled using a uniform, rectangular grid with a resolution of $400 \times 400$, $800 \times 800$, or $1600 \times 1600$. Variable resolution was used to improve statistics for high values of $\varepsilon$, for which only few orbits remain bound to the system, soon after the beginning of the experiments. These initial conditions were

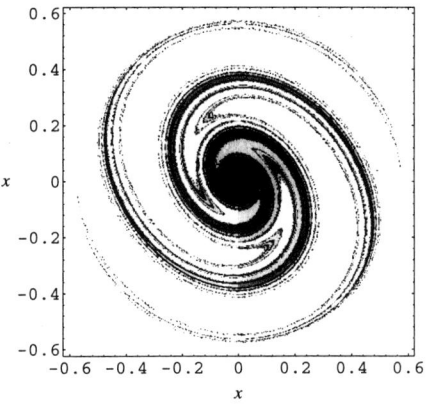

**FIGURE 2.** The plane $(x - \dot{x})$ from which the initial conditions were sampled, for orbits characterized by the Hamiltonian $H_3$ for $\varepsilon = 1.3$. *Darker shading* implies longer escape times. Resolution is $400 \times 400$.

then evolved in time, for a fixed value for $\varepsilon$. If the particle escaped from the system, the escape time was recorded; otherwise the orbit was followed for up to a maximum time. This procedure was repeated for several values of $\varepsilon$, namely 1.13, 1.16, 1.18, 1.20, 1.30, 1.40, 1.50, 1.60, 1.70, 1.80, 1.90, 2.00, 2.10, 2.20, 2.30, 2.40, 2.50, 2.60, and 2.70. The number of crossings of the positive $y$-axis was used as a measure of time.

It should be noted that, for this work, the time-series integrator used in the preceding papers was replaced by a Lie-integration method.[11] With this new, improved code, it was possible to run more cases and use significantly higher grid resolutions than in the earlier work. This was essential in order to improve the statistics.

### The Short-Term Behavior

Following the statistical analysis performed for the other two potentials, the results of the numerical simulations were reduced to extract the relative escape probability per unit time, defined as

$$p(\varepsilon, t) \pm \Delta p(\varepsilon, t) = \frac{N_{esc} \pm \sqrt{N_{esc}}}{N_{tot}}. \quad (7)$$

Here $N_{esc}$ is the number of trajectories from a given cell that escape between $t - 1$ and $t$; $N_{tot}$ is the total number of trajectories still present at $t - 1$; and $\Delta p(\varepsilon, t)$ represents an estimate of the uncertainty in the computed escape probability.

For values of $\varepsilon$ larger than $\varepsilon_{esc} = 1$, but lower than $\varepsilon_2 \approx 1.10$, the escape probability for late times tends toward zero (FIG. 3). However, when $\varepsilon > \varepsilon_2$, it asymptotes instead toward a constant, nonzero value, $p_\infty(\varepsilon)$, which depends on $\varepsilon$ but is independent of the initial phase-space region that is sampled. Moreover, $p_\infty(\varepsilon)$ is well fit by a power-law function of $\varepsilon$ similar to (3), with $\alpha = 0.45 \pm 0.05$ (FIG. 4). The time, $T(\varepsilon)$, required for convergence to $p_\infty(\varepsilon)$ is also found to be a power-law function of $\varepsilon$, of the form of (4), with $\beta = 0.37^{+0.10}_{-0.07}$ (FIG. 5).

It therefore appears that the power-law scaling observed for the cubic and quartic potentials persists in the Hénon–Heiles case as well. Moreover, given the substantial statistical uncertainties, the data are consistent with the conclusion that the values of the exponents $\alpha$ and $\beta$ are the same for all three potentials. In other words, it would seem that the scaling and, possibly, the values of the exponents, are universal.

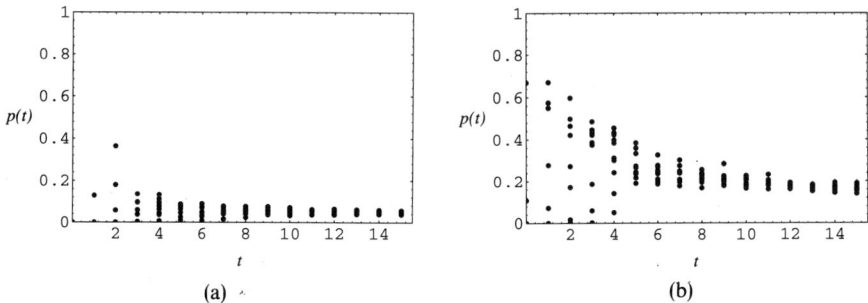

**FIGURE 3.** The escape probability as a function of time for $H_3$, for (a) $\varepsilon = 1.06$, and (b) $\varepsilon = 1.3$ (the *error bars* are too small compared to the *dot size*). Results of experiments from ten different cells of initial conditions with size $0.05 \times 0.05$ have been superimposed to produce these graphs.

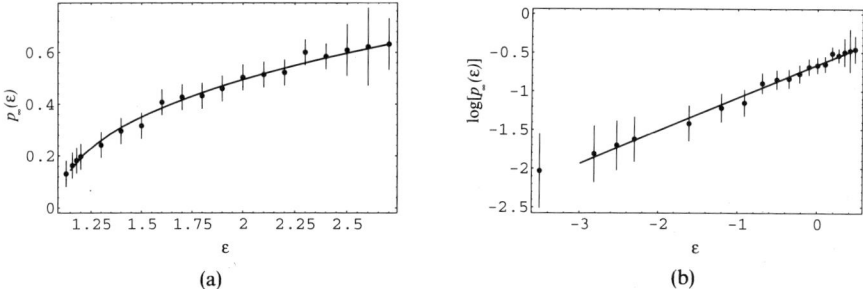

**FIGURE 4.** The asymptotic escape probability $p_\infty(\varepsilon)$ as a function of $\varepsilon$, and the corresponding power-law fitting, using data from ten different cells with size $0.05 \times 0.05$.

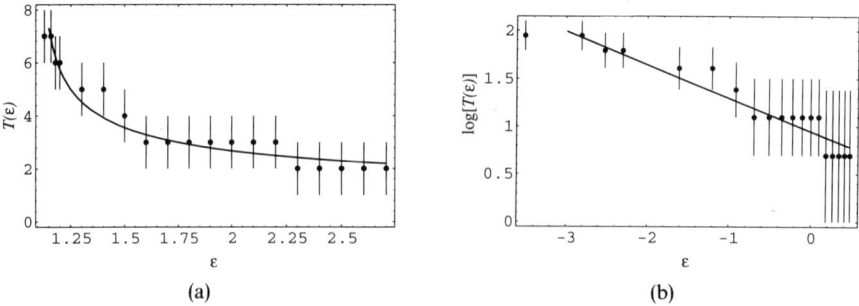

**FIGURE 5.** The convergence time $T(\varepsilon)$ as a function of $\varepsilon$, and the power-law fitting for ten cells with size $0.05 \times 0.05$.

## The Long-Term Decay

In the preceding discussion, it was assumed that, for $\varepsilon > \varepsilon_2$, the escape probability asymptotes toward a constant, nonzero value. Using this as a starting point, it was discovered that the asymptotic escape probability and the convergence time follow a simple, power-law scaling and that the exponents may be universal. However, it is possible that the escape probability is not a constant, but slowly decreases at later times.

In order to test this hypothesis, one must follow the orbits long enough for the decay to manifest itself. All the numerical experiments in this paper were performed for a maximum time equal to 100 consequents with the positive $y$-axis. The long-term decay is only apparent for $t \gtrsim 15$. FIGURE 6 shows the difference between short-term and long-term behavior. On short time scales, the escape probability seemingly assumes a constant, nonzero asymptotic value. However, if one waits long enough, it is obvious that this probability slowly decays toward zero, over time scales about an order of magnitude longer than the time scales required for the short-term convergence toward $p_\infty$.

An obvious question is what causes this behavior. Using qualitative arguments and previous experience with chaotic systems, one might give the following interpretation. At the beginning of the simulations, some of the orbits can escape immediately, leading to the observed irregular pattern for $p(\varepsilon, t)$ for early times. This transient phase is rather short, and soon the system reaches a quasi equilibrium, where some orbits appear to be trapped around KAM "islands" of stability, while others can move anywhere in phase space—outside the islands—and therefore, are allowed to escape.

One way to explain this behavior is to make the assumption that there are two clearly distinct populations of orbits, those that can escape and those that cannot, because they are trapped inside islands of stability. It is then easily shown[3] that the total escape probability $\mathscr{P}(t_k)$ at late times is given by

$$\mathscr{P}(t_k) = p_\infty \left[ 1 + \left( \frac{n_N}{1 - n_N} \right) \exp\left( \lambda t_{k-1} \right) \right]^{-1}, \tag{8}$$

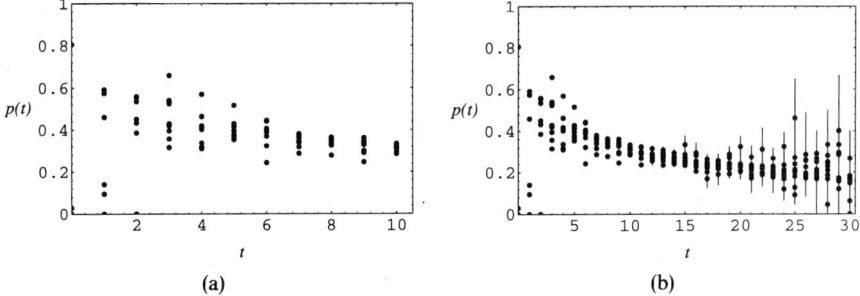

**FIGURE 6.** Short-term versus long-term behavior. Both graphs were created using data from the same numerical experiment. The only difference is the time scale.

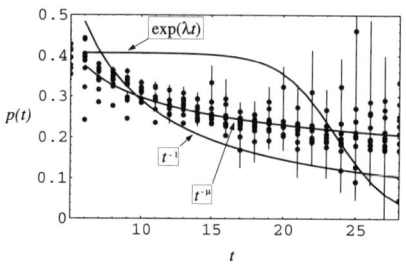

FIGURE 7. Exponential, inverse time, and power-law fittings for the long-term escape probability.

where $\lambda = -\ln(1 - p_\infty)$ and $n_N$ denotes the fraction of the initial trajectories that never escape. After a long time $t_k$, the total escape probability decreases exponentially with $\mathcal{P}(t_k) \propto \exp(-\lambda t_k)$.

An alternative interpretation is that the orbits are only temporarily trapped close to small KAM islands, due to a "stickiness" effect. This phenomenon was first observed numerically[12,13] and later proved mathematically,[14,15] and seems to be quite general. If one models the structure of the cantori around the islands as self-similar fractals, it follows [14,16,17] that the total escape probability varies inversely with time at late times:

$$\mathcal{P}(t_k) \sim t_k^{-1}. \tag{9}$$

Data from the numerical experiments make it is possible to check the validity of these predictions. The results for one particular case ($\varepsilon = 1.6$) are shown in FIGURE 7(a) and 7(b). Other values of $\varepsilon$ yield similar results. It is quite obvious that neither the exponential nor the power-law fitting is satisfactory. However, it was found that an empirical relation

$$\mathcal{P}(t_k) \sim t_k^{-\mu}, \tag{10}$$

with $\mu \neq 1$, will in general yield a reasonable fit. For example, the fit for $\varepsilon = 1.6$, shown in FIGURE 7(c), is quite satisfactory. Experiments for several values of $\varepsilon$ indicate that $\mu$ may be independent of the value of $\varepsilon$. This is demonstrated in FIGURE 8. The best value for $\mu$ is $0.31 \pm 0.05$.

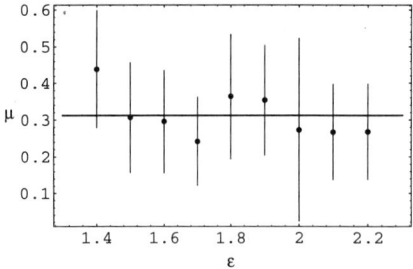

FIGURE 8. Best line fitting for the values of $\mu$ as a function of $\varepsilon$.

Integrating (8)–(10) leads to the following simple expressions for the number of particles remaining at late times $t$:

$$N(t) \sim \exp(-p_\infty t) \tag{11}$$

$$N(t) \sim t^{-q}, \quad q > 0 \tag{12}$$

$$N(t) \sim \exp(-qt^{1-\mu}), \quad \mu < 1. \tag{13}$$

The fact that (10) provides a good fit to the data indicates that the number of remaining particles is intermediate between a power law and an exponential.

## CONCLUSIONS

The object of this paper was to check the genericity of the results obtained for two Hamiltonians, $H_1$ and $H_2$, that represent quartic and cubic perturbations of two harmonic oscillators, allowing, respectively, for four and two channels of escape. This was done here by studying a third Hamiltonian, $H_3$, a modified version of the Hénon–Heiles Hamiltonian that allows for three channels of escape. An analysis of all three Hamiltonians leads to the following conclusions:

1. There is a critical $\varepsilon_2$ larger than the escape coupling $\varepsilon_{esc}$ such that, for $\varepsilon > \varepsilon_2$, the escape probability asymptotes to a constant, nonzero value, instead of tending to zero. This transition is rather abrupt and strongly reminiscent of a phase transition. Viewed macroscopically, one sees a qualitative change near $\varepsilon_2$, even though, at the level of individual orbits, there does not seem to be any abrupt change.
2. In all three cases, the asymptotic escape probability, $p_\infty(\varepsilon)$ and the time $T(\varepsilon)$ required to converge toward $p_\infty(\varepsilon)$ both exhibit simple power-law scalings in terms of an "order parameter" $\varepsilon - \varepsilon_2$.
3. Within statistical uncertainties, the exponents of the power laws are the same in all three cases. The existence of universal scaling exponents is a much stronger property than scaling itself. If generally true, this universality implies a fundamental similarity in the basic structure of dynamical systems.
4. The escape probability at late times does not remain constant, but decays with time. The number of remaining particles decreases at a rate intermediate between a power law and an exponential. This means that the simple models proposed hitherto cannot explain the observed late-time escape behavior. Therefore, a more satisfactory theory is required.

## ACKNOWLEDGMENTS

C. V. S. and R. D. would like to thank Mr. Karl Wodnar for useful discussions and help during the implementation of the Lie integrator code.

## REFERENCES

1. CONTOPOULOS, G. 1990. Astron. Astrophys. **231**: 41.
2. CONTOPOULOS, G. & D. E. KAUFMANN. 1992. Astron. Astrophys. **253**: 379.

3. CONTOPOULOS, G., H. E. KANDRUP & D. E. KAUFMANN. 1993. Physica D **64:** 310.
4. SIOPIS, C. V., G. CONTOPOULOS & H. E. KANDRUP. 1993. *In* Three-Dimensional Systems, H. E. Kandrup, S. T. Gottesman, J. R. Ipser, Eds.: 205. N. Y. Acad. Sci. New York.
5. ECKHARDT, B. & H. AREF. 1988. Philos. Trans. R. Soc. London **326:** 655.
6. RANKIN, C. C. & W. H. MILLER. 1971. J. Chem. Phys. **55:** 3150.
7. BLÜMEL, R. & U. SMILANSKY. 1988. Phys. Rev. Lett. **60:** 5971.
8. WERSINGER, J. M., E. OTT & J. M. FINN. 1978. Phys. Fluids **21:** 2263.
9. CHURCHILL, R. C., G. PECELLI & D. L. ROD. 1979. *In* Como Conf. Proc. on Stochastic Behavior in Classical and Quantum Hamiltonian Systems, G. Casati and J. Ford, Eds.: 76. Springer Lecture Notes in Physics, Vol. 93. Springer-Verlag. Berlin.
10. HÉNON, M. & C. HEILES. 1964. Astron. J. **69:** 73.
11. HANSELMEIER, A. & R. DVORAK. 1984. Astron. Astrophys. **132:** 203.
12. CONTOPOULOS, G. 1971. Astron. J. **76:** 147.
13. SHIRTS, R. S. & W. P. REINHART. 1982. J. Chem. Phys. **77:** 5204.
14. CHIRIKOV, B. V. & D. L. SHEPELYANSKY. 1984. Physica D **13:** 395.
15. MATHERS, J. N. 1982. Topology **21:** 457.
16. HANSON, J. D., J. R. CARY & J. D. MEISS. 1985. J. Stat. Phys. **39:** 327.
17. BLEHER, S., C. GREBOGI & E. OTT. 1990. Physica D **46:** 87.

# Counterrotating Bars[a]

CHAD L. DAVIES[b,d] AND JAMES H. HUNTER, JR.[c]

[b]*Department of Physics*
*University of Florida*
*Gainesville, Florida 32611*

[c]*Department of Astronomy*
*University of Florida*
*Gainesville, Florida 32611*

## INTRODUCTION

In 1993, the authors undertook the task of investigating the effects of the inclusion of a significant percentage of counterrotating angular momentum in two-dimensional disks with constantly rising rotation curves[1] (this work will hereafter be referred to as Paper I). In the present communication, we continue reporting on this study for disk systems in which the rotation curve turns over and falls beyond some radius, $R_{v_{max}}$. The simulations reported herein have used consistently truncated Toomre $n = 1$ discs[2] as described by Hunter et al.[3] The surface density of such a disc is given by:

$$\sigma_1(b, r) = \frac{C_1^2}{\pi^2 G} \left\{ [r^2 + b^2)^{-3/2}] \tan^{-1}\left(\frac{\sqrt{R_D^2 - r^2}}{\sqrt{r^2 + b^2}}\right) + \frac{\sqrt{R_D^2 - r^2}}{(R_D^2 + b^2)(r^2 + b^2)} \right\} \quad r \leq R_D$$

$$\sigma_1(b, r) = 0, \quad r > R_D,$$

and the cold rotation velocity (as shown in FIG. 1) is

$$V_1^2(r, b) = \frac{C_1^2 b r^2}{(r^2 + b^2)^{3/2}} \quad r \leq R_D,$$

where $C_1$ would be the local circular (cold rotational) velocity when $r$ and $R_D$ go to infinity. The maximum of $V_1^2(r, b)$ occurs at $r_m^2 = 2b^2$. Given this, it can be shown that

$$C_1 = \sqrt{\frac{3\sqrt{3}}{2}} V_1.$$

To parametrize the amounts of kinetic energy stored in the random and circular motions of the disc system, we express the (hydrodynamic) circular velocity of a

---

[a]This work was supported in part by National Science Foundation Grant AST9022827.
[d]Current address: Division of Natural Sciences, Cloud County Community College, 2221 Campus Drive, Concordia, Kansas 66901.

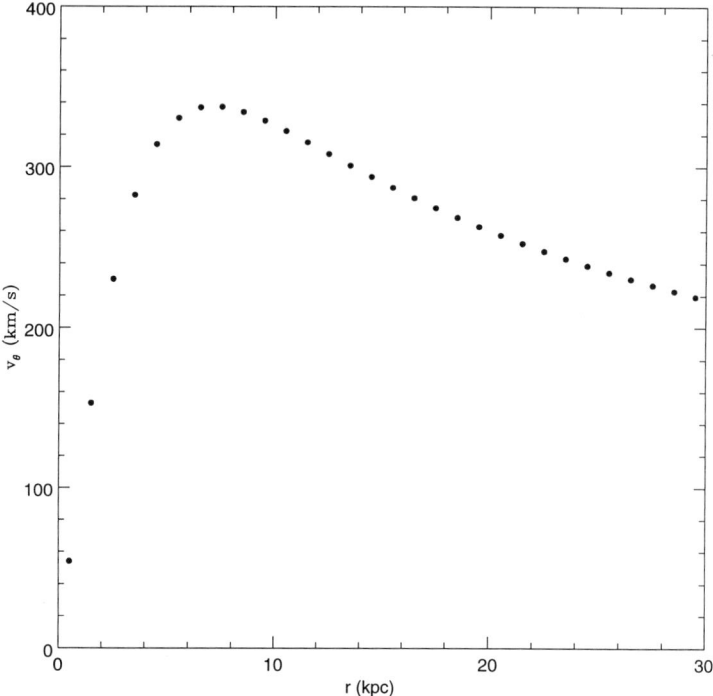

**FIGURE 1.** The rotation curve of a Toomre $n = 1$ disk.

stellar particle with the relation:

$$V(r) = kV_1(r, b),$$

where $k$, when assumed to be constant throughout the disc, may be related to the ratio of the total rotational kinetic energy to the absolute value of the gravitational potential energy as given by Ostriker and Peebles[4] by using the virial theorem to obtain:

$$t \equiv \frac{KE_{\text{rot}}}{|W|} = \frac{k^2}{2}$$

By forming the moments of the collisionless Boltzmann equation and assuming a truncated Gaussian distribution for the velocity dispersions, we are able to calculate $\sigma_r(r)$ and $\sigma_\phi(r)$. As a general rule, for a fully direct disc, $k \leq 0.53$ ($t \leq 0.14$) will ensure that the disc is stable agaist the formation of an $m = 2$ (bar) mode.

The simulations reported here have been run using two similar hierarchical tree algorithms, TREECOD as developed by Hernquist and Katz,[5,6] and FTM (Fast Tree Method), developed by Heller.[7] The results reported have been taken from the FTM runs, but have been checked for accuracy against runs done using TREECOD, and have been found to agree very closely in terms of the global parameters such as conservation of energy and angular momentum as well as global development.

The disc parameters that have been used are as follows: $N = 32000$; $M_D = 2.6 \times 10^{11}\ M_\odot$; $R_D = 25$ kpc; and $b = 5$ kpc. Machine units are chosen such that $t_{comp} \simeq t_{dyn}$ ($t_{dyn} = 4.713 \times 10^7$ yr) and will be reported as such. This is the free-fall time at a radius of 10 kpc and as such, the rotation period at this radius will be roughly $2\pi t_{dyn}$.

## THE MODELS

The models examined may be classified into two groups. The first group has its initial angular momentum inside some radius, $R_C$, reversed so that the inner portion of the initial disc counterrotates with respect to the outer disk (see FIG. 2). This is referred to as a step-function model. A variation of this is a slope-function model, where the disk's central region is fully counterrotating, its intermediate region has mixed angular momentum, and its outer region is rotating in a fully direct sense. While these models have only been of academic interest until recently, observations of NGC4826 (M64)[8,9] have revealed that the system has two segregated, counterrotating gas flows similar to these models. The second class of models have counterrotating angular momentum fully mixed throughout the disk. Numerous models of both types were run and examined in detail.

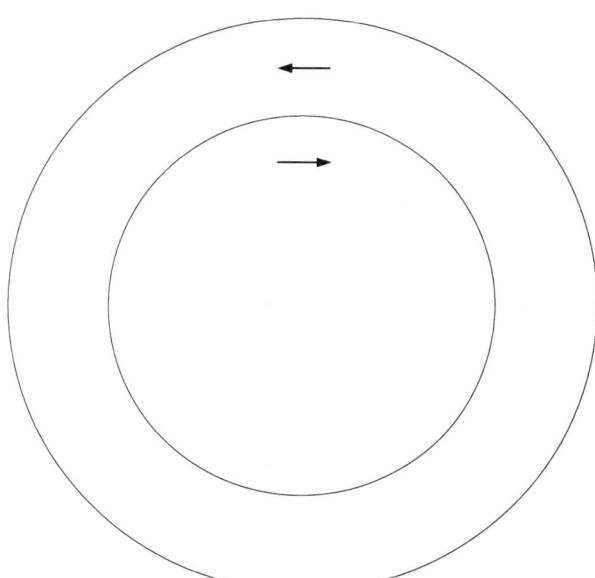

**FIGURE 2.** A schematic representation of the step function distribution.

TABLE 1. The effect of the Inclusion of Counterrotating Angular Momentum into a Bar Unstable Disk

| Percent Counterrotating | Bar Strength and Pattern Speed (in rad/$t_{dyn}$) after 10 $t_{dyn}$ |
|---|---|
| 0 | 0.35, 1.60 |
| 5 | 0.35, 1.55 |
| 10 | 0.33, 1.65 |
| 15 | 0.26, 1.54 |
| 20 | 0.26, 1.66 |
| 25 | 0.20, 1.60 |
| 30 | 0.05, — |

In Paper I, we reported that the inclusion of counterrotating angular momentum in a fully mixed sense was able to stabilize a Toomre $n = 0$ disc against the formation of an $m = 2$ mode. As can be seen in TABLE 1, th same is also true for the $n = 1$ disk with a value of $k = 0.7$, but only if the percentage of angular momentum exceeds a certain amount. Note that the pattern speed of the $m = 2$ mode is unaffected by the inclusion of the counterrotating angular momentum and that the decreased amplitude of the mode is not a linear function of the percentage of counterrotating angular momentum.

TABLE 2 shows the results of a series of step/slope function models. The most interesting of these, Model 1, develops an inner and outer bar that initially counterrotate with respect to an inertial frame. The inner bar forms at $t \sim 8.5\ t_{dyn}$ and the outer bar forms at $t \sim 13.5\ t_{dyn}$. At $t \sim 29\ t_{dyn}$, the inner bar reverses direction and then overtakes the outer bar. After this time the inner bar oscillates around the outer bar with a diminishing amplitude until it finally matches the pattern speed of the outer bar and the two rotate in unison. Also of interest, is Model 3 in which the central region develops a stable core that is not destroyed by the ensuing bar formation. FIGURES 3, 4, and 5 show the behavior of these systems over an extended time.

## CONCLUSIONS

From the models we have considered both here and in Paper I, we draw the following conclusions. The inclusion of counterrotating angular momentum, fully mixed, into a disk system will stablize the system against nonaxisymmetric modes

TABLE 2. Step Function Models and Resulting Global Development

| M Number | Model Type | Inner CR Bar | Outer Bar | Comments | $t_{dyn}$ |
|---|---|---|---|---|---|
| 1 | Step function | Yes | Yes | Inner bar reverses | 40 |
| 2 | Steep slope | Yes | Yes | Inner bar destroyed | 40 |
| 3 | Shallow slope | Develops core | Yes | Core persists | 40 |

*Note:* For all models, $k = 0.7$ and $J_{CR} = 0.1\ J_T$.

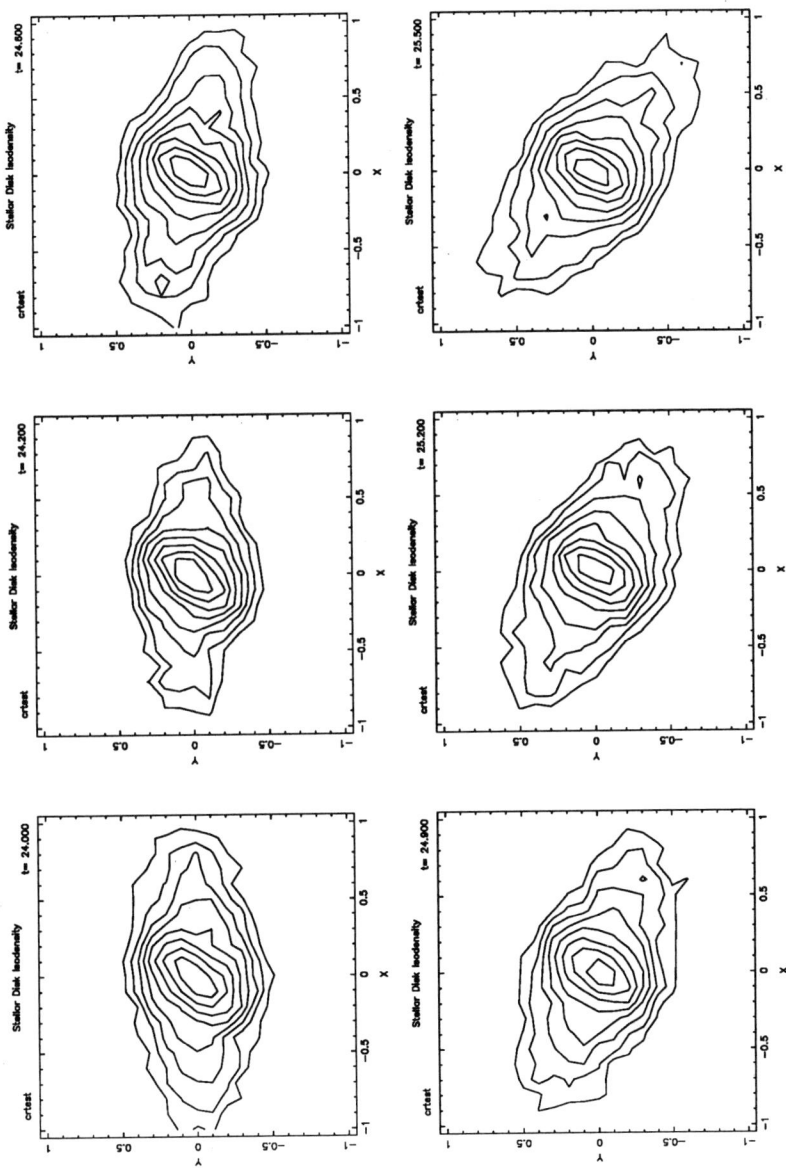

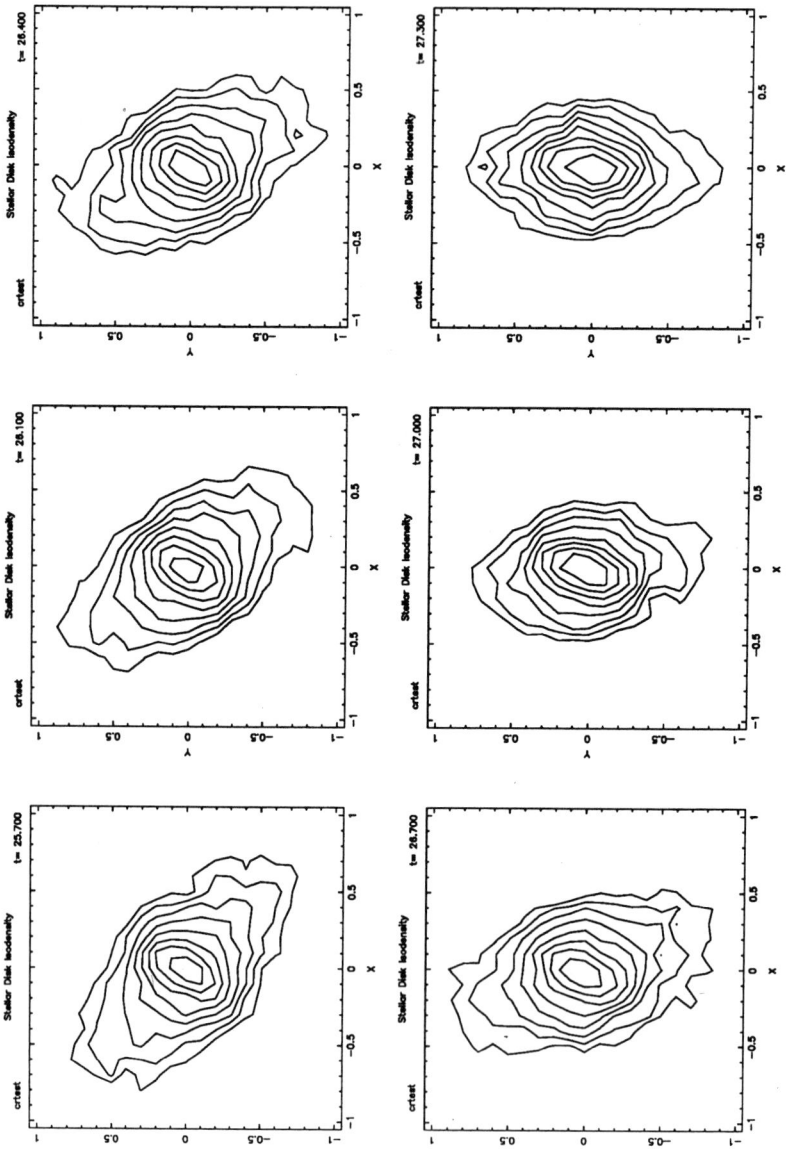

**FIGURE 3.** Isodensity plots for a Toomre $n = 1$ consistently truncated disk with 10 percent of its angular momentum counterrotating in a step function distribution type. This model corresponds to model 1 in TABLE 2.

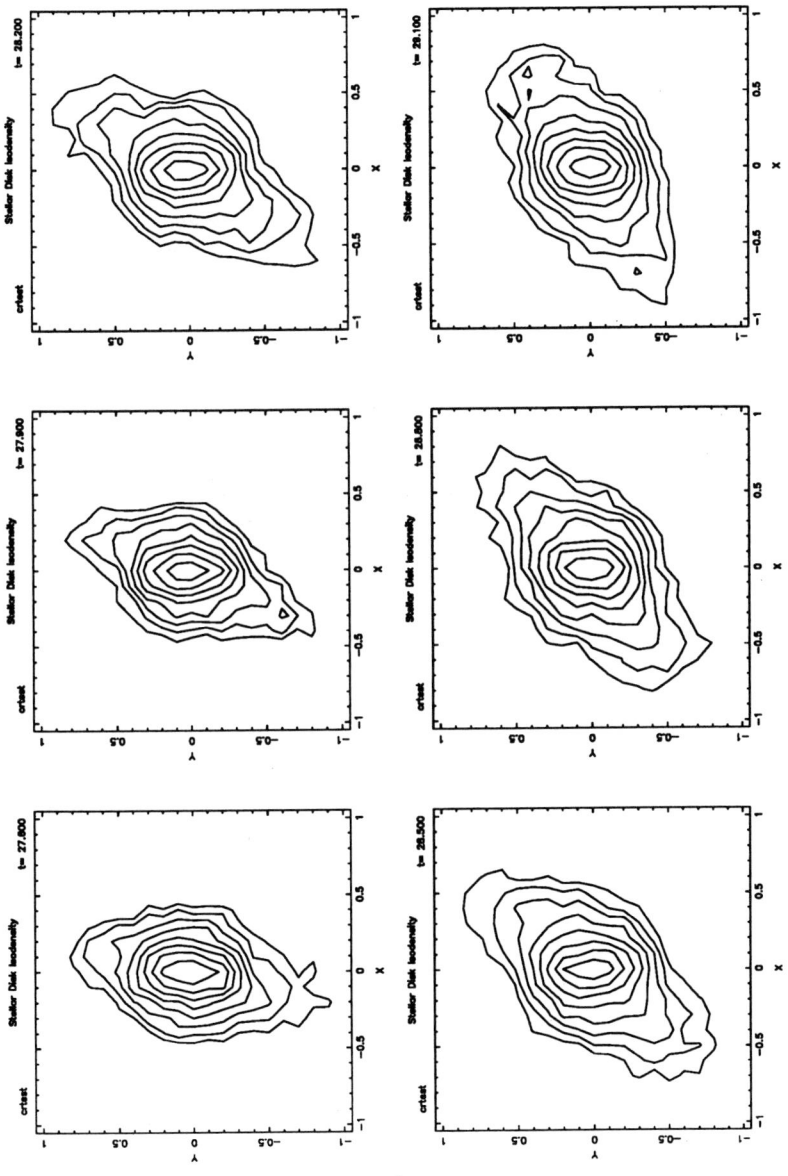

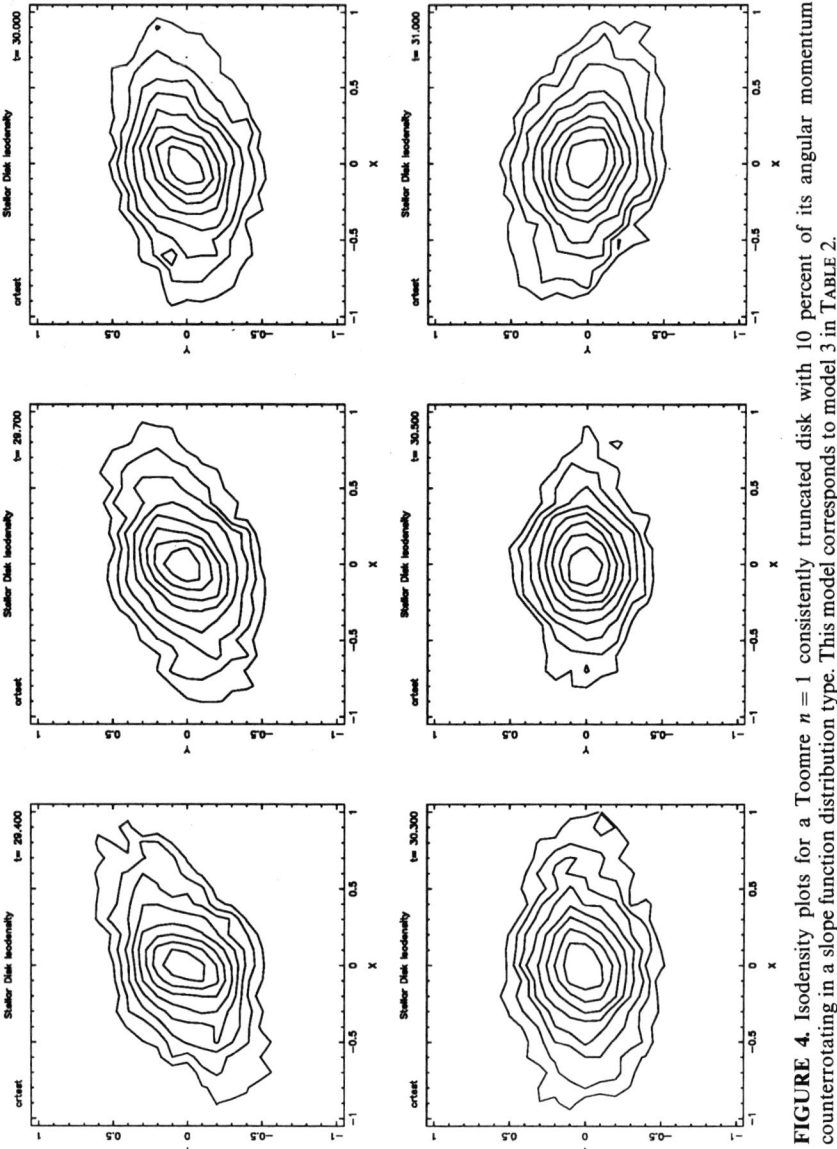

**FIGURE 4.** Isodensity plots for a Toomre $n = 1$ consistently truncated disk with 10 percent of its angular momentum counterrotating in a slope function distribution type. This model corresponds to model 3 in TABLE 2.

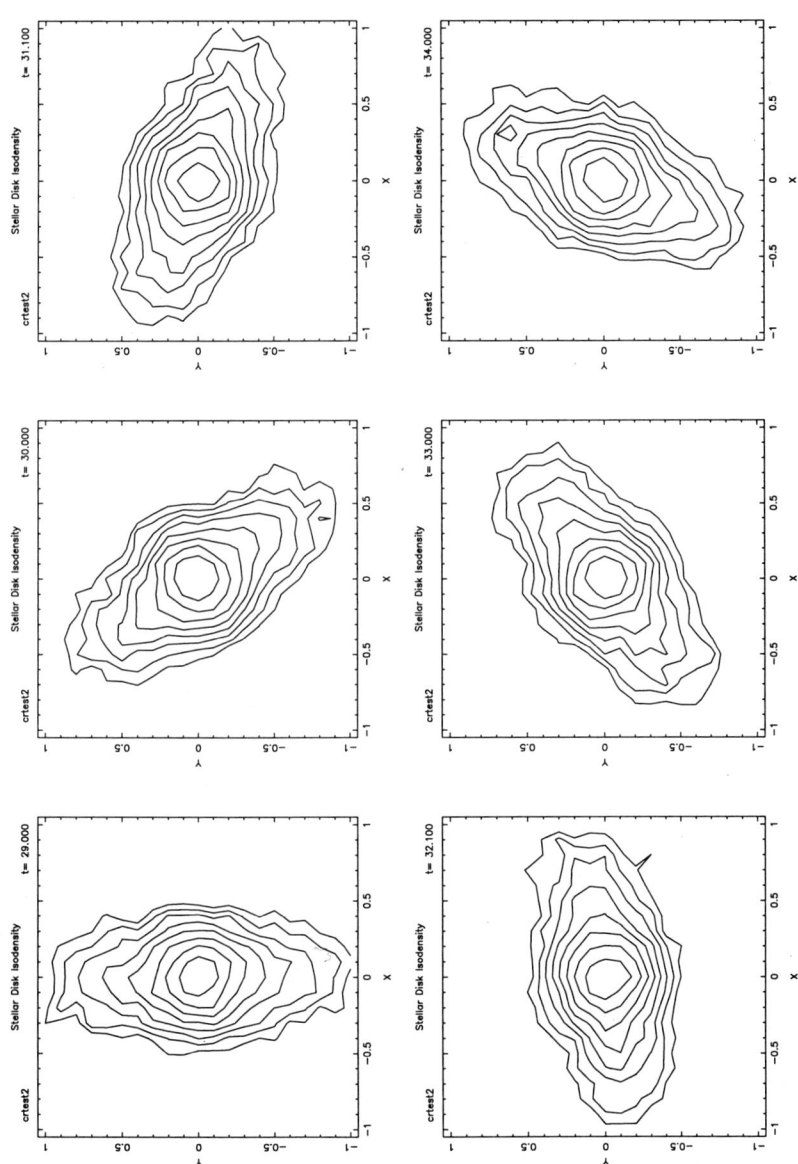

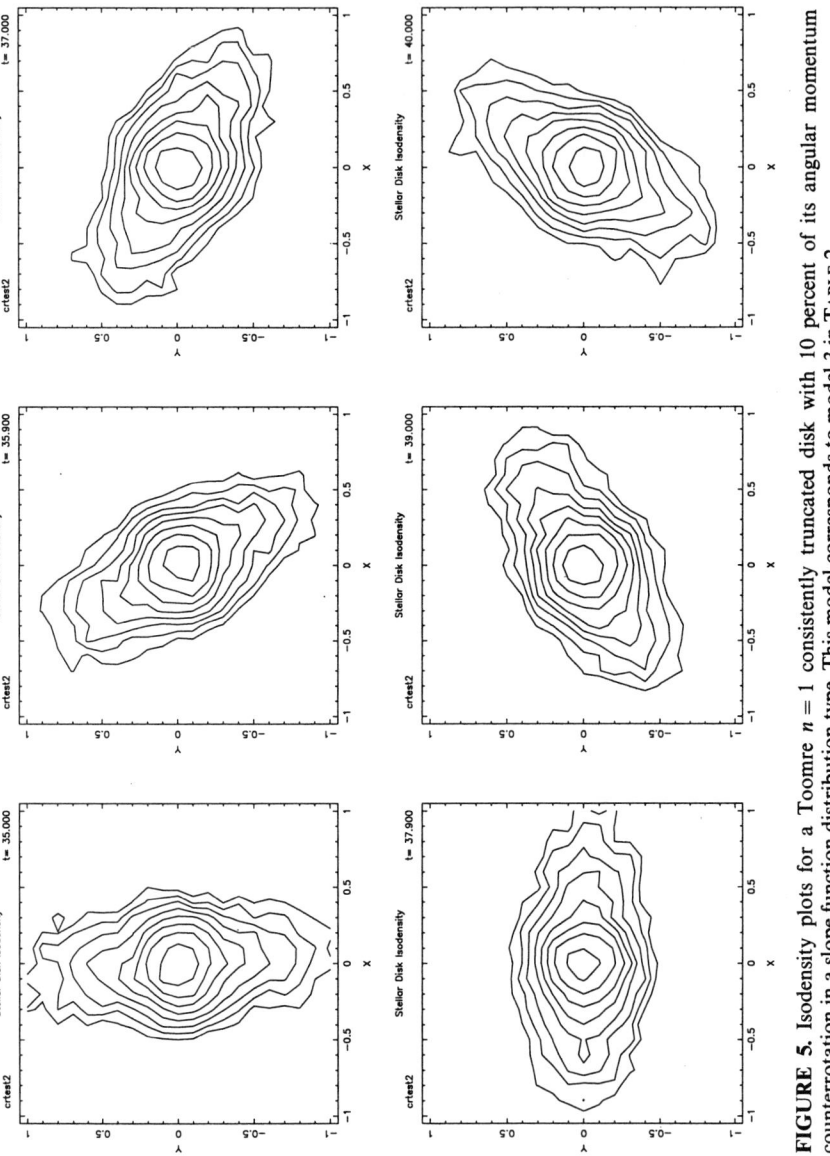

**FIGURE 5.** Isodensity plots for a Toomre $n = 1$ consistently truncated disk with 10 percent of its angular momentum counterrotation in a slope function distribution type. This model corresponds to model 3 in TABLE 2.

for both systems with constantly rising rotation curves and for those whose rotation curve reaches a maximum and then turn over. This concurs with earlier work done by Zang and Hohl[10] and Christodoulou et al.[11] Given the observations of NGC 4550 by Rubin et al.,[12] wherein they find two fully mixed counterrotating stellar components over the inner third of the disk, these results may imply a larger role for counterrotating angular momentum in galactic dynamics than was heretofore thought to be the case. We also find that models in which the direct and counter-rotating angular momentum are fully segregated into distinct regions are able to produce bar/disk (see Paper I) or multiple bar configurations in cases where fully prograde models cannot. Recently, Voglis et al.,[13] have shown that it is possible in the early universe for protogalaxies to form with the angular momentum of their inner regions counterrotating with respect to their outer regions.

As can be seen from these studies, the initial distribution of angular momentum in a disk sysem can have permanent effects upon its evolution. While these models have been confined to two dimensions, three-dimensional simulations have produced similar results, as was found in Paper I. Thus far, none of our merger simulations have resulted in "initial" conditions resembling our step/slope function models.

## ACKNOWLEDGMENTS

We wish to thank L. Hernquist for making the TREECOD available to us and C. Heller for providing us with a version of the FTM code and also for useful discussions during this study. We also thank G. Contopoulos and N. Hiotelis for their helpful comments.

## REFERENCES

1. DAVIES, C. L. & J. H. HUNTER, JR. 1994. Three-Dimensional Systems, H. E. Kandrup, S. T. Gottesmann, J. R. Ipser, Eds.: **751**: 222. N.Y. Acad. Sci. New York.
2. TOOMRE, A. 1963. Astrophys. J. **138**: 385.
3. HUNTER, J. H., JR., R. BALL & S. T. GOTTESMANN. 1984. Mon. Not. R. Astron. Soc. **208**: 1.
4. OSTRIKER, J. P. & P. J. E. PEEBLES. 1973. Astrophys. J. **186**: 467.
5. HERNQUIST, L. 1987. Astrophys. J., Suppl. Ser. **64**: 175.
6. HERNQUIST, L. & N. KATZ. 1989. Astrophys. J., Suppl. Ser **70**: 419.
7. HELLER, C. 1992. Ph.D. dissertation, Yale University, New Haven, Conn.
8. BRAUN, R., R. A. M. WALTERBOS & R. C. KENNICUTT, JR. 1992. Nature **360**: 442.
9. RUBIN, V. C. 1994. Astron. J. **104**: 173.
10. ZANG, T. A. & F. HOHL. 1978. Astrophys. J. **226**: 521.
11. CHRISTIDOULOU, D. M., I. SCHLOSMAN & J. TOHLINE. 1994. Astrophys J. (Submitted)
12. RUBIN, V. C., J. A. GRAHAM & J. D. P. KENNEY. 1992. Astrophys. J., Lett. **394**: L9–L12.
13. VOGLIS, N., N. HIOTELIS & P. HOFLICH. 1991. Astron. Astrophys. **249**: 5.

# Shadowing and Noise in Nonhyperbolic Systems

DAVID E. WILLMES

*Department of Physics*
*University of Florida*
*Gainesville, Florida 32611*

## INTRODUCTION

Chaotic dynamical systems exhibit sensitive dependence on initial conditions. Thus, the presence of noise in a chaotic dynamical system radically alters the details of individual orbits; noisy orbits quickly diverge from purely deterministic orbits with the same initial conditions. Undesirable noise may be introduced into physical experiments via errors associated with the measurement process, and into computer experiments via roundoff and truncation errors. Noise due to measurement errors, that is, observational noise, can generally be considered *additive noise*, in which noise introduced at one instant does not affect the orbit at the next instant. However, computer errors are often the cause of *dynamical noise*. Roundoff errors are due to the finite floating-point arithmetic that occurs whenever one does a computer simulation involving irrational numbers, whereas truncation errors occur whenever attempting to integrate a differential equation numerically, and depend on the details of the integration scheme used. Dynamical noise introduced at one instant alters the dynamics of the orbit in the future, and in a chaotic system causes a divergence from the purely deterministic orbit.

For both additive noise and dynamical noise, it is common to attempt some form of *noise reduction* on the system;[1,2] however, the problem is somewhat more difficult for the case of dynamical noise. Even so, the effect of computer errors on the behavior of particle orbits is often ignored by the galactic dynamicist doing numerical simulations. It is frequently assumed that, even if the details of the noisy orbits differ drastically from the deterministic orbits, the properties of the orbits are the same, in some statistical sense. Statistical tests seem to indicate that this is indeed true for computer simulations of $N$-body gravitational systems.[3,4]

Moreover, in some cases the neglect of computer errors in a simulation can perhaps be justified by the possibility that they may actually mimic the random perturbations that influence real, physical systems. In the case of galactic dynamics, these perturbations may be assumed to be due to close encounters of nearby stars. While it may be true that the computer noise will not exactly reproduce the physical noise, both types of noise are dynamical, and the exact form of the noise may not really matter. If this is true, one should not necessarily be concerned with the use of noise reduction methods to remove the impact of noise on a simulation. Instead, one should be concerned whether the amplitude of the noise has any physical relevance. The existence of computer errors, or any dynamical noise for that matter, should not necessarily be considered destructive in a computer simulation, but may actually be useful for modeling a given physical system.

Nevertheless, the galactic dynamicist should still be concerned as to the qualitative effect a given amount of noise will have on the properties of orbits in a chaotic dynamical system. A potentially useful technique for determining the qualitative effect of noise on a dynamical system may be to study the *shadowability* of the system. Shadowable systems possess the property that a deterministic orbit remains arbitrarily close to a noisy orbit. Hyperbolic systems have been proved to be shadowable by Anosov[5] and Bowen[6] for all time. The important quality of a hyperbolic system is the separability of the stable and unstable manifolds; at each point in a hyperbolic system, the dynamics can be factored into contracting directions, defined by the stable manifold, and expanding directions, given by the unstable manifold. The angle between the stable and unstable manifolds must remain bounded away from zero.[7]

THEOREM (Anosov–Bowen, 1967): For hyperbolic systems, given a shadow distance $\varepsilon > 0$, there exists a $\delta > 0$ such that any $\delta$-pseudo-orbit can be $\varepsilon$-shadowed by a true orbit for all time.

Here, $\delta$ is the noise amplitude, so defined that the $\delta$-pseudo-orbit is a noisy orbit that, at any given time step, remains within $\delta$ of the deterministic orbit that the system would have followed at that time step without any noise. The true orbit (or shadow orbit) is the purely deterministic orbit that remains within $\varepsilon$ of the noisy orbit. The hyperbolic shadowing theorem states that a deterministic orbit and a noisy orbit remain within a distance $\varepsilon$ as long as they are separated by no more than $\delta$ initially. Thus, the shadow orbit and the noisy orbit do not, in general, have the same initial conditions, and, as such, the initial conditions of an orbit in a hyperbolic dynamical system need not be known exactly to determine the qualitative behavior of the orbit. This is shown in FIGURE 1.

It is important to note that, in the standard interpretation, the true orbit in the Anosov–Bowen theorem is only a true orbit of the purely deterministic dynamical system. However, if one wishes to consider the noise to be of some physical origin, then the $\delta$-pseudo-orbit should more appropriately be considered a true orbit of the noisy dynamical system. Given this new interpretation, the hyperbolic shadowing theorem states that a true, noisy orbit with noise amplitude $\delta$ would remain within $\varepsilon$ of some orbit of the purely deterministic system.

However, most problems of physical interest are *not* hyperbolic, and the Anosov–Bowen theorem is not applicable to the majority of systems that the physicist or galactic dynamicist would like to represent. Two representative systems that are considered paradigms of chaotic systems are the Hénon map and the standard map, neither of which is hyperbolic. Other commonly studied systems of physical interest that may well be nonhyperbolic are the Hénon–Heiles system, often used as a simple galactic model, and the complex Ginsburg–Landau equation, which has important applications in condensed-matter physics. Hence, a revised shadowing theorem is necessary to determine the extent of the shadowability of the system.

The difficulty of shadowing nonhyperbolic systems lies in the existence of homoclinic tangencies and near-tangencies. These tangencies are points along the orbit where the stable and unstable manifolds cannot be factored into separate directions,

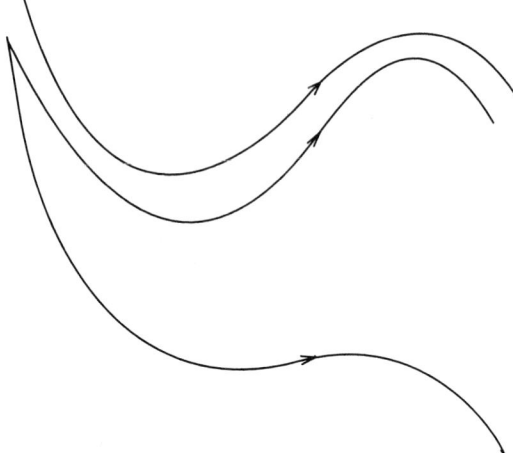

**FIGURE 1.** Deterministic and noisy orbits of a dynamical system. The lowest curve is the purely deterministic orbit for a given initial condition. The middle curve is the noisy orbit with the same initial condition. The upper curve is the shadow orbit that remains within a distance ε of the noisy orbit.

because at these points the angle between the manifolds tends to zero. Orbits in nonhyperbolic systems can be shadowed as long as they do not encounter regions where this angle approaches zero—a homoclinic near-tangency. One can then ask how long a noisy orbit will remain in such a region, where it is "sufficiently hyperbolic" to be shadowed by a deterministic orbit. The *shadowing time* may be defined as the length of time $N$ such that a shadow orbit exists given a noisy orbit for all times $n = 0, 1, \ldots, N$, but such an orbit does not exist for $n = 0, 1, \ldots, N + 1$. An algorithm that may be used to calculate this shadowing time was derived by Sauer and Yorke.[8]

THEOREM (Sauer–Yorke, 1991): Assume $\delta < 1/20m^2$ and let $B$ be a bound on the first and second partial derivatives of the map $f$ and $f^{-1}$. If

$$\max \{C_n, D_n\} \leq \frac{1}{m^{5/2} B^2 \sqrt{\delta}}$$

for all $n = 0, \ldots, N$, then there exists an orbit $\{y_n\}$ of $f$ such that $|x_n - y_n| < \sqrt{\delta}$ for $n = 0, \ldots, N$.

Here, $m$ is the dimension of the system, $\delta$ is the noise amplitude, $x_n$ is the noisy orbit, while $y_n$ is the shadow orbit. The quantities $C_n$ and $D_n$ are obtained through recursion relations involving the evolution of the stable and unstable manifolds along the orbit at each time step $n$. At the time step for which either $C_n$ or $D_n$ gets too large, a homoclinic near-tangency has been encountered and the system is no longer shadowable.

## THE HÉNON MAP

As a first example, the theorem of Sauer and Yorke was used to determine shadowing times of the Hénon map:[9]

$$x_{n+1} = by_n + a - x_n^2$$
$$y_{n+1} = x_n.$$

This is a dissipative map as long as $b$ is chosen to be less than one. In this study, the parameters $a$ and $b$ were chosen to be 1.4 and 0.3, respectively, so that iteration of the map yields the familiar strange attractor shown in FIGURE 2.

Numerical experiments were done to determine the "shadowability" of this map, as a function of the initial conditions and the noise amplitude. These experiments involved computing the quantities associated with the nonhyperbolic shadowing theorem until a "glitch" is encountered. A *glitch* is a point in phase space where the system is no longer shadowable, that is, a homoclinic near-tangency. For each noise level $\delta$, the location of the glitch and the shadowing time were determined for an ensemble of 1000 initial conditions, randomly chosen within varying ranges of the origin. The shadowing time is the number of iterations the orbit is computed before a glitch is encountered. Noise levels were chosen to be from $\delta = 10^{-14}$ to $\delta = 10^{-7}$.

The locations of the glitches for one representative ensemble of initial conditions with $\delta = 10^{-14}$ are shown in FIGURE 3. It should be apparent that, if one were to continue computing the locations of these near-tangencies, the resulting figure would not be distinguishable from the Hénon attractor itself. Thus, it appears that these homoclinic near-tangencies are dense within the attractor. Similar results are found for other choices of the noise level.

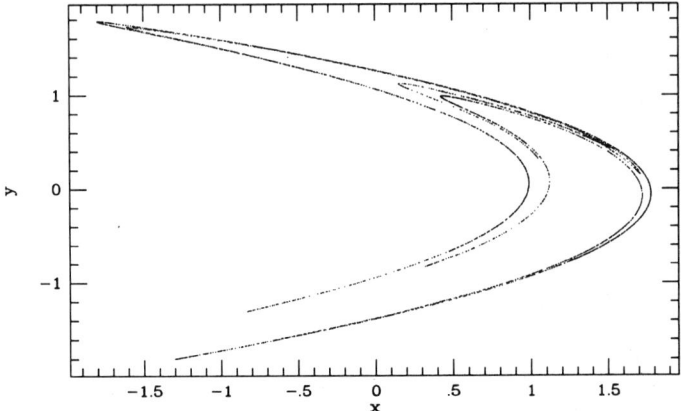

**FIGURE 2.** The Hénon map, with $a = 1.4$ and $b = 0.3$.

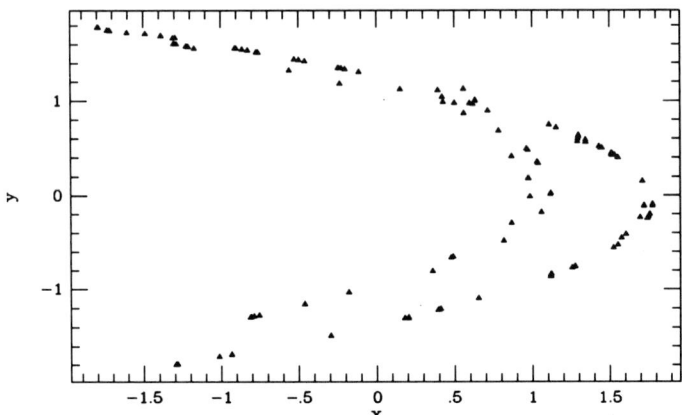

**FIGURE 3.** Location of glitches for the Hénon map, for an ensemble of initial conditions chosen to lie within $10^{-10}$ of the origin, with $\delta = 10^{-14}$.

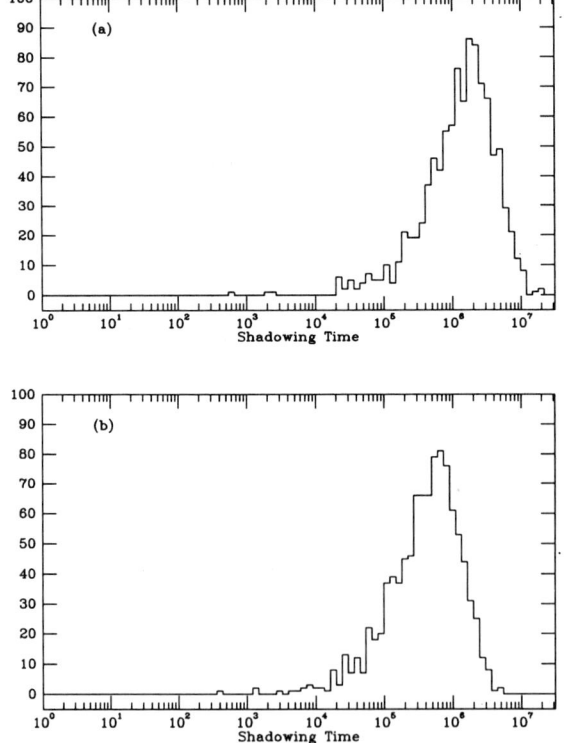

**FIGURE 4.** Histograms showing the distribution of shadowing times for randomly chosen ensembles of 1000 initial conditions, chosen to lie within $10^{-5}$ of the origin, for the Hénon map. **(a)** $\delta = 10^{-14}$; **(b)** $\delta = 10^{-13}$; **(c)** $\delta = 10^{-12}$; **(d)** $\delta = 10^{-11}$; **(e)** $\delta = 10^{-10}$; **(f)** $\delta = 10^{-8}$.

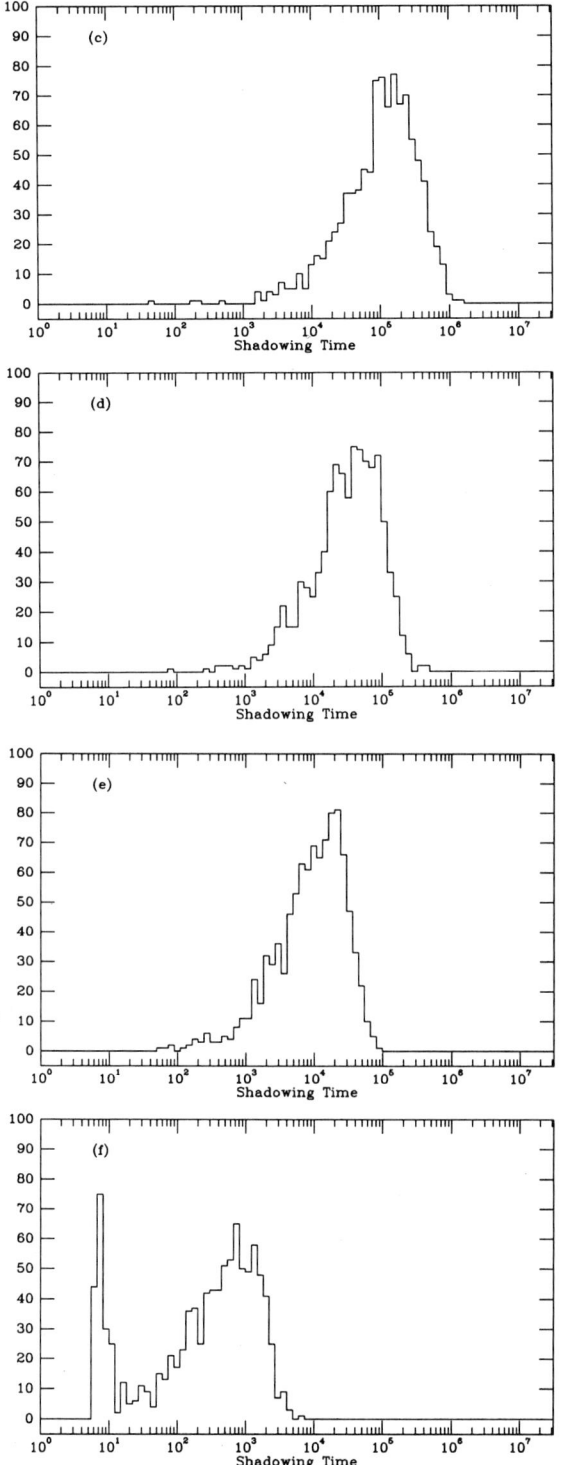

**FIGURE 4.** Histograms showing the distribution of shadowing times for randomly chosen ensembles of 1000 initial conditions, chosen to lie within $10^{-5}$ of the origin, for the Hénon map. (a) $\delta = 10^{-14}$; (b) $\delta = 10^{-13}$; (c) $\delta = 10^{-12}$; (d) $\delta = 10^{-11}$; (e) $\delta = 10^{-10}$; (f) $\delta = 10^{-8}$.

The shadowing time was computed for each of the 1000 randomly chosen initial conditions for each noise level. FIGURE 4 shows the distribution of shadowing times for representative ensembles of initial conditions, chosen to lie within a certain distance to the origin. As can easily be seen, the shadowing time varies significantly with the initial conditions chosen. Even though the mean shadowing time can be long, there is an appreciable number of orbits with relatively short shadowing times, extending down to about two orders of magnitude below the mean. Also, these distributions remain essentially identical if one chooses a different ensemble of initial conditions, for example, by grouping the initial conditions around a randomly chosen location instead of the origin, or changing the range around the location about which the initial conditions are chosen to lie. This last observation seems to imply that the dependence of shadowing time on initial conditions is fractal, that is, the distributions have similar structure on all scales. This conclusion is supported by results of the Kolmogorov–Smirnoff test, which shows that these sample distributions have about a 95 percent chance of being drawn from the same distribution.

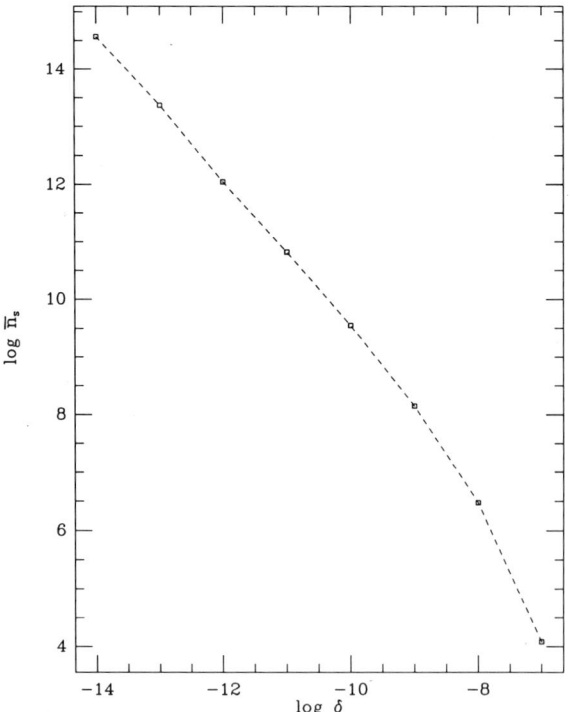

**FIGURE 5.** Log $\bar{n}_s$ versus log $\delta$ for the Hénon map, where $\bar{n}_s$ is the mean shadowing time and $\delta$ is the noise level. The *error bars*, obtained by averaging over five distributions, are smaller than the *squares*.

By comparing the sample distributions in FIGURE 4, it is apparent that the mean shadowing time decreases as the noise level is increased. This can easily be understood—increasing the noise level increases the range of angles that are close enough to zero to create a homoclinic near-tangency, making the orbit experience these glitches more frequently. The relationship between the mean shadowing time, $\bar{n}_s$, and the noise level $\delta$ is shown in FIGURE 5. The data points were obtained by averaging $\bar{n}_s$ over five different ensembles of 1000 initial conditions each for each noise level. However, the sample distributions are essentially identical, so any error caused by using only one ensemble of initial conditions would be minimal. The data points for $\delta = 10^{-8}$ and $\delta = 10^{-7}$ lie somewhat below the line obtained by fitting the other six data points. This appears to be due to the large spikes at low shadowing times for these distributions, which can be seen in FIGURE 4(f).

A conjecture by Hammell et al.[10] implies effectively that $\bar{n}_s \sim 1/\sqrt{\delta}$ for dissipative maps. The six data points for $\delta = 10^{-14}$ through $\delta = 10^{-9}$ are well fit by a power law $\bar{n}_s \sim \delta^p$, where $p = -0.556 \pm 0.006$. Since the standard deviation for each point is insignificant, the error estimate for the exponent $p$ is simply the formal error associated with the least squares fit of the average values plotted in FIGURE 5. If one were to include the data points for $\delta = 10^{-8}$ and $\delta = 10^{-7}$, the exponent would be $p = -0.624 \pm 0.030$.

## THE STANDARD MAP

As an example of shadowing a Hamiltonian system, the shadowing algorithm was applied to the standard map, which, in action-angle variables, has the form

$$J_{n+1} = J_n + (K/2\pi) \sin 2\pi\theta_n \quad (\text{mod } 1),$$

$$\theta_{n+1} = \theta_n + J_{n+1} \quad (\text{mod } 1).$$

This map is nonhyperbolic for all positive values of the stochasticity parameter $K$.[11] As $K$ increases, the measure of the stochastic region of phase space also increases.

Numerical experiments, similar to those done with the Hénon map, have been done with the standard map for various values of $K$. FIGURE 6(a) shows a single stochastic orbit of the standard map for $K = 3$. The location of the glitches associated with homoclinic near-tangencies are shown in FIGURE 6(b) for an ensemble of 1000 randomly chosen initial conditions within the stochastic region, given a noise level of $\delta = 10^{-14}$. As can be seen, these glitches are not randomly distributed throughout the stochastic region of phase space, but appear to lie along distinct curves within the region. It can also be seen that a number of these glitches occur at the boundary of the stochastic and regular regions. A similar distribution of glitches occurs for $K = 7$ as well, as shown in FIGURE 7(b), even though, as shown in FIGURE 7(a), practically the entire phase space appears stochastic.

It should be noted that, even though shadowing may break down at these points, the statistical properties of the orbits may remain unaffected. It is certainly

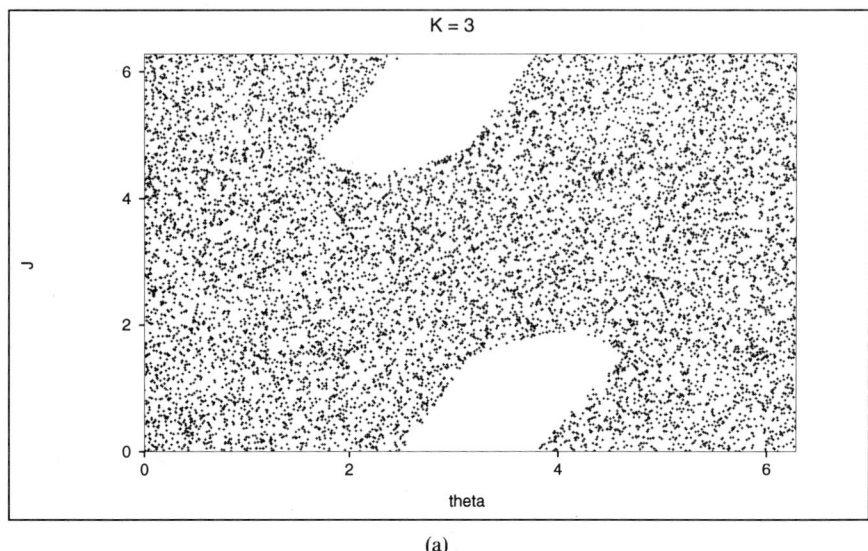

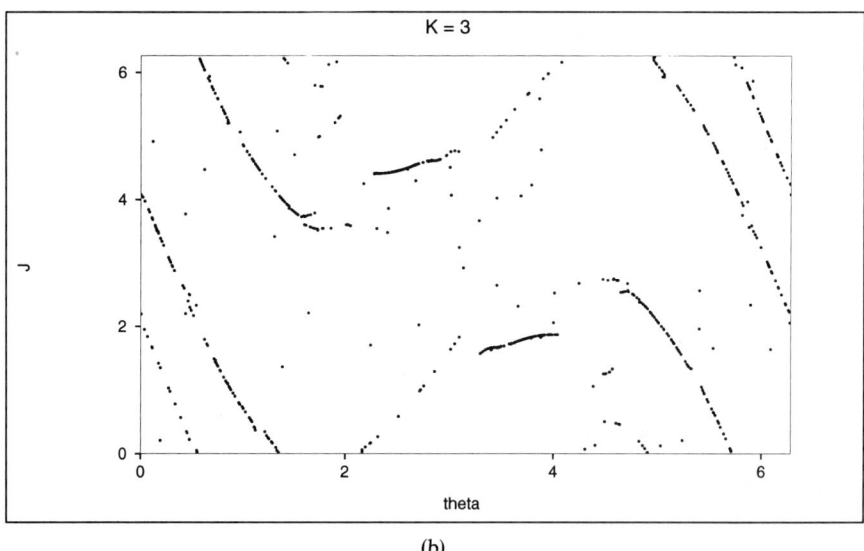

**FIGURE 6.** (a) Many iterates of a single stochastic orbit of the standard map for $K = 3$. (b) Location of glitches for the standard map, for an ensemble of initial conditions chosen to lie within $10^{-10}$ of the origin, with $\delta = 10^{-14}$.

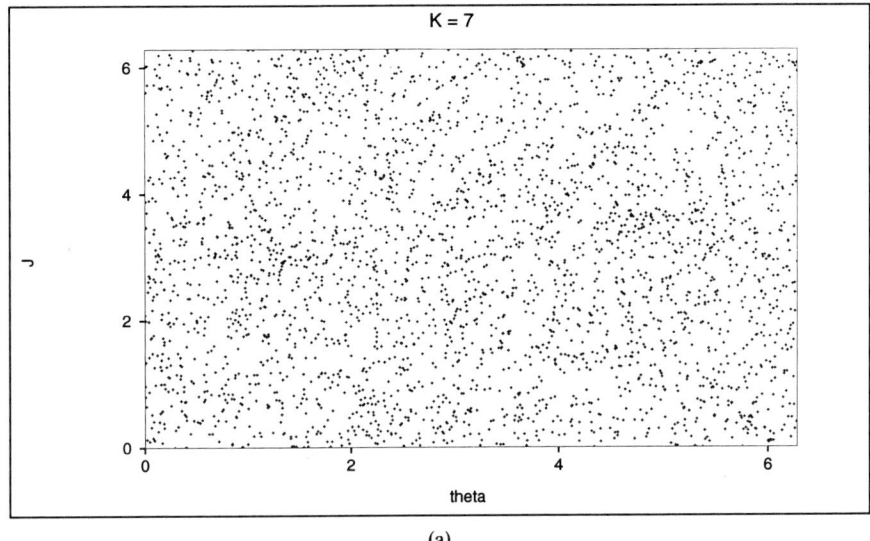

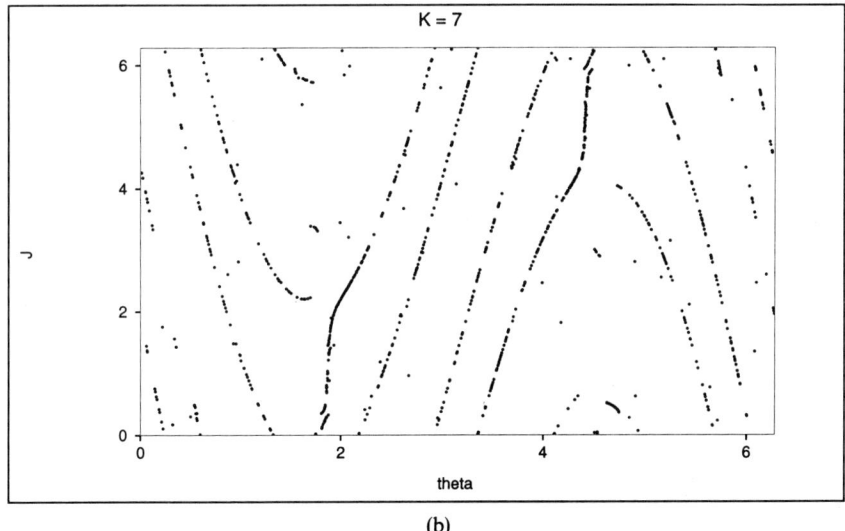

**FIGURE 7.** Same as FIGURE 6, for $K = 7$.

true that a noisy orbit experiencing a glitch may be "thrown" away from its shadowing orbit; however, it most likely will "land" nearby another deterministic orbit that will shadow it for some length of time. Thus, such a noisy orbit does not correspond to any *single* deterministic orbit, but as long as one is not interested in the behavior

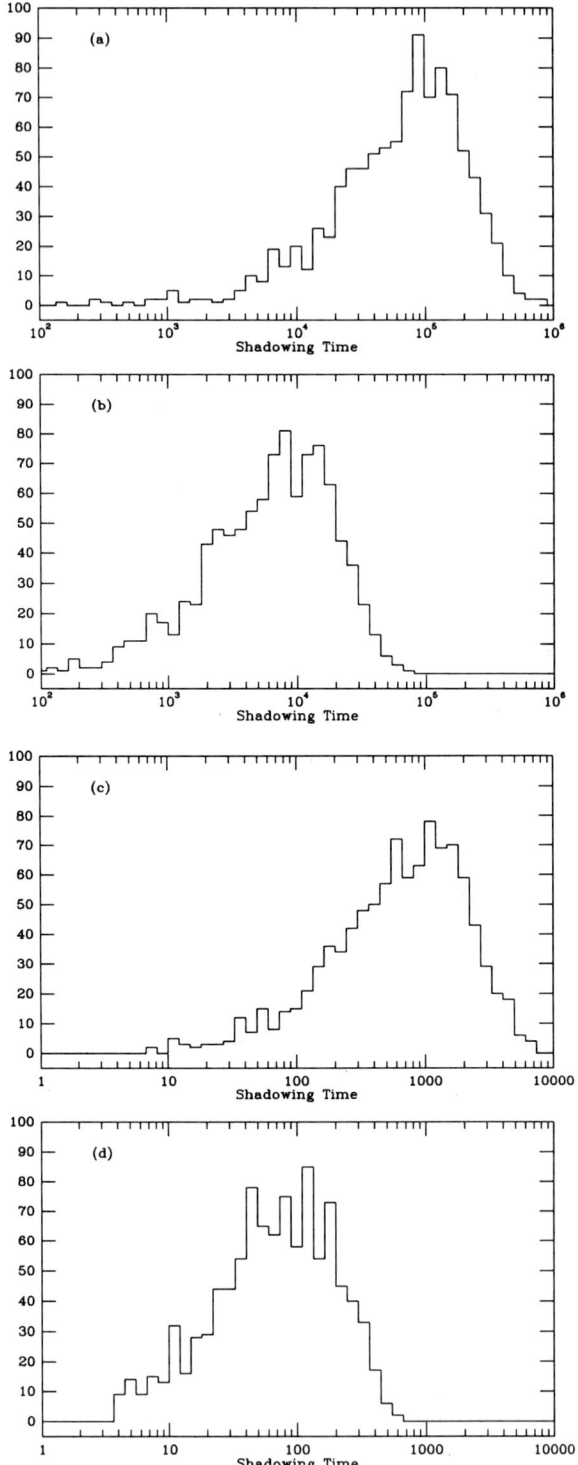

**FIGURE 8.** Histograms showing the distribution of shadowing times for randomly chosen ensembles of 1000 initial conditions, chosen to lie within $10^{-5}$ of the origin, for the standard map. (a) $\delta = 10^{-14}$; (b) $\delta = 10^{-12}$; (c) $\delta = 10^{-10}$; (d) $\delta = 10^{-8}$.

of this orbit specifically, the results should not really matter. Indeed, for the case of the standard map, it appears that the statistical behavior of a typical noisy orbit is similar before and after experiencing a glitch. Despite this, the existence of these glitches may well be important, considering that many of these glitches occur at regional boundaries.

The sample distributions for 1000 randomly chosen initial conditions are shown in FIGURE 8, for the value $K = 7$. This value of $K$ was chosen to ensure that all orbits would be within the stochastic sea, and that there would be no effect due to choosing initial conditions close to regular regions of phase space. Again, these distributions are independent of the range and location of the chosen initial conditions, implying a fractal dependence here as well. These results are supported by the Kolmogorov–Smirnoff test, which verifies that these sample distributions have a probability of within 90 percent of being drawn from the same distribution.

The relationship between the mean shadowing time, $\bar{n}_s$, and the noise level $\delta$ is shown in FIGURE 9. Again, the data points were obtained by averaging $\bar{n}_s$ over five different ensembles of 1000 initial conditions each for each noise level. Fitting the data points to a straight line yields $\bar{n}_s \sim \delta^p$, where $p = -0.498 \pm 0.002$. This compares well with a conjecture of Grebogi et al.[11] similar to the previous conjecture, that effectively $\bar{n}_s \sim 1/\sqrt{\delta}$ for two-dimensional Hamiltonian maps.

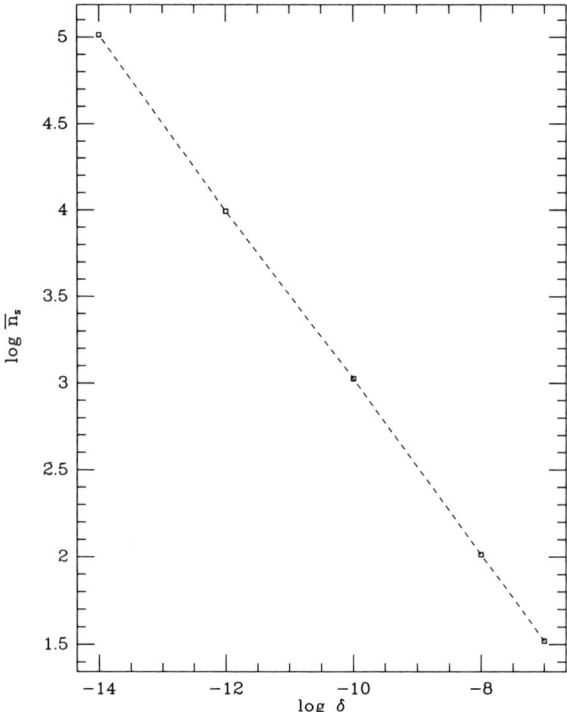

**FIGURE 9.** Log $\bar{n}_s$ versus log $\delta$ for the standard map. The *error bars*, obtained by averaging over five distributions, are smaller than the *squares*.

## CONCLUSIONS

Results from both the Hénon map and the standard map support the claim that systems that include dynamical noise can be shadowed for long times for *most* orbits, provided that the noise amplitude $\delta$ is small enough. However, the existence of orbits with appreciably *shorter* shadowing times should cause concern when integrating chaotic dynamical systems for long times. The existence of these "unshadowable" orbits makes conclusions about the "shadowability" of orbits in a nonhyperbolic system unclear, even though the mean shadowing time may be quite large. This is especially true considering the fractal nature of the sample distributions of shadowing times. If one is interested in the shadowing time for an orbit with a certain initial condition, one has to realize that there are orbits with initial conditions *arbitrarily close* to the desired initial condition with drastically different shadowing times.

However, nonhyperbolic shadowing may be used to determine the fraction of orbits may be qualitatively changed by the presence of noise. Even given the fractal nature of the shadowing times, the sample distributions yield information as to how many orbits are "unshadowable" given the number of iterations of the map.

It remains to be seen whether "unshadowability" of orbits actually affects the statistical properties of the orbits involved, however. It is certainly true that if a system is *shadowable*, then it is structurally stable with respect to noise. In fact, Anosov[12] proved hyperbolic systems to be structurally stable. The converse of this statement, however, is not necessarily true. In fact, for the standard map with large stochasticity ($K = 7$), the existence of homoclinic near-tangencies does not appear to alter the statistical properties of the orbits. This would seem to imply that the standard map for $K = 7$ is *structurally stable* in the sense that the statistical properties of the map are unaffected by the presence of noise. It remains to be seen whether this behavior is universal. For example, it has been observed that even weak noise can affect the near-invariant measure of Hamiltonian potentials.[13] One would expect that this behavior is due to the existence of homoclinic near-tangencies for the system.

The fact that the glitches in the standard map are not randomly distributed would seem to imply that the location of the homoclinic near-tangencies may be related to some other property of the dynamical system, for example, the Lyapunov exponents. The existence of glitches adjacent to KAM tori may be associated with "sticky" orbits with smaller local Lyapunov exponents than are found in the stochastic sea. These sticky orbits have been observed for the standard map[14] and for several two-dimensional potentials as well.[15] This may have important consequences; orbits that experience a glitch adjacent to invariant tori or cantori may perhaps "jump" across the barrier, even if a purely deterministic orbit would not. These glitches would then serve as the mechanism for diffusion across these barriers in noisy dynamical systems.

## ACKNOWLEDGMENTS

The author thanks Henry Kandrup for many useful discussions.

## REFERENCES

1. FARMER, J. D. & J. J. SIDOROWICH. 1991. Physica D **47**: 373.
2. HAMMEL, S. M. 1990. Phys. Lett. **A148**: 421.
3. SMITH, H. 1977. Astron. Astrophys. **61**: 305.
4. HEGGIE, D. C. 1991. In Predictability, Stability, and Chaos in N-body Systems, P. Hut and S. L. W. McMillan, Eds.: 47. Plenum Press. New York.
5. ANOSOV, D. V. 1967. Proc. Steklov Inst. Math. **90**: 1.
6. BOWEN, R. 1975. J. Differential Equations **18**: 333.
7. EUBANK, S. & J. D. FARMER. 1990. In 1989 Lectures in Complex Systems, E. Jen, Ed.: 75. Addison-Wesley. Redwood City, Calif.
8. SAUER, T. & J. A. YORKE. 1991. Nonlinearity **4**: 961.
9. HÉNON, M. 1976. Commun. Math. Phys. **50**: 69.
10. HAMMEL S. M., J. A. YORKE & C. GREBOGI. 1988. Bull. Amer. Math. Soc. **19**: 465.
11. GREBOGI, C., S. M. HAMMEL, J. A. YORKE & T. SAUER. 1990. Phys. Rev. Lett. **65**: 1527.
12. ANOSOV, D. V. 1962. Dokl. Akad. Nauk. **145**: 707.
13. HABIB, S., H. E. KANDRUP & M. E. MAHON. 1995. Phys. Rev. Lett. **275**: 443.
14. SEPÚLVEDA, M. A., R. BADII & E. POLLACK. 1989. Phys. Rev. Lett. **63**: 1226.
15. MAHON, M. E., R. A. ABERNATHY, B. O. BRADLEY & H. E. KANDRUP. 1995. Mon. Not. R. Astron. Soc. In press.

# Low-Frequency Modes of Pulsation of Relativistic Accretion Disks

## JAMES R. IPSER

*Department of Physics
University of Florida
Gainesville, Florida 32611*

### INTRODUCTION

Variability on a variety of timescales is a commonly observed feature of astronomical systems ranging from compact binaries to active galatic nuclei. In some cases, such as those involving the x-ray emissions of certain black-hole candidates[1] and the Z-sources among low-mass x-ray binaries,[2] quasi-periodic oscillations are observed to exhibit periods of oscillation that are much longer than the dynamical timescale of the underlying system. In an attempt to understand these long-timescale phonomena, several analyses[3-5] have focused on the possibility that they are associated with low-frequency oscillations in the inner regions of accretion disks associated with the observed systems. These analyses found that a variety of disk structures admit low-frequency nonaxisymmetric bending modes that are trapped within the inner disk regions where most of the radiation originates. This led to speculation that these modes are the origin of the quasi-periodic x-ray phenomena.

The preceding analyses are compromised by their use of a pseudo-Newtonian approximation in which the Newtonian gravitational potential is modified in a particular way in order to simulate certain effects of general relativity. Our purpose here is to point out and demonstrate that this approximation can be replaced by a fully relativistic treatment based on a relativistic formalism that has recently become available.[6] In what follows we describe the first applications of the relativistic formalism to the problem of low-frequency modes in accretion disks around compact objects.

### THE RELATIVISTIC FORMALISM

The formalism[6] enables one to describe the adiabatic pulsations of rotating relativistic fluid configurations. The equilibrium configuration is described by a perfect-fluid stress-energy tensor of the form ($G = c = 1$)

$$T^{ab} = \rho u^a u^b + pq^{ab}, \quad q^{ab} = g^{ab} + u^a u^b. \tag{1}$$

Here $\rho$ and $p$ are the density and pressure, $u^a$ is the fluid four-velocity, $g^{ab}$ is the metric, and $q^{ab}$ projects orthogonal to $u^a$. The velocity $u^a$ is a Killing trajectory and is given by

$$u^a = \gamma(t^a + \Omega\phi^a), \tag{2}$$

where $\gamma$ is a redshift factor, $t^a$ and $\varphi^a$ are the time and rotational Killing vectors, and $\Omega$ is the angular velocity. The equilibrium geometry is of the form

$$ds^2 = -e^{2\nu} dt^2 + B^2 e^{-2\nu}\varpi^2(d\varphi - \alpha \, dt)^2 + e^{2(\zeta-\nu)}(d\varpi^2 + dz^2), \tag{3}$$

where the metric functions depend only on the cylindrical coordinates $\varpi$ and $z$.

In this paper our interest lies with a non-self-gravitating barotropic disk that is governed by the passive gravitational field of the central source about which it rotates. In this case the structure of the equilibrium disk is essentially determined by the Euler equations

$$p_{,\varpi}/(\rho + p) = -\tfrac{1}{2}\gamma^2[(\Omega_K^2 - \Omega^2)g_{\varphi\varphi,\varpi} + 2(\Omega_K - \Omega)g_{t\varphi,\varpi}], \tag{4}$$

$$p_{,z}/(\rho + p) = -\gamma^2 e^{2(\zeta-\nu)}\Omega_\perp^2 z$$
$$\equiv -\tfrac{1}{2}\gamma^2(g_{tt,z} + 2\Omega g_{t\varphi,z} + \Omega^2 g_{\varphi\varphi,z}), \tag{5}$$

and by the choices of rotation law $\Omega(\varpi)$ and speed of sound $c_s(\varpi)$. A comma denotes partial derivative, $\Omega_K$ is the Keplerian angular velocity, and $\Omega_\perp$ is the perpendicular oscillation frequency of circular orbits.

When a non-self-gravitating disk is perturbed adiabatically, the resulting oscillations are described by the relativistic analogs of the perturbed Euler equations and continuity equation. In fact, the formalism of Ipser and Lindblom[6] reveals that for normal-mode perturbations $\propto e^{i\omega t + im\varphi}$ the Euler equations can be solved for the Eulerian perturbation $\delta u^a$ of the four-velocity in terms of the scalar function $\delta U \equiv \delta p/(\rho + p)$, where $\delta p$ is the Eulerian pressure perturbation. This yields

$$q_b^a \delta u^b = Q^{ab}\delta U_{,b}, \tag{6}$$

where the matrix $Q^{ab}$ depends only on the equilibrium configuration. The relativistic generalization of the continuity equation then yields a pulsation equation for $\delta U$. For disks of constant effective thickness $c_s/[(1 + c_s^2)^{1/2}\Omega_\perp]$, the pulsation equation can be solved by separation of variables; and in a special case it yields a radial eigenequation of the form:

$$\frac{1}{(\rho+p)}\left[\frac{(\rho+p)\varpi B \sigma^2 \gamma^2}{(\sigma^2\gamma^2 - \kappa^2)} F_{,\varpi}\right]_{,\varpi} + \frac{\varpi B}{c_s^2} e^{2(\zeta-\nu)}\gamma^2[\sigma^2 - (1 + c_s^2)\Omega_\perp^2]F$$
$$+ e^{2(\zeta-\nu)}\left\{\sigma\gamma\left(\frac{\rho_{,\varpi}}{\rho+p}\mathscr{F} + \mathscr{F}_{,\varpi}\right) + \sigma\gamma[\varpi B \Lambda e^{-2(\zeta-\nu)}(\sigma\gamma)_{,\varpi}]\right.$$
$$\left.+ \frac{(\rho+p)_{,\varpi}}{(\rho+p)}\frac{\varpi B \Lambda}{\sigma\gamma} e^{-2(\zeta-\nu)}(\sigma\gamma)_{,\varpi} - \frac{\Lambda(m + \sigma\gamma u_\varphi)^2}{\varpi B \gamma^2}\right\}F. \tag{7}$$

Here $\sigma = \omega + m\Omega$, $\kappa$ is the epicyclic frequency, $\Lambda = \sigma^2\gamma^2/(\sigma^2\gamma^2 - \kappa^2)$, and

$$\mathscr{F} = \frac{\Lambda(m + \sigma\gamma u_\varphi)e^{-2(\zeta-\nu)}}{\sigma^2\gamma^5\varpi B}[(\gamma u_\varphi)_{,\varpi} - (\gamma u_\varphi)^2\Omega_{,\varpi} - (\gamma^2\varpi B)^2\Omega_{,\varpi}]. \tag{8}$$

The radial eigenfunction $F(\varpi)$ is subject to the boundary conditions that the Lagrangian pressure perturbation vanish at the disk's inner edge and that the locally measured perturbation energy die out at large radii in the disk's outer regions (trapped mode). Also, it is consistent to expand (7) in powers of $c_s^2$ and to keep only the lowest order terms.

## RESULTS

We have searched for low-frequency trapped modes satisfying (7) in the case of thin disks around Kerr black holes. Analysis of that equation reveals that the best case for low-frequency modes arises for the value $m = 1$ of the aximuthal index. Mathematically, this is due essentially to the fact that the quantity $\Omega_K - \Omega_\perp$ vanishes in the nonrotating limit (Kerr parameter $a = 0$).

In this paper we focus on disks with the rotation law

$$\Omega^2 = \Omega_K^2 \left[ 1 - \frac{\beta}{2M^2} (r_{BL} - r_{BL,\,min})(r_{BL} - r_2)(r_{BL}/r_2)^{-4} \right], \qquad (9)$$

where $\beta$ and $r_2$ are constants, $r_{BL}$ is the Boyer–Lindquist radial coordinate, with $r_{BL,\,min}$ its value at the inner edge of the disk. The inner edge is taken to lie at the position of the last stable circular orbit.

Our results in the nonrotating limit for $c_s^2 = 10^{-3}$ are exhibited in TABLE 1, and they provide the first available comparison of relativistic results with those obtained in the pseudo-Newtonian theory. The column label $N$ denotes the number of nodes of the radial eigenfunction $F$. The label $r_E/2M$ denotes the approximate radius, or value of $\varpi$ in the equatorial plane, at which a mode begins to die out. The unit $2M$ is twice the mass of the central source. The last two columns yield the frequencies $f = \omega/2\pi$ of relativistic modes in the nonrotating limit and of the corresponding pseudo-Newtonian modes. The frequencies are all positive, corresponding to modes that are retrograde (points with a given displacement rotate in the direction opposite to the hole's rotation). A key result exhibited by these columns is that the relativistic frequencies, in the nonrotating limit, are consistently smaller than their pseudo-Newtonian counterparts by about a factor of 2, which can be attributed to redshift effects. Otherwise, the pseudo-Newtonian results mimic the nonrotating-hole relativistic results fairly well.

TABLE 1. Corresponding Frequencies of Pseudo-Newtonian and Relativistic Modes ($a = 0$)

| | | | | $f \times (M/M_\odot)$(Hz) | |
|---|---|---|---|---|---|
| $r_2/2M$ | $\beta/c_s^2$ | $N$ | $r_E/2M$ | Relativistic ($a = 0$) | Pseudo-Newtonian |
| 3.5 | 100 | 1 | 4 | 0.46 | 0.92 |
| 3.5 | 100 | 2 | 4.5 | 0.94 | 1.9 |
| 3.5 | 10 | | | No Trapped Mode | |
| 4.5 | 100 | 1 | 5 | 0.96 | 1.8 |
| 4.5 | 100 | 2 | 5 | 1.8 | 3.6 |
| 4.5 | 10 | 1 | 5.5 | 0.086 | 0.18 |
| 4.5 | 10 | 2 | 6 | 0.18 | 0.34 |
| 5.0 | 100 | 1 | 5.5 | 1.3 | 2.5 |

*Note:* The column labels are defined in the text.

This is perhaps somewhat misleading, however, because significant differences, associated with the dragging of inertial frames, begin to emerge when the central hole is allowed to rotate. That frame-dragging effects should strongly affect the mode frequencies is suggested by the dispersion relation

$$(\sigma^2 - \kappa^2)(\sigma^2 - \Omega_\perp^2) = \sigma^2 c_s^2 k_\varpi^2/\gamma^2, \tag{10}$$

which can be obtained from (7). Here $k_\varpi$ is the effective radial wave number. Setting $m = 1$ and assuming that the mode frequency $\omega$ is small compared to $\Omega$, one obtains

$$\omega \approx -\frac{(\Omega_K - \Omega_\perp)(\Omega + \Omega_\perp)}{2\Omega} + \frac{(\Omega_K - \Omega)(\Omega + \Omega_\perp)}{2\Omega} + \frac{\sigma^2 c_s^2 k_\varpi^2}{2\Omega(\sigma^2 - \kappa^2)\gamma^2}. \tag{11}$$

It can be shown that $(\Omega_K - \Omega_\perp)/\Omega_K$ increases as the Kerr parameter increases. This is because the perpendicular oscillation frequency $\Omega_\perp$ of an orbiting particle is, in rough terms, given by the rotation rate as sensed locally; and, due to frame-dragging effects, that local rate becomes progressively less than the rate $\Omega_K$ seen at infinity, as the hole is spun up. Thus one expects that frame-dragging should decrease $\omega$ below its nonrotating value by an amount

$$\frac{\bar{\delta}\omega}{2\pi}\frac{M}{M_\odot} \sim -\frac{2aM}{r^3}\frac{M/M_\odot}{2\pi} \sim -10^2\frac{a}{M}\text{ Hz}. \tag{12}$$

Equation (12) suggests that frame-dragging, even for small amounts of rotation, can produce large-percentage decreases in mode frequencies, and can in fact drag them through zero so that they become prograde. FIGURE 1 shows that this is exactly what happens. FIGURE 1 exhibits the lowest three modes associated with the

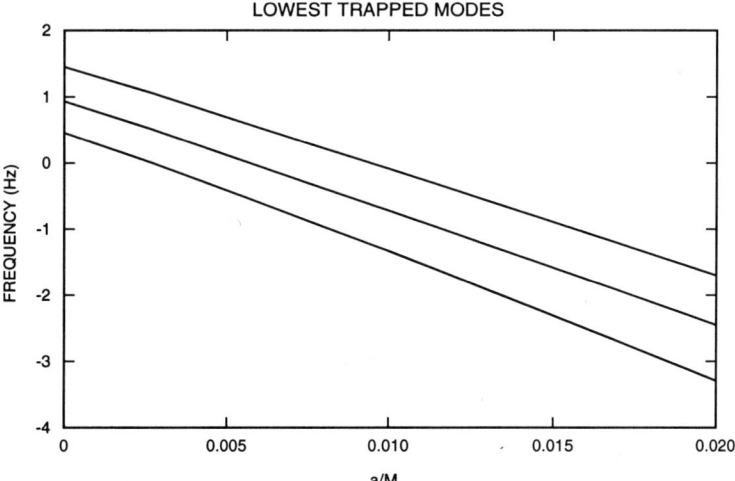

**FIGURE 1.** The frequencies of the three lowest modes associated with $c_s = 10^{-3}$ and the rotation law parameters $r_2/2M = 3.5$, $\beta/c_s^2 = 10^2$. The *vertical axis* is the frequency in Hz multiplied by $M/M_\odot$, where $M$ is the mass of the central black hole. The *horizontal axis* is the ratio of the Kerr parameter to $M$.

rotation-law parameters $r_s/2M = 3.5$, $\beta/c_s^2 = 100$. It is evident that the lowest mode is dragged through zero frequency and becomes prograde at $a/M \approx 0.005$, which is in good agreement with the order-of-magnitude equation 12. FIGURE 1 also indicates that each increase of $a/M$ by an amount $\sim 0.005$ causes another mode to be dragged through zero frequency. This is borne out by the fact that as $a/M$ increases, typically more than a hundred modes behave in this way, with all existing modes being prograde for $a/M \gtrsim 0.6$.

The general conclusion is that relativistic accretion disks around black holes can support precession/tilt modes with very low frequencies. For solar-mass objects with disk temperature $\sim 10^7$ K, the frequencies range from of the order of 10 Hz down to 0. Arbitrarily small values of the frequency can in principle be obtained, due to effects associated with the dragging of inertial frames. One is thus led to the suggestion that these modes are associated with at least some of the quasi-periodic phenomena observed in black-hole candidates. It will be important, in this connection, to investigate further the way in which these modes can modulate disk emissions and the extent to which they are influenced by the accretion process and viscous effects.

## REFERENCES

1. MIYAMOTO, S., K. KIMURA & S. KITAMOTO. 1991. Astrophys. J. **383**: 784.
2. VAN DER KLIS, M. 1989. Annu. Rev. Astron. Astrophys. **27**: 517.
3. KATO, S. 1989. Publ. Astron. Soc. Japan **41**: 745.
4. KATO, S. & F. HONNA. 1991. Publ. Astron. Soc. Japan **43**: 95.
5. IPSER, J. R. 1994. Astrophys. J. **435**: 767.
6. IPSER, J. R. & L. LINDBLOM, 1990. Astrophys. J. **335**: 226.

# Density Waves and Warps Generated by Tidal Perturbation of a Gaseous Disk[a]

J. C. B. PAPALOIZOU, D. G. KORYCANSKY,
AND C. TERQUEM

*Astronomy Unit*
*School of Mathematics*
*Queen Mary and Westfield College*
*Mile End Road, London E1 4NS*

## 1. INTRODUCTION

Recently much progress has been made in our knowledge of pre-main-sequence binary and multiple stars that apparently form the majority of young stellar systems. Some recent observational results can be found in Reipurth and Zinnecker[1] and Ghez et al.[2] The presence of a binary companion can affect the disks that are observed to exist around young stars.[3] Such a companion provides external forcing that can result in the excitation of waves and angular momentum transport between it and the disk. Under favorable conditions, the waves propagate toward the center of the disk where they become nonlinear. Subsequent dissipation causes some mass accretion to occur.[4] Furthermore, tidal interactions can control the location of the disk boundaries and so have direct observational consequences.

Disk-companion interactions leading to tidal truncation and sculpting of the disk have previously been studied in the context of close binary systems and the early solar system.[5] In these cases, circular or near-circular coplanar orbits were assumed for the binary. Artymowicz and Lubow[6] have considered the tidal effect of a binary companion in a coplanar eccentric orbit on a viscous circumstellar disk. They used smoothed particle hydrodynamics to simulate the disk and considered eccentricities less than or equal to 0.7.

In this paper we consider an inviscid disk undergoing tidal interaction as a result of a perturber flying by on a coplanar parabolic orbit. A sequence of such encounters also approximates the situation appropriate to a companion in a highly eccentric orbit. Korycansky and Papaloizou[7] considered the case when the perturbing and central mass were equal. Here, we extend the calculations to consider the case when the perturber's mass is one-tenth of the central mass. Later we consider the disk response when the perturber's orbit is not coplanar, but circular.

The problem of calculating the linear response of a disk to a perturber on a parabolic orbit has been considered, using asymptotic methods, in a galactic context by Palmer and Papaloizou.[8] Ostriker[9] similarly examined the effect of an encounter of a star-plus-disk system with a point mass on a parabolic orbit. The primary interest was in calculating the change in energy and angular momentum suffered by

---

[a]This work was supported by PPARC Grant GR/H/09454.

the perturbing mass, the motivation being whether such encounters could provide a mechanism of binary formation, as suggested by Clarke and Pringle.[10]

We have performed fully nonlinear calculations for coplanar encounters. However, we have concentrated on cases where the interactions are relatively mild, with the net angular momentum transfer being from the disk. Very close encounters that disrupt the disk have been considered by Clarke and Pringle[11] and Heller.[12]

We find that the interaction occurs in an impulsive manner near to pericenter passage. An excited wave propagates inwards toward the disk center with a speed that asymptotically is close to the sound speed. In the central regions, weak shocks form. After the passage of the wave, the outer regions of the disk remain relatively unperturbed. This indicates that a sequence of encounters of the type we consider can be used to model the interaction of a disk with a companion in a highly eccentric orbit. We are thus able to estimate conditions fo the tidal interaction to control the location of the disk edge.

The work just cited assumes the disk and binary orbit to be coplanar. However, it is possible that the plane of the disk is inclined to that of the orbit. Then tidal effects may induce the presence of large-scale warps in circumstellar disks, which are important for reprocessing radiation originating from the disk center.[13]

When the perturber is in an inclined orbit, warps and bending waves, for which the density perturbation is antisymmetric with respect to reflection in the midplane, may be excited. Papaloizou and Terquem[14] have discussed the excitation of such waves with $m = 1$ by a perturber in a circular orbit. They considered components of the perturbing potential that have nonzero frequency in a frame that comoves with the central star but does not rotate. Such waves are of interest because they may result in long wavelength global perturbation of the disk as well as provide angular momentum transport.

Here we consider the zero-frequency perturbation that provides a net torque on the disk and that would cause it, if rigid, to precess as a rigid body. We evaluate the linear response of a disk with varying thickness when the perturbing and central mass are equal. If the binary separation is great enough, the response is found to be similar to that of a rigid body.

## 2. BASIC EQUATIONS

We work in an initially nonrotating cylindrical coordinate system $(r, \varphi, z)$ defined with the origin at the primary mass $M$, at the disk center, with $z$ being the vertical coordinate. In the limit of a small viscosity, the disk is described by the equation of motion

$$\rho \frac{\partial v}{\partial t} + \rho(v \cdot \nabla)v = -\nabla P - \rho \nabla(\psi + \Psi_p) \tag{1}$$

and the equation of continuity

$$\frac{\partial \rho}{\partial t} + \nabla \cdot (\rho v) = 0, \tag{2}$$

where $P$ is the pressure, $\rho$ the density, and $v$ the flow velocity. The gravitational potentials due to the primary point mass $M$ and the perturbing companion mass are $\psi$ and $\Psi_p$, respectively.

### 2.1. Coplanar Encounters

To describe the response of a thin disk to coplanar encounters with a perturbing mass, we adopt a thin-disk approximation in which there is assumed to be no vertical motion and no dependence of the horizontal velocities on $z$. The governing equations are obtained by integratig (1) and (2) with respect to $z$ from $+\infty$ to $-\infty$.

The radial component of the equation of motion for the disk becomes

$$\frac{\partial v_r}{\partial t} + v_r \frac{\partial v_r}{\partial r} + \frac{v_\phi}{r} \frac{\partial v_r}{\partial \phi} - \frac{v_\phi^2}{r} = -\frac{1}{\Sigma} \frac{\partial p}{\partial r} - \frac{\partial(\Psi_p + \psi)}{\partial r}, \qquad (3)$$

where $v_r$ and $v_\phi = r\Omega$ are the radial and azimuthal components of velocity and $\Omega$ is the angular velocity. Here, $\Sigma$ and $p(\equiv \int_{-\infty}^{\infty} P\, dz)$ are the surface density and vertically integrated pressure, respectively.

The azimuthal component of the equation of motion similarly becomes

$$\frac{\partial v_\phi}{\partial t} + \frac{v_\phi}{r} \frac{\partial v_\phi}{\partial \phi} + v_r \frac{\partial v_\phi}{\partial r} + \frac{v_\phi v_r}{r} = -\frac{1}{r\Sigma} \frac{\partial p}{\partial \phi} - \frac{1}{r} \frac{\partial \Psi_p}{\partial \phi}. \qquad (4)$$

Finally, the continuity equation becomes

$$\frac{\partial \Sigma}{\partial t} + \frac{1}{r} \frac{\partial}{\partial r} (\Sigma r v_r) + \frac{1}{r^2} \frac{\partial}{\partial \phi} (\Sigma r v_\phi) = 0. \qquad (5)$$

We adopt a two-dimensional barotropic equation of state such that $p = p(\Sigma)$, with the local sound speed, $c$, defined by $c^2 \equiv dp/d\Sigma$. In all calculations that we consider, the disk is assumed to be initially unperturbed and axisymmetric. The surface density, angular velocity, and sound speed of the disk are then $\Sigma(r)$, $\Omega(r)$, and $c(r)$.

## 3. LINEAR THEORY

In order to gain understanding of the processes involved and to provide a reference point for nonlinear calculations, we first consider the case when the perturbation to the disk is small so that it may be calculated using linear perturbation theory.

We suppose that the encounter produces infinitesimal velocity perturbations $v'_r$, $v'_\phi$, and surface density perturbation $\Sigma' \equiv w'\Sigma/c^2$. The perturbing potential $\Psi_p$ can be written as a sum of azimuthal Fourier components with mode number $m$ such that

$$\Psi_p = \sum_{m=-\infty}^{\infty} \Psi_m \exp(im\phi). \qquad (6)$$

When the perturbing mass is unbound it is convenient to take $t = 0$ to correspond to pericenter passage. Then the $\Psi_m$ are functions of $r$ and $t$ that vanish for $t = -\infty$ and $t = \infty$. It is therefore convenient to work in terms of their Fourier transforms, $\Psi_{m\omega}(\omega, r)$ defined through

$$\Psi_m(r, t) = \int_{-\infty}^{\infty} \Psi_{m\omega}(\omega, r) \exp(-i\omega t) \, d\omega. \tag{7}$$

In a similar way we define azimuthal Fourier components and their Fourier transforms for each of the disk perturbations $v_r'$, $v_\phi'$, $w'$, and we denote these by $V_r$, $V_\phi$, and $W$, respectively, where the dependence on $m$ and $\omega$ has been suppressed. We shall also suppress this dependence in $\Psi_{m\omega}$, dropping the subscripts so that $\Psi$ will from now on denote the Fourier transform.

We obtain equations for $V_r$, $V_\phi$, and $W$ by linearizing the basic equations and taking their Fourier transforms. This procedure results in a set of equations of identical form to those obtained for a disk being forced by a binary in circular orbit with frequency $\omega$, for a particular value of $m$.[15–17]

As these have been well studied, we merely quote them here:

$$i\tilde{\sigma} V_r - 2\Omega V_\phi + \frac{dW}{dr} + \frac{d\Psi}{dr} = 0, \tag{8}$$

$$i\tilde{\sigma} V_\phi + \frac{\kappa^2}{2\Omega} V_r + \frac{imW}{r} + \frac{im\Psi}{r} = 0, \tag{9}$$

$$\frac{i\tilde{\sigma} W}{c^2} + \frac{dV_r}{dr} + V_r \left( \frac{1}{r} + \frac{1}{\Sigma} \frac{d\Sigma}{dr} \right) + \frac{imV_\phi}{r} = 0. \tag{10}$$

Here $\kappa^2 = (2\Omega/r)d(r^2\Omega)/dr$ denotes the square of the epicyclic frequency, $\tilde{\sigma} = m\Omega - \omega$, and we define a pattern speed $\Omega_p \equiv \omega/m$. Just as in the circular orbit case these equations have a corotation singularity in the disk where $\Omega = \Omega_p$.[16,18]

### 3.1. Wave Excitation and Angular Momentum Transport

The interaction between the disk and companion results in wave excitation and an angular momentum exchange that occur via the tidal torque. The unforced ($\Psi = 0$) solutions of (8)–(10) consist of propagating disturbanes that behave like sound waves. These are also in general present in the forced solution.

Disturbances propagating toward small radii have a conserved wave action and are expected to become nonlinear and dissipate.[19] The inner boundary condition on the response is thus that the disturbances should take the form of ingoing waves. At small radii, the solutions can be matched to Wentzel-Kramers-Brillouin (WKB) ingoing waves[7,17] which, because their pattern speed is slower than the angular velocity of the disk, carry negative angular momentum. When the waves become nonlinear and dissipate, the negative wave angular momentum is transferred to the disk material, which accordingly loses angular momentum.[20] We remark that this process of angular momentum loss by the inner disk material will occur simply as a result of the inner boundary condition and is independent of whether the net angular momentum transfer between disk and companion is positive or negative. In addition to the inner boundary condition an outer boundary condition is specified by the requirement that the solution be nonsingular.[5] The solution is thus completely determined.

The local dispersion relation for linear waves propagating in a non-self-gravitating disk is[21]

$$\tilde{\sigma}^2 = \kappa^2 + c^2 k_r^2,$$

where $k_r$ is the radial wavenumber. Because $k_r$ tends to $\infty$ as $r \to 0$, the WKB approximation is valid for small $r$.

For an inward propagating wave this gives, to within an arbitrary constant amplitude,

$$W = \frac{\tilde{\sigma}}{(r\Sigma c)^{1/2}(\tilde{\sigma}^2 - \kappa^2)^{1/4}} \exp\left[i \int_{r_1}^{r} \frac{dr}{c} (\tilde{\sigma}^2 - \kappa^2)^{1/2}\right]. \quad (11)$$

At any time the advective flux $F$ due to angular momentum transport by waves, or the rate of outward passage of angular momentum through a cylindrical shell of radius $r$, is[18]

$$F = r^2 \Sigma \int_0^{2\pi} \Re(v'_r)\Re(v'_\phi) \, d\phi, \quad (12)$$

where $\Re$ indicates that the real part is to be taken. It is possible to consider positive values of $m$ only, taking into account negative values by multiplying the formula for $F$ by a factor of 4. Thus we need consider $m$ to be positive only.

We integrate $F$ with respect to $t$ between $t = -\infty$ and $t = \infty$. The result is the total angular momentum associated with a single $m$ that has flowed:

$$\Delta F_A = 8\pi^2 r^2 \Sigma \int_{-\infty}^{\infty} \Re(V_r V_\phi^*) \, d\omega. \quad (13)$$

We note that the integrand $F_A = 8\pi^2 r^2 \Sigma \Re(V_r V_\phi^*)$, considered as a function of $\omega$, tends to a positive constant in the WKB limit as $r \to 0$.

### 3.2. Total Angular Momentum Exchange

The total angular momentum that is exchanged between the disk and perturber, $\Delta J$, can be found from the external torque

$$\frac{d\Delta J}{dt} = -\int_{\text{disk}} \Sigma' \frac{\partial \Psi_p}{\partial \phi} r \, dr \, d\phi. \quad (14)$$

Integrating with respect to $t$ and using Fourier transforms gives for each $m$

$$\Delta J = 8\pi^2 \int_{-\infty}^{\infty} \int_{r_i}^{r_o} \Re\left(\frac{im\Sigma W \Psi^*}{c^2}\right) r \, dr \, d\omega \equiv \int_{-\infty}^{\infty} \Gamma_T \, d\omega. \quad (15)$$

Here $r_i$ and $r_o$ denote the radii of the disk inner and outer boundaries, respectively. If there were no sources or sinks of angular momentum in the disk, angular momentum conservation would imply that at each frequency $\Gamma_T = -F_A(r_i)$. However, the corotation singularity provides a location where angular momentum may be emitted or absorbed. This is handled automatically by the Landau prescription in any integrations. By conservation of angular momentum, the total corotation torque is given, for a particular $m$, by

$$\Delta J_C = \Delta J + \Delta F_A(r_i). \tag{16}$$

For a more detailed discussion of coration torques in the context described here, see Korycansky and Papaloizou.[7]

We comment that the preceding description of the interaction of a disk with a perturber involving the excitation of inwardly propagating negative angular momentum waves and angular momentum transport at corotation is quite general. This would also be expected to apply to a self-gravitating disk. In particular the effects of an encounter can be regarded as the superposition of the effects of circular orbit tidal interactions expressed through a Fourier integral.

Linear torque calculations of non self-gravitating disks subject to encounters with perturbers on parabolic orbits have been performed by Palmer and Papaloizou[8] and Ostriker[9] who used asymptotic approximations throughout. Korycansky and Papaloizou[7] solved (8)–(10) directly for the cases when the perturbing mass was either equal to $M$ or was very small.

In the equal mass case it was found that typically about one percent of the disk angular momentum may be lost to the perturber before strong nonlinear effects occur.

## 4. NONLINEAR CALCULATIONS

We adopt a polytropic equation of state for the disk material so that the sound speed $c$ is given by $c^2 = c_0^2(\Sigma/\Sigma_0)^{1/n}$, where $n$ is the polytropic index and $c_0$ is the sound speed for surface density $\Sigma_0$. We are free to choose a form for $\Sigma(r)$ in the unperturbed disk. Our choice, based on that used by Papaloizou and Lin,[15] is

$$\Sigma = \Sigma_0(1 - b) \tanh^n\left(\frac{r_o - r}{\Delta}\right) + \Sigma_0 b, \tag{17}$$

where $\Delta$ is a scale length associated with the edge and $b$, which prevents the surface density from vanishing, is a small positive number taken to be 0.001.

We have performed fully nonlinear numerical calculations of disks undergoing encounters with perturbers in parabolic orbits. The method we used solves the equations by explicit finite difference techiques using operator splitting. The advection terms are handled using Van Leer's[22] monotonic transport scheme for advection. This method is similar to that described by Kley.[23]

The calculations were performed with 200 equally spaced zones in radius and 64 zones in azimuth. For convenience we adopt units such that $r_o = GM = \Sigma_0 = 1$. Rigid boundaries were adopted at $r = r_o \equiv 1$ and $r = r_i = 0.2$. We remark that the torques calculated here should scale as the fiducial constant interior surface density

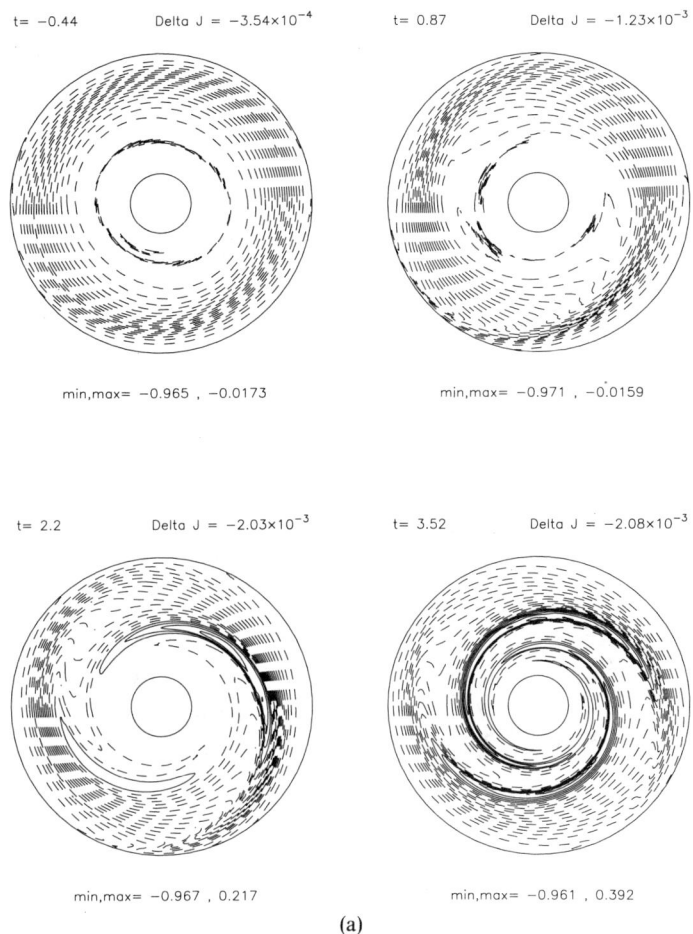

(a)

$\Sigma_0$ as long as the effects of disk self-gravity can be neglected. The disk self-gravity becomes important once the ratio of disk to central mass becomes comparable to $c_0$.

Results for encounters in which the perturbing mass was either very small or equal to the central mass are given in Korycansky and Papaloizou.[7] Here, we present results for an encounter with a perturbing mass equal to $0.1M$. The disk parameters were $n = 2.5$, $c_0 = 0.1$, nd $\Delta = 0.2$. We consider encounters for which the pericenter distance was 2.0 (model 1) and 1.8 (model 2). Surface density contour plots at different times for these cases are presented in FIGURE 1(a) and 1(b), and FIGURE 2, respectively.

Very little change in the disk occurs until just before pericenter passage, and nearly all the torque on the disk is applied impulsively within one orbital period at the disk edge.[7] At the outer edge of the disk, a disturbance forms that develops into

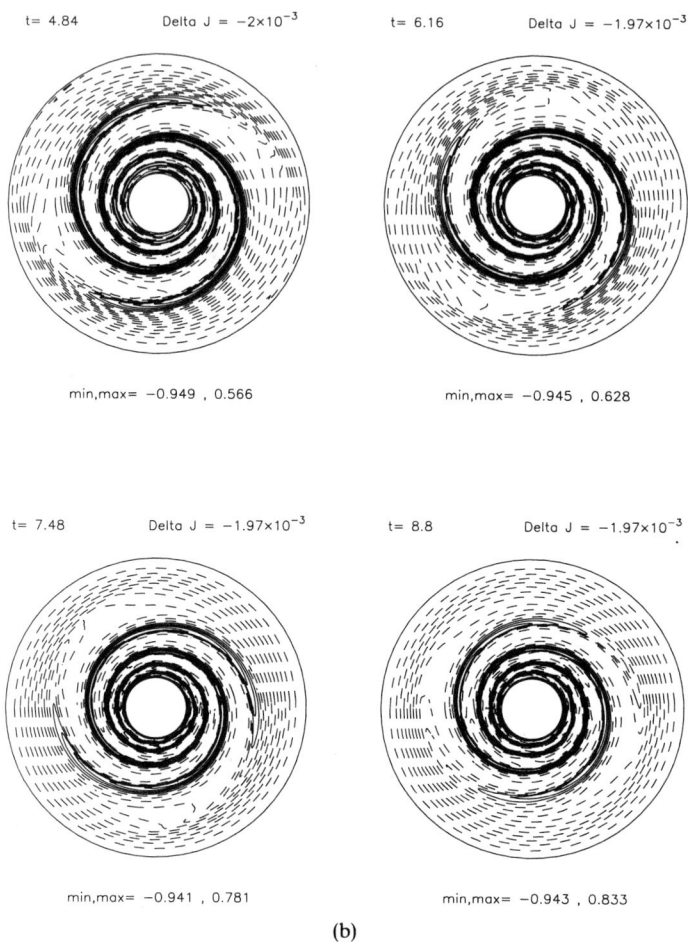

**FIGURE 1. (a)** A sequence of surface density contour plots for model 1. The maximum and minimum levels indicated are for $(\Sigma - \Sigma_0)/\Sigma_0$. Contour levels are equally spaced, those being indicated by *dashed lines* corresponding to negative values. The sense of rotation of the disk and perturber is anticlockwise. The time of pericenter is $t = 0$, at which time the line joining the disk center to the perturber points due west. The angular momentum gained by the disk since the beginning of the encounter, *DeltaJ*, is given for each plot. **(b)** Surface density contour plots for the same encounter illustrated in **(a)** at later times.

a spiral wave that decreases its wavelength as it propagates inwards. The propagation speed is close to the sound speed. Because of the short wavelength, the waves develop into weak shocks as they propagate inward causing matter to flow inward (strong shocks would dissipate after propagation through one wavelength). The situation is similar to that found in circular orbit calculations.[4,17] But note that radial wave propagation may be easier here because of the neglect of vertical structure. In

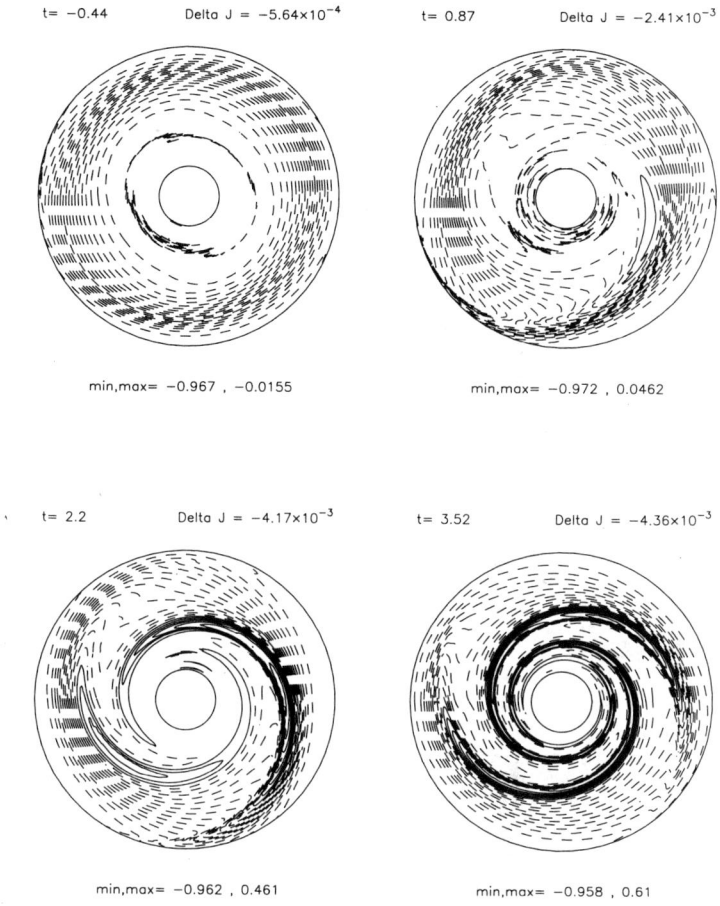

**FIGURE 2.** As in FIGURE 1(a) but for model 2.

reality, in a nonisothermal disk, propagating wavefronts may tilt upwards as they propagate, resulting in vertical losses.[24]

If the pericenter distance is decreased further, large density enhancements occur near the disk outer boundary as material is pulled outward toward the companion. This is noticeable for model 2 and can be seen to occur just prior to pericenter passage in that case. It is likely that this material would form a bridge and be captured by the perturber,[25] a situation that we were unable to model. In our calculations nonlinear effects start to become noticeable when, at pericenter passage, the disk edge is about 80 percent of the mean Roche lobe radius that would apply if the perturber was in a circular orbit.

The total angular momentum lost from the disk for model 1 was $2.0 \times 10^{-3}$ in our units and for model 2, $4.4 \times 10^{-3}$. The result for model 1 agrees to within 15 percent of the linear result of Korycansky and Papaloizou[7] for very small mass perturbers scaled up to a perturbing mass of $0.1M$.

### 4.1. Tidal Interactions with Companions on Highly Eccentric Orbits

For encounters with perturbers on parabolic orbits most of the interaction occurs when the perturber is close to pericenter. We found that the disk becomes relatively undisturbed after the inward propagating wave has passed. This suggests that the results can be used to model tidal interactions with perturbers on highly eccentric orbits. This should be reasonable provided successive pericenter passages of the perturber occur with a time interval that is longer than the time for sound to cross the disk. The calculations presented here taken together with those of Korycansky and Papaloizou[7] indicate that $2 \times 10^{-3}$ of the disk angular momentum could be transferred to the companion every orbital period, for perturbers with mass $> 0.1M$, without disrupting the disk.

This effect may control the edge of a weakly viscous disk. For this to occur the mean tidal torque $F_J = 2.0 \times 10^{-3}/T$, where $T$ is the orbital period, should exceed the outward viscous angular momentum flux $F_v = -2\pi v \Sigma r^3 \, d\Omega/dr$, $v$ being the kinematic viscosity. For an orbit with eccentricity of 0.7 and pericenter at two disk radii, $T = 108$, $F_J \sim 2 \times 10^{-5}$, and $F_v = 9.4v$. Thus the edge might be controlled by tidal effects if $v < 2 \times 10^{-6}$ corresponding to a Reynolds number of $5 \times 10^5$. This Reynolds number corresponds to an accretion timescale $\sim 10^{7-8}$ years for a protostellar disk of radius $\sim 60$ astronomical units centered on a $1 \, M_\odot$ star. This suggests the possible importance for protostellar disks of tidal effects due to perturbers on highly eccentric orbits. Because such companions spend most time at apocenter, they may be observed to be far from the disk edge they may control.

## 5. WARPED DISKS AND BENDING WAVES

When the perturber's orbit is inclined to the disk plane, warps and bending waves may be excited. For bending waves, the density perturbation is antisymmetric with respect to reflection in the midplane. Here we describe some recent results on the excitation of such disturbances with $m = 1$. They are of interest because they can result in long wavelength global perturbation of the disk. It has been proposed that large-scale warps may exist in protostellar disks where they are important in radiation reprocessing.[13]

Here, for simplicity, we consider the disk response to a companion in an inclined circular orbit. But note that, as before, the effects of encounters can be regarded as arising from the superposition of such responses at different frequencies.

We consider a binary system in which the stars have comparable masses. As before the primary has mass $M$ and the perturbing companion mass is $M_s$. The orbital separation is $D$.

We take the orbit of the performing star to be circular and in a plane that has an inclination angle $\delta$ with respect to the $(x, y)$ plane. For a disk of negligible mass, as adopted here, the secondary describes a Keplerian orbit with angular velocity $\omega$ about the primary such that the plane of the orbit does not precess.

We adopt an orientation of coordinates and an origin of time such that the line of nodes coincides with, and the secondary is on, the $x$-axis at $t = 0$. The perturbing potential may be expanded in powers of $1/D$. The lowest order terms that excite density perturbations that are odd in $z$ and have $m = 1$ can be written[14]

$$\Psi_p = -6GM_s(2D)^{-3}rz[(1 - \cos \delta) \sin \delta \sin (\varphi + 2\omega t)$$
$$- (1 + \cos \delta) \sin \delta \sin (\varphi - 2\omega t)$$
$$+ \sin 2\delta \sin (\varphi)]. \quad (18)$$

Because the principle of linear superposition can be used, the general problem may be reduced to calculating the response due to a complex potential of the form

$$\Psi_p = frze^{i(\varphi - \Omega_p t)}.$$

Here the pattern frequency $\Omega_p$ of the perturber is one of $0$, $2\omega$, or $-2\omega$, and $f$ is an appropriate complex amplitude. The magnitude of the tidal perturbation acting on the disk is measured by the dimensionless quantity $GM_s/(\Omega^2 D^3)$. For comparable primary and secondary masses this is of order $\omega^2/\Omega^2$.

In addition, if the disk were composed of noninteracting particles, the axisymmetric component of the tidal potential would induce a precession of disk particle orbital planes at a rate $\omega_z$, where

$$2\Omega\omega_z = -\frac{3GM_s}{4D^3}(3 \cos^2 \delta - 1).$$

When the tidal perturbation is considered to be small, the precession frequency is also small.

### 5.1. Unperturbed Disk Model

To examine density perturbations that are odd with respect to reflection in the disk midplane, we work with the full (1) and (2). For simplicity we assume a polytropic equation of state $P = K\rho^{1 + 1/n}$, where $K$ is the polytropic constant and $n$ is the polytropic index. The sound speed is given by $c_s^2 = dP/d\rho$. In the absence of the secondary the disk is axisymmetric and in hydrostatic equilibrium. The barotropic equation of state implies that the angular velocity is a function of $r$ alone.

For a thin disk under the influence of a point mass, integration of the hydrosatic equilibrium equations gives[14]

$$\rho(r, z) = [2K(1 + n)C_n^2]^{-n/(2n+1)}(\Sigma\Omega_K)^{2n/(2n+1)}\left(1 - \frac{z^2}{H^2}\right)^n \quad (19)$$

$$H(r) = \left(\frac{[2K(1 + n)]^n}{C_n}\right)^{1/(2n+1)}(\Sigma\Omega_K^{-2n})^{1/(2n+1)}. \quad (20)$$

Here $\Omega_K$ is the Keplerian angular velocity given by $\Omega_K^2 = GM/r^3$, $H$ is the semi-thickness of the disk, and

$$C_n = \frac{\Gamma(\frac{1}{2})\Gamma(n + 1)}{\Gamma(\frac{1}{2} + n + 1)},$$

where $\Gamma$ denotes the gamma function.

The angular velocity is close to the Keplerian value and is given by

$$\Omega^2 = \Omega_K^2 + \frac{1}{r}\left(\frac{[K(1+n)]^{2n}}{2C_n^2}\right)^{1/(2n+1)} \frac{\partial}{\partial r}[(\Sigma\Omega_K)^{2/(2n+1)}].$$

In order to perform response calculations an unperturbed surface density profile must be specified. As in the coplanar case, we specify a function that is constant in the main body of the disk with a taper to make it vanish at the outer boundary:

$$\Sigma(r) = \Sigma_0 \frac{\sigma(r)}{(1+\sigma(r)^s)^{1/s}}, \tag{21}$$

which approximates $\Sigma(r) = \Sigma_0 \min(1, \sigma(r))$ with

$$\sigma(r) = \left[1 - \left(\frac{r - (r_o - \Delta_r)}{\Delta_r}\right)^{p_r}\right]^N. \tag{22}$$

Here $\Delta_r$ is the distance interior to the outer boundary occupied by the taper, $N = (2n+1)/2$ and $p_r$ and $s$ are constants. The greater $s$ is, the more abrupt is the decrease of the density near $r_o - \Delta$. The value of $p_r$ controls the rate of decrease of the surface density at the outer edge of the disk. Parameters must be chosen to avoid $\kappa^2$ becoming negative in the region of the taper.

The polytropic constant $K$ may be fixed by specifying a maximum value of the relative thickness, $H/r$ in the disk, and the fiducial surface density $\Sigma_0$ is fixed by specifying the total mass of the disk.

## 6. THE DISK RESPONSE

### 6.1. Linear Perturbations

In this case one assumes that the perturbation is small, so that the basic equations, (1) and (2), can be linearized. Because the relevant perturbing potential, $\Psi_p$, is proportional to $\exp(i(\varphi - \Omega_p t))$, the $\varphi$ and $t$ dependence of the induced perturbations is the same; the factor will be henceforth taken as read.

As the relevant linearization of the equations has been discussed elsewhere,[26] we provide an abbreviated discussion here.

Denoting perturbations with a prime, we define a quantity, $g$, through

$$\frac{P'}{\rho} + \Psi_p = -ir z \Omega^2 g.$$

This formulation is convenient because a constant value of $g$ corresponds to a rigid tilt of the entire disk. Because there is no preferred direction, this is always a possible solution, with zero pattern speed, of the linear equations without a forcing term. Using the linearized motion, the velocity perturbations can be expressed in terms of $g$ in the form

$$v'_z = \frac{\Omega^2 r}{\Omega - \Omega_p} \frac{\partial(gz)}{\partial z} \tag{23}$$

$$\frac{v_r'}{z} = -g(\Omega - \Omega_p) + \frac{r\Omega^2(\Omega - \Omega_p)(dg/dr) + g\Omega_p^2(3\Omega - \Omega_p)}{(\Omega - \Omega_p)^2 - \kappa^2} \quad (24)$$

$$\frac{v_\varphi'}{z} = \frac{ig\Omega^2 + i(d(r^2\Omega)/dr)v_r'/(rz)}{\Omega - \Omega_p}. \quad (25)$$

We consider perturbations with radial wavelength significantly longer than the disk thickness for which we can use a vertical averaging approximation.

Papaloizou and Lin[26] do this by assuming $g$ to be independent of $z$. This means that $g$ is the inclination associated with the warped disk. Each of $v_r'$ and $v_\varphi'$ is then proportional to $z$, with $v_z$ being independent of $z$.

Linearization of the continuity equation followed by multiplication by $z$ and vertical averaging then gives

$$-i(\Omega - \Omega_p)\mathscr{I} + \frac{\Omega^2 g}{\Omega - \Omega_p}\left[r((\Omega - \Omega_p)^2 - \Omega^2)\mathscr{J} + \Omega^2)\mathscr{J} + \frac{d\mu}{dr}\right] = -\frac{1}{r}\left[\frac{\partial}{\partial r}\left(r\mu\frac{v_r'}{z}\right) + i\mu\frac{v_\varphi'}{z}\right] \quad (26)$$

in which

$$\mathscr{J} = \int_{-\infty}^{+\infty} \frac{\rho z^2}{c_s^2} dz, \quad \mu = \int_{-\infty}^{+\infty} \rho z^2\, dz, \quad \text{and} \quad \mathscr{I} = \int_{-\infty}^{+\infty} \frac{\rho z}{c_s^2} \Psi_p\, dz. \quad (27)$$

It is possible to eliminate $(v_z', v_r', v_\varphi')$ from the preceding using (23)–(25) and so obtain a single second-order ordinary differential equation for $g(r)$. This is in general quite lengthy, so we do not give it here. However, when $\Omega_p$ is zero, it takes on the relatively simple form

$$\frac{d}{dr}\left(\frac{\mu\Omega^2}{(\Omega^2 - \kappa^2)}\frac{dg}{dr}\right) = \frac{i\mathscr{I}}{r}. \quad (28)$$

For a perturbing potential $\propto z$, we have

$$\mathscr{I} = \mathscr{J}\left(\frac{\partial \Psi_p}{\partial z}\right)_{z=0}. \quad (29)$$

### 6.2. Angular Momentum Flux

Just as in the coplanar case the perturbing potential components with nonzero frequency excite inward propagating waves that result in angular momentum being transported from the disk to the perturber. In the WKB limit these waves are described for a strictly Keplerian disk in which $\omega_z = 0$, $\kappa = \Omega$, by

$$4g\Omega\Omega_p^2 \mathscr{J} + \frac{d}{dr}\left(\mu\Omega\frac{dg}{dr}\right) = 0. \quad (30)$$

The preceding equation shows that localized free disturbances for which $g \propto \exp(ik_r r)$, have $\Omega_p = \bar{c}_B k_r$, where the propagation speed $\bar{c}_B = (1/2)\sqrt{\mu/(\mathscr{J})}$. From (27) this is an appropriate mean value of the sound speed. The waves thus behave like sound waves propagating with no dispersion. Papaloizou and Lin[26] found that a similar situation occurs for waves in a disk in which self-gravity is included. Papaloizou and Terquem[14] calculated the angular momentum flux associated with waves generated by the nonzero frequency components in the perturbing potential. They found that this could be significant over the lifetime of inclined protostellar disks.

Here we present some results for the zero frequency response. This is associated with the possible precession of the disk.

### 6.3. Zero Frequency Response

To calculate the zero frequency response, we need to solve (28) with the appropriate boundary conditions. However, this is not possible with $\mathscr{I}$ in its present form. For example, suppose that $\mu$ vanishes at the disk boundaries. Then integrating (28) with respect to $r$ over the disk gives the integrability condition

$$\int_{r_i}^{r_o} \frac{\mathscr{I}}{r} dr = \int_{-\infty}^{+\infty} \int_{r_i}^{r_o} \frac{\rho z}{c_s^2} \frac{\Psi_p}{r} dr\, dz = 0.$$

This will in general not be satisfied for a thin Keplerian disk, as it can then be shown to be equivalent to the requirement that the net external torque vanish.

This can be dealt with by supposing our coordinate system rotates slowly about the orbital rotation axis with angular velocity $\omega_p$. We suppose that $\omega_p$ is of the same order as the external perturbation, namely $\Omega_p^2/\Omega^2$. Then we need consider only the addition of the coriolis force, the centrifugal force being a second-order quantity. This procedure is equivalent to supposing that the disk precesses about the orbital rotation axis with angular velocity $\omega_p$. For responses of the symmetry type we consider, only the $z$ component of the coriolis force contributes and this is of magnitude $-2\omega_p r\Omega \sin(\delta) \sin(\varphi)$. It can be regarded as providing an additional forcing term with zero pattern speed. For a thin Keplerian disk, this can be taken into account by modifying the zero pattern speed perturbing potential such that

$$\Psi_p \to \Psi_p + 2i\omega_p rz\Omega \sin(\delta).$$

The integrability condition is accordingly modified to become

$$\int_{-\infty}^{+\infty} \int_{r_i}^{r_o} \frac{\rho z}{c_s^2} \frac{(\Psi_p + 2i\omega_p rz\Omega \sin(\delta))}{r} dr\, dz = 0. \tag{31}$$

The precession frequency $\omega_p$ may now be chosen so that (31) is satisfied. Using the fact that our tidal perturbing potential is $\propto z$ and the fact that for a thin Keplerian disk, hydrostatic equilibrium in the $z$ direction gives $\Omega^2 \mathscr{I} = \Sigma$, we obtain

$$\omega_p = -\int_{r_i}^{r_o} \frac{\Sigma}{r\Omega^2} \left(\frac{\partial \Psi_p}{\partial z}\right)_{z=0} dr \bigg/ \int_{r_i}^{r_o} \frac{2ir\Sigma \sin(\delta)}{\Omega r} dr. \tag{32}$$

After finding $\omega_p$ from (32), (28) may be integrated to give $g$ to within an arbitrary constant inclination that may be eliminated by choosing the coordinate system so $g = 0$ at the disk inner boundary. Then if elsewhere $g$ is small, the disk approximately precesses like a rigid body. We remark that $\omega_p$ is found to be of the same order as $\omega_z$, the precession frequency of a free particle orbit in the disk.

We calculate $g$ in the preceding manner for disks with $M_s/M = 1$, $D/r_o = 7$, $n = 1.5$, $p_r = 1$, $s = 4$, and inclination angle $\delta = \pi/4$. But note that the response scales as $\sin(2\delta)$ for other inclinations and that it can be found for other values of $D$ by noting that it is $\propto D^{-3}$. For model 3, $\Delta_r = 0.1r_o$ and the maximum value of $H/r$ was 0.1. For model 4, $\Delta_r = 0.04r_o$ and the maximum value of $H/r$ was 1/30. For model 3 the integrability condition gave $\omega_p = -8.7 \times 10^{-4}$, while for model 4,

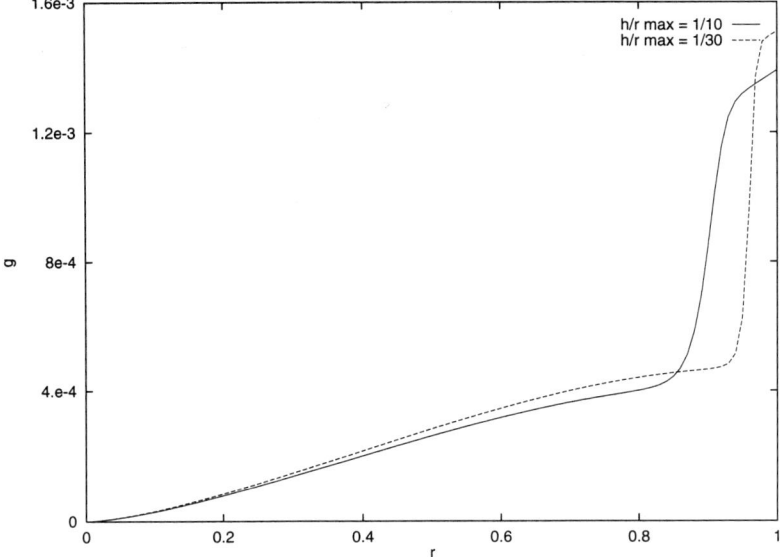

**FIGURE 3.** A plot of the inclination $g$ as a function of $r$ measured in units of the disk outer radius, $r_o$, for model 3 (*full line*) and model 4 (*dashed line*). In these calculations $r_i = 2 \times 10^{-4} r_o$.

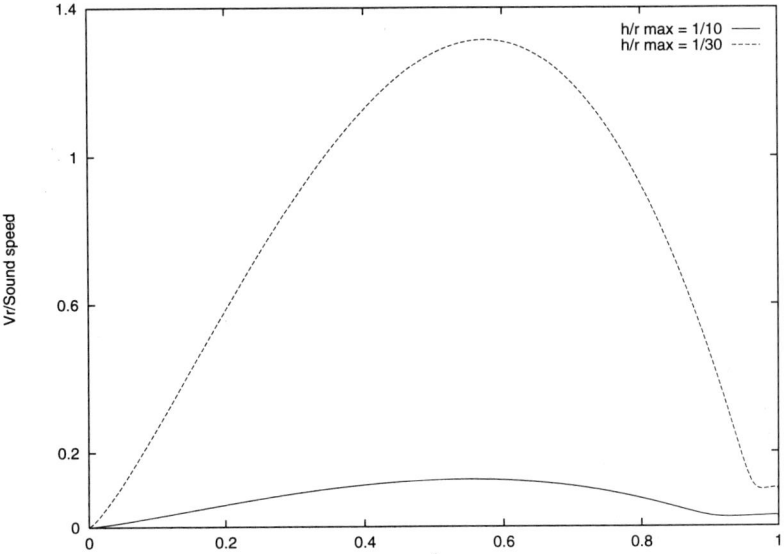

**FIGURE 4.** A plot of $v'_r H/(zc_s)$, as a function of $r$ measured in units of the disk outer radius, $r_o$, for model 3 (*full line*) and model 4 (*dashed line*). In these calculations $r_i = 2 \times 10^{-4} r_o$ and $c_s$ is evaluated in the midplane.

$\omega_p = -9.2 \times 10^{-4}$. Thus $\omega_p$ is insensitive to the disk thickness and taper, as expected.

From (28), for a fixed surface density profile, the response of a thin disk is independent of $H$. However, the neglect of the effect of the perturber on $\kappa^2$ requires $\omega_p < (H/r)^2$. This is only marginally satisfied for model 4. We present the forms of $g$ for models 3 and 4 in FIGURE 3. The inclination is everywhere small in these cases with $g \sim -\omega_p$, indicating near rigid body precession, but it varies quite rapidly near the outer boundary. We present the forms of $v'_r H/(zc_s)$, with the sound speed $c_s$ evaluated in the midplane, in FIGURE 4. This is found to be a maximum in the central regions of the disk reaching a value of about unity for model 4.

The fact that these models approximately precess as a rigid body is not unreasonable on physical grounds, as the sound crossing time is shorter than the precession time, enabling the disk to communicate within a precession period.

## REFERENCES

1. REIPURTH, B. & H. ZINNECKER. 1993. Astron. Astrophys. **278:** 81.
2. GHEZ, A. M., G. NEUGEBAUER & K. MATTHEWS. 1993. Astrophys. J. **106:** 2005.
3. MATHIEU, R. D., F. C. ADAMS & D. W. LATHAM. 1991. Astrophys. J. **101:** 2184.
4. SPRUIT, H. C., T. MATSUDA, M. INOUE & K. SAWADA. 1987. Mon. Not. R. Astron. Soc. **229:** 517.
5. LIN, D. N. C. & J. C. B. PAPALOIZOU. 1993. In Protostars and Protoplanets III, E. Levy, J. Lunine, and M. S. Mathews, Eds. Univ. of Arizona Press. Fayetteville.
6. ARTYMOWICZ, P. & S. H. LUBOW. 1994. Astrophys. J. **421:** 651.
7. KORYCANSKY, D. K. & J. C. B. PAPALOIZOU. 1995. Mon. Not. R. Astron. Soc. **274:** 85.
8. PALMER, P. L. & J. C. B. PAPALOIZOU. 1982. Mon. Not. R. Astron. Soc. **199:** 869.
9. OSTRIKER, E. C. 1994. Astrophys. J. **424:** 292.
10. CLARKE, C. J. & J. E. PRINGLE. 1991. Mon. Not. R. Astron. Soc. **249:** 564.
11. CLARKE, C. J. & J. E. PRINGLE. 1991. Mon. Not. R. Astron. Soc. **261:** 190.
12. HELLER, C. 1993. Astrophys. J. **408:** 337.
13. TERQUEM, C. & C. BERTOUT. 1993. Astron. Astrophys. **274:** 291.
14. PAPALOIZOU, J. C. B. & C. TERQUEM. 1995. Mon. Not. R. Astron. Soc. **274:** 987.
15. PAPALOIZOU, J. C. B. & D. N. C. LIN. 1984. Astrophys. J. **285:** 818.
16. KORYCANSKY, D. G. & J. B. POLLACK. 1993. Icarus **102:** 150.
17. SAVONIJE, G. J., J. B. P. PAPALOIZOU & D. N. C. LIN. 1994. Mon. Not. R. Astron. Soc. **268:** 13.
18. GOLDREICH, P. & S. TREMAINE. 1979. Astrophys. J. **233:** 857.
19. GOLDREICH, P. & S. TREMAINE. 1978. Icarus **34:** 227.
20. GOLDREICH, P. & P. D. NICHOLSON. 1989. Astrophys. J. **342:** 1075.
21. BINNEY, J. J. & S. TREMAINE. 1987. Galactic Dynamics. Princeton Univ. Press. Princeton, N.J.
22. VAN LEER, B. 1982. Lect. Notes Phys. **170:** 505.
23. KLEY, W. 1989. Astron. Astrophys. **222:** 141.
24. LIN, D. N. C., J. C. B. PAPALOIZOU & G. J. SAVONIJE. 1990. Astrophys. J. **364:** 326.
25. TOOMRE, A. & J. TOOMRE. 1972. Astrophys. J. **178:** 623.
26. PAPALOIZOU, J. C. B. & D. N. C. LIN. 1995. Astrophys. J. **438:** 841.

# Nonlinear Waves in Magnetized Accretion Disks[a]

R. V. E. LOVELACE[b] AND M. M. ROMANOVA[c]

[b]*Department of Applied Physics*
*Cornell University*
*Ithaca, New York 14853*

[c]*Space Research Institute*
*Russian Academy of Sciences*
*Profsoyuznaja 84/32,*
*Moscow 117810, Russia*

## 1. INTRODUCTION

Very-long-baseline interferometer (VLBI) radio observations of the brightest quasars have shown that the jets (0.1–10 pc) typically consist of one or more bright components moving with apparent "superluminal" speed with respect to the nuclear component. The extrapolated zero-point time of their formation often coincides with the brightness amplification of the continuum radiation in the infrared, optical, and X-ray wavebands. This correlation was noticed for several well-studied quasars, such as 3C 273, 3C 345, and 3C 120.[1–5] VLBI monitoring of the quasar 3C 273 after an optical/infrared flux density outburst in 1988 revealed the appearance of a new component in the jet at the predicted time.[6] The data appear to confirm the earlier hypothesis of Kinman[7] that both observed features, jet formation, and flux-density outburst occur simultaneously and reflect the same physical process.

Section 2 discusses the theory of outbursts from an accretion disk threaded by an ordered magnetic field and is based in part on the work of Lovelace *et al.*[8] The twisting of this field acts to drive jets from the disk surfaces that carry away angular momentum and energy and thereby increasing the accretion speed as depicted schematically in FIGURE 1. The increased accretion speed can in turn amplify the **B** field and lead to runaway or implosive accretion and explosive wind or jet formation. Section 3 gives conclusions from this work.

## 2. TIME-DEPENDENT ACCRETION WITH *B* FIELD

The basic equations are

$$\frac{\partial \rho}{\partial t} + \nabla \cdot (\rho \mathbf{v}) = 0, \tag{1a}$$

$$\nabla \times \mathbf{B} = \frac{4\pi}{c} \mathbf{J}, \tag{1b}$$

---

[a]This work was supported in part by grants from the Russian Fundamental Research Foundation, the European Southern Observatory, the Soros Foundation, NASA, and the National Science Foundation.

$$J = \sigma_e(E + v \times B/c), \quad (1c)$$

$$\nabla \times E = -\frac{1}{c}\frac{\partial B}{\partial t}, \quad (1d)$$

$$\rho \frac{dv}{dt} = -\nabla p + \rho g + \frac{1}{c} J \times B + F^{vis}. \quad (1e)$$

Here, $v$ is the flow velocity in the disk, $\rho$ is the density, $\sigma_e$ is the effective electrical conductivity, $F^{vis}$ is the viscous force density, $p = \rho k_B T/\mu + aT^4/3$ is the total, gas plus radiation, pressure (with $k_B$ the Boltzmann constant, $\mu$ the mean particle mass, and $a$ the usual radiation constant), and $g$ is the gravitational acceleration. Outside of the disk, dissipative effects are considered to be negligible ($\sigma_e \to \infty$, $F^{vis} = 0$, etc.). We assume a low mass disk and neglect general relativistic effects so that $g = -\nabla \Phi_g$ with $\Phi_g = -GM/(r^2 + z^2)^{1/2}$, where $M$ is the mass of the central object. Equations (1) are supplemented later by an equation for the conservation of energy in the disk. Even field symmetry[9] is assumed so that $B_r(r, z, t) = -B_r(r, -z, t)$, $B_\phi(r, z, t) = -B_\phi(r, -z, t)$, and $B_z(r, z, t) = +B_z(r, -z, t)$. Previous treatments of disks with ordered B fields have assumed stationary solutions.

The plasma flow in a thin disk can be described approximately by $v = -u(r, t) \hat{r} + v_\phi(r, t)\hat{\phi} + (z/h)(dh/dt)\hat{z}$, where $u$ is the accretion speed, $2h(r, t) \ll r$ is disk thickness, and $d/dt = \partial/\partial t - u\partial/\partial r$ is the convective time derivative following a ring of disk matter. Integration of the continuity equation, (1a), over the vertical ($z$) thick-

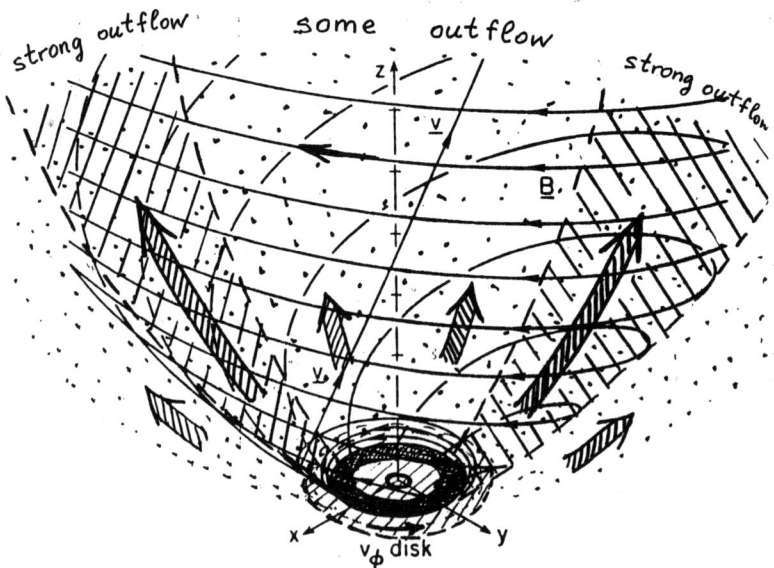

**FIGURE 1.** The figure shows field lines and a streamline for the outflow from the $+z$ surface of a magnetized disk.

ness of the disk gives

$$\frac{\partial}{\partial t}(\sigma) - \frac{1}{r}\frac{\partial}{\partial r}(ru\sigma) = 2(\rho v_z)_h, \qquad (2)$$

where $\sigma(r, t) = \int_{-h}^{h} dz\, \rho(r, z, t)$ is the disk's surface mass density, and $2(\rho v_z)_h$ is the mass flux-density from the $\pm z$ surfaces of the disk. In principle $(\rho v_z)_h$ can be obtained from the magnetohydrodynamic (MHD) outflow solutions external to the disk.

Integration of the $r$-component of the Navier–Stokes equation, (1e), over $z$ gives

$$\sigma \frac{du}{dt} = \frac{\partial}{\partial r} P + \frac{\sigma(v_K^2 - v_\phi^2)}{r} - \int_{-h}^{h} dz\, F_r^{\text{mag}},$$

or

$$\frac{\partial}{\partial t}(u\sigma) = \frac{1}{r}\frac{\partial}{\partial r}[r\sigma(u^2 + P/\sigma)] + \frac{\sigma}{r}[v_K^2 - v_\phi^2 - (P/\sigma)] - \int_{-h}^{h} dz\, F_r^{\text{mag}}, \qquad (3)$$

where $P = \sigma k_B T/\mu + 2haT^4/3$ is the height integrated *total* pressure, $T(r, t)$ is the midplane temperature of the disk, and $v_K \equiv (GM/r)^{1/2}$ is the Keplerian velocity at $r$. Assuming even field symmetry, the radial magnetic force can be written as $\int_{-h}^{h} dz\, F_r^{\text{mag}} = (1/2\pi)(B_r B_z)_h$, where the $h$ subscript indicates that the quantity is evaluated at $z = h$. We have dropped terms of order $h/r$ in (3). Also, the viscous force is negligible in (3).

For a broad range of conditions, the *quasi-static approximation* holds where $\sigma|du/dt|$ is negligible compared with the other terms in (3). In this case the accretion flow is subsonic and there is hydrostatic balance in the $r$ direction. Also for a broad range of conditions there is approximate balance of the centrifugal and gravitational forces so that $v_\phi^2 \approx v_K^2$.

Integration over $z$ of the $\phi$ component of the Navier–Stokes equation, (1e), gives

$$\frac{\partial}{\partial t}(\sigma l) - \frac{1}{r}\frac{\partial}{\partial r}\left[u\sigma l + r^3 \sigma v_t \frac{d\omega}{dr}\right] = \frac{1}{2\pi}(rB_\phi B_z)_h + 2(\rho v_z)_h l, \qquad (4)$$

where $l \equiv rv_\phi$ is the specific angular momentum, $\omega \equiv v_\phi/r$, and $v_t$ is the turbulent viscosity discussed below. The term $(1/2\pi)(rB_\phi B_z)_h$ represents the torque on the disk due to the external magnetic field associated with the outflows from the disk's surfaces. The term $2(\rho v_z)_h l$ represents the angular momentum carried by the matter outflowing from the surfaces of the disk.

For the quasi-static conditions mentioned previously, we have $l \approx rv_K$ so that $\partial l/\partial t \approx 0$. Thus, combining (2) and (4) gives the accretion speed

$$u(r, t) = u_v + u_B, \qquad (5a)$$

where

$$u_v = \frac{3}{r\sigma v_K}\frac{\partial}{\partial r}(r\sigma v_K v_t) \qquad (5b)$$

is the viscous accretion speed, which is of the order of $v_t/r$, and

$$u_B = \frac{4r(\rho v_z)_j}{\sigma}(R_A^2 - 1) \tag{5c}$$

is the magnetic accretion speed. Here,

$$R_A^2 - 1 \equiv \frac{(-B_\phi B_z)/4\pi}{(\rho v_z)_j v_K} > 0$$

is the ratio of the angular momentum carried by the **B** field to that carried by the matter of the outflow. The radius (in cylindrical coordinates), $R_A$, of the Alfvén point of the outflow (at $z_A \gg h$) is measured in units of the disk radius $r_d$ where the flow originates.[10] Thus in general $R_A = R_A(r_d)$. For a magnetically dominated flow, $R_A \gg 1$ and the $(\rho v_z)_h$ terms in (2) and (4) are negligible.

The magnitude of the turbulent viscosity $v_t$ is assumed to be given by an alpha model[11,12] based on the gas pressure of the disk; namely, $v_t = (2/3)\alpha(p_{gas}/\rho)(r/v_K)$, where $\alpha$ is a dimensionless constant less than or of order unity. Sakimoto and Coroniti[13,14] give physical arguments for this form of the viscosity law for active galactic nuclei (AGN) disks, and show that for this law the inner part of the disk is thermally stable under typical conditions where radiation pressure is dominant. The possible contribution to the turbulent momentum flux due to small-scale magnetic field fluctuations is assumed included in $\alpha$.[14-16]

The $z$-component of the Navier–Stokes equation, (1e), gives the condition for vertical hydrostatic balance. This can be written as

$$\left(\frac{h}{r}\right)^2 + b\left(\frac{h}{r}\right) - \frac{p}{\rho v_K^2} = 0, \tag{6}$$

where $c_s \equiv (p/\rho)^{1/2}$ is an effective sound speed based on the total (gas plus radiation) pressure at the midplane of the disk, and $b \equiv r[(B_r)_h^2 + (B_\phi)_h^2]/(4\pi\sigma v_K^2)$.[17] For $b \ll 2c_s/v_K$, this equation gives the well-known relation $h/r = c_s/v_K$,[12] while for $b \gg 2c_s/v_K$ it gives $h/r = b^{-1}(c_s/v_K)^2$, which is smaller than $c_s/v_K$ owing to the compressive effect of the magnetic field external to the disk.[17] It is useful to write $b = \varepsilon\beta^2$, where $\varepsilon \equiv rB_z^2(r,0)/(2\pi\sigma v_K^2)$ and $\beta^2 \equiv [(B_r)_h^2 + (B_\phi)_h^2]/(2B_z^2)$. As we discuss below, $\beta = O(1)$ in order to have outflows from the disk, and $\beta \leq 1$ with no outflows. Thus, the Alfvén speed in the midplane of the disk is $v_A = v_K \varepsilon^{1/2}(h/r)^{1/2}$. We may term the magnetic field as weak if $\varepsilon < c_s/v_K \approx h/r$. In this limit, $v_A/c_s \approx (\varepsilon r/h)^{1/2} \leq 1$, and the magnetic compression of the disk is always small. In the opposite, strong field limit, $\varepsilon > c_s/v_K$. If at the same time $\beta = O(1)$, then the disk is magnetically compressed, and for $\varepsilon \gg c_s/v_K$ and $v_A/c_s = 1/\beta$.

The solution for the magnetic field **B** outside of the disk ($|z| > h$) is matched to the field solution inside the disk at $z = \pm h$ as discussed by Lovelace et al.[9] We first treat the internal field solution, which is described by (1b), (1c), and (1d). These can be combined to give the "induction equation,"

$$\frac{\partial \mathbf{B}}{\partial t} = \nabla \times (\mathbf{v} \times \mathbf{B}) - \nabla \times (\eta \nabla \times \mathbf{B}), \tag{7}$$

where the magnetic diffusivity $\eta \equiv c^2/(4\pi\sigma_e)$ has the same units as kinematic viscosity. We assume that the magnitude of $\eta$ is comparable to that of the turbulent viscosity $v_t$,[18,19] that is, we let $\eta = Dv_t$ with $D = O(1)$.

For the even field symmetry assumed, $B_z$ is an even function of $z$. Because $\nabla \cdot \mathbf{B} = 0$, this implies $\Delta B_r/r \approx \Delta B_z/h$. Then, the variation of $B_z$ from $z = 0$ to $z = h$ is $\Delta B_z \approx (h/r)(B_r)_h$. Assuming $(B_r)_h \lesssim B_z$ as discussed below, it follows that $\Delta B_z \ll B_z$; that is, the variation of $B_z$ with $z$ within the disk is negligible. The $z$-component of (7) is

$$\frac{\partial B_z}{\partial t} = \frac{1}{r}\frac{\partial}{\partial r}(urB_z) + \frac{1}{r}\frac{\partial}{\partial r}\left[\eta r\left(\frac{\partial B_z}{\partial r} - \frac{\partial B_r}{\partial z}\right)\right].$$

For even field symmetry, $B_r$ is an odd function of $z$, and consequently $\partial B_r/\partial z \approx (B_r)_h/h$. The condition for validity of this equality is $h^2|\partial^3 B_r/\partial z^3| \ll |\partial B_r/\partial z|$. Thus,

$$\frac{\partial}{\partial t}(rB_z) = \frac{\partial}{\partial r}\left[urB_z - \frac{\eta r(B_r)_h}{h} + \eta r\frac{\partial B_z}{\partial r}\right]. \quad (8)$$

Inside the square brackets, the term proportional to $u$ describes the inward advection of the poloidal field with the accretion flow, the term proportional to $\eta(B_r)_h$ describes the diffusive outward drift of the poloidal field, and the term proportional to $\eta(\partial B_z/\partial r)$ represents the radial diffusion of the poloidal field.

In contrast with (8) for $B_z(r, t)$, the values of $(B_r)_h$ and $(B_\phi)_h$ are determined by the field solution external to the disk. External MHD outflow solutions are discussed by Blandford and Payne[20] and Lovelace et al.[10] The timescale for setting up or changing the outflow is the Alfvén propagation time over distances of order $r$ in the low-density external plasma. Therefore, this timescale is much shorter than any timescales for changes in the disk. The external field solutions for magnetically driven outflows obey the relation

$$(B_r)_h = \beta_r B_z(r, t), \quad (9a)$$

where $\beta_r$ is a constant of order unity. Blandford and Payne[20] give the condition $\beta_r \gtrsim 1/\sqrt{3} \approx 0.577$ for MHD outflows. On the other hand, the MHD solutions of Lovelace et al.[10] and Lovelace et al.[8] give $\beta_r \approx 1$ for outflows. Consistent with the detailed models, we assume

$$(B_\phi(r, t))_h = -\beta_\phi B_z(r, t), \quad (9b)$$

with $\beta_\phi$ a positive constant of order unity.

Conservation of energy in the disk can be written as an equation for the midplane temperature of the disk $T(r, t)$,

$$c_v\sigma\frac{dT}{dt} = c_v(\gamma - 1)Th\frac{d}{dt}\left(\frac{\sigma}{h}\right) + D_{v\phi} + D_{vr} + D_O - 2F_R - \frac{1}{r}\frac{\partial(rh\mathscr{F}_r)}{\partial r}. \quad (10a)$$

Here, $c_v \equiv k_B/[\mu(\gamma - 1)]$ is the specific heat; $\gamma$ is the specific heat ratio;

$$D_{v\phi} = \sigma v_t\left[r\left(\frac{\partial \omega}{\partial r}\right)\right]^2 \quad (10b)$$

is the viscous dissipation (per unit area of the disk) due to the shear in the azimuthal motion; $D_{vr}$ is the viscous dissipation due to the radial motion and is given by Chen and Taam,[21] and

$$D_O = \frac{4\pi}{c^2} \int_{-h}^{h} dz\, \eta |\mathbf{J}|^2 = \frac{\eta}{4\pi h}\left[2(B_r)_h^2 + \frac{3}{5}(B_\phi)_h^2\right] \quad (10c)$$

is the Ohmic dissipation in the disk;

$$F_R = \frac{2acT^4}{3\kappa\sigma} = \sigma_B T_{\text{eff}}^4 \quad (10d)$$

is the radiative energy flux per unit area from the top or bottom surface of the disk which is assumed optically thick; $a$ and $\sigma_B$, the usual radiation constants; $\kappa$ is the opacity; $T_{\text{eff}}$ is the effective temperature of the surface of the disk; and $\mathscr{F}_r = -[(8acT^3)/(3\kappa\rho)](\partial T/\partial r)$ is the radial energy flux of the radiation in the disk.

In the quasi-static approximation previously mentioned, the radial accretion speed is $u = u_v + u_B$, where $u_v$ is the contribution due to viscosity, and $u_B =$

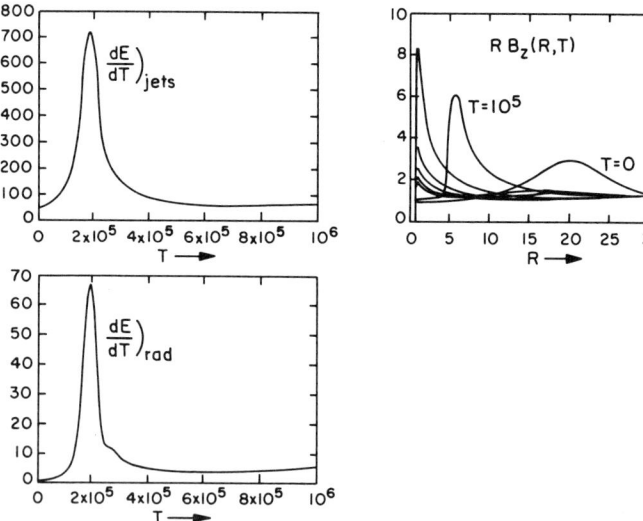

**FIGURE 2.** The figure shows the disk behavior for a case where the initial field consists of a background component $B_z = B_0/R$, a bump in $B_z$ centered at $R = 20$, of full width to $1/e$ of $\Delta R = 8$, and of peak amplitude three times the background. Here, $R$ is the radius measured in units of the inner radius of the disk, $r_i$, and the time $T$ is measured in units of $r_i/v_{Ki}$, where the $i$ subscript indicates that the quantity is evaluated at $r = r_i$. For this case, $\varepsilon_i = r_i(B_0)^2/(2\pi\sigma_i v_{Ki}^2) = 3 \times 10^{-6}$, $\alpha = 0.1$, $D = 1$ for the magnetic diffusivity, and $R_A \gg 1$ for the Alfvén radius. The luminosity of the disk if it were nonmagnetized is small compared with the Eddington luminosity (0.038). The power in the jets and in the disk radiation are measured in units of the radiative power of the nonmagnetized disk. The **B** field is measured in units of $B_0$. The results shown are obtained using a conservative second-order Eulerian method to integrate (2) and (5)–(10) (see reference 8). The outer radius is $R_0 = 31$ and $\Delta R = 0.1$. The vertical lines at $R = 1$ and $R = 31$ are retrace artifacts of the plotting routine.

$-r(B_\phi)_h B_z/(\pi v_K \sigma) \approx \beta_\phi B_z^2 r^{3/2}/(\pi\sigma\sqrt{GM})$ is the accretion speed that results from the removal of angular momentum by the outflows. Thus, the dynamics of a magnetized disk is controlled by the evolution of $B_z(r, t)$ and $\sigma(r, t)$. In the $B_z$ equation, (8), the inward advection ($\propto u$) is counteracted by the outward diffusive drift ($\propto \eta$), while the diffusion term is generally smaller. The nonlinear dependence of $u_B$ on $B_z$ suggests that an enhancement or bump in $B_z$ at, say, $r_0$ can make the inward advection of $B_z$ dominant over the outward drift. Consequently, advection will cause both $\sigma$ and $B_z$ to increase with time. This can give rise to what we term *implosive accretion*.

A disk initially with a distributed $B_z$ field, for example, $B_z = B_0 R^{-5/4}$, suggested by Blandford and Payne,[20] remains approximately steady for a long time ($>10^6$ orbits of the inner part of the disk) with an approximately constant ratio of the power in the $\pm z$ jets to the radiated power of the disk.

In contrast, a disk initially with both a distributed $B_z$ field and an enhancement or "bump" in $B_z$ in the vicinity of some large radius $r_0$ evolves with the bump moving inward and diffusively spreading. For a larger amplitude of the initial bump in $B_z$ and/or a larger background field, the bump moves inward more rapidly, it narrows in radial width, and it forms a soliton-like disturbance. The disturbance propagates inward while growing in strength up to the time when it is "swallowed," that is, when it passes within $r_i$. The disturbance produces almost simultaneous bursts in the power output in the winds and in the disk radiation as shown in FIGURE 2.

## 3. DISCUSSION

Time-dependent accretion for a magnetized disk can exhibit short timescale outbursts of high-power jets associated with bursts of disk radiation. The burst of energy in jets can produce outward propagating disturbances in preexisting ($\pm z$) jet flows. The disturbances are strong and can be expected to steepen and form strong shock waves as assumed in the model of Blandford and Königl[22] and Hughes *et al*.[23] The strong linear polarization (up to $\sim 30$ percent) and large polarization position angle changes ($90°$ and more) seen in the optical outbursts of some quasars[24-28] clearly indicate synchrotron radiation, and this component of the radiation is probably associated with the burst in jet energy in the present model.

## ACKNOWLEDGMENTS

One of the authors (R. L.) thanks Drs. G. Contopoulos, J. Hunter, and R. Wilson for the hospitality of the Tenth Florida Workshop.

## REFERENCES

1. BABADZHANYANTS, M. K. & E. T. BELOKON. 1986. Astrophysics **23**: 639.
2. BELOKON, E. T. 1988. Astrophysics **27**: 588.
3. BELOKON, E. T. 1991. Sov. Astron. **35**: 1.
4. BREGMAN, J. N., *et al.* 1986. Astrophys. J. **301**: 708.

5. ZENSUS, J. A., S. C. UNWIN, M. H. COHEN & J. A. BIRETTA. 1990. Astron. J. **100**: 1777.
6. KRICHBAUM, T. P., et al. 1990. Astron. Astrophys. **237**: 3.
7. KINMAN, T. D. 1977. Nature **267**: 798.
8. LOVELACE, R. V. E., M. M. ROMANOVA & W. I. NEWMAN. 1994. Astrophys. J. **437**: 136.
9. LOVELACE, R. V. E., J. C. L. WANG & M. SULKANEN. 1987. Astrophys. J. **315**: 504.
10. LOVELACE, R. V. E., H. L. BERK & J. CONTOPOULOS. 1991. Astrophys. J. **379**: 696.
11. SHAKURA, N. I. 1973. Sov. Astron. **16**: 756.
12. SHAKURA, N. I. & R. A. SUNYAEV. 1973. Astron. Astrophys. **24**: 337.
13. SAKIMOTO, P. J. & F. V. CORONITI. 1981. Astrophys. J. **247**: 19.
14. CORONITI, F. V. 1981. Astrophys. J. **244**: 587.
15. BALBUS, S. A. & J. F. HAWLEY. 1992. Astrophys. J. **400**: 610.
16. KAISIG, M., T. TAJIMA & R. V. E. LOVELACE. 1992. Astrophys. J. **386**: 83.
17. WANG, J. C. L., R. V. E. LOVELACE & M. E. SULKANEN. 1990. Astrophys. J. **355**: 38.
18. BISNOVATYI-KOGAN, G. S. & A. A. RUZMAIKIN. 1976. Astrophys. Space Sci. **42**: 401.
19. PARKER, E. N. 1979. Cosmical Magnetic Fields, Chap. 17. Clarendon Press. Oxford.
20. BLANDFORD, R. D. & D. G. PAYNE. 1982. Mon. Nat. R. Astron. Soc. **199**: 883.
21. CHEN, X. & R. E. TAAM. 1992. Mon. Not. R. Astron. Soc. **255**: 51.
22. BLANDFORD, R. D. & A. KÖNIGL. 1979. Astrophys. J. **232**: 34.
23. HUGHES, P. A., H. D. ALLER & M. F. ALLER. 1989. Astrophys. J. **341**: 54 and 68.
24. KINMAN, T. D. 1967. Astrophys. J., Lett. **148**: L53.
25. KINMAN, T. D., E. LAMLA, T. CIURLA, E. HARLAN & C. A. WIRTANEN. 1968. Astrophys. J. **152**: 357.
26. BABADZHANYANTS, M. K. & V. A. HAGEN-THORN. 1975. Astrophysics **11**: 259.
27. MOORE, A. L. & H. S. STOCKMAN. 1981. Astrophys. J. **243**: 60.
28. MOORE, A. L. & H. S. STOCKMAN. 1984. Astrophys. J. **279**: 465.

# Structural Stability and $\lambda$-Transitions[a]

DIMITRIS M. CHRISTODOULOU,[b]
DEMOSTHENES KAZANAS,[c] ISAAC SHLOSMAN,[d,e]
AND JOEL E. TOHLINE[b]

[b]*Department of Physics and Astronomy*
*Louisiana State University*
*Baton Rouge, Louisiana 70803*

[c]*NASA Goddard Space Flight Center*
*Code 665*
*Greenbelt, Maryland 20771*

[d]*Department of Physics and Astronomy*
*University of Kentucky*
*Lexington, Kentucky 40506*

## INTRODUCTION

In this work, we review some new discoveries and physical interpretations that concern the secular and dynamical instabilities and the associated structural changes in rotating, self-gravitating, incompressible fluids.[1] For such fluids, we have been able to examine in depth and to understand physically a variety of classic results, including the analytical investigations of Poincaré,[2] Jeans,[3] Cartan,[4] Chandrasekhar,[5] Lebovitz,[6-8] and Bertin and Radicati,[9] as well as the numerical computations of Eriguchi *et al.*[10-13] A complete list of the relevant literature on which we build and details of our analysis can be found in a recent series of four papers[1] entitled "Phase-Transition Theory of Instabilities."

An extension of the analysis to compressible polytropic fluids has not been completed yet, although the results are expected to be equally interesting and significant in this case as well. Compressible, rotating, self-gravitating fluid binaries have been recently studied both analytically and numerically by Lai *et al.*[14,15] and a wealth of important information coming out of these studies is currently under investigation using our theory of "phase transitions" and additional numerical calculations.[16]

The results obtained by all the just cited investigators are directly relevant to the physics of star formation, the structure and evolution of single and binary stars, and the formation of disks, rings, and tori around protostellar objects.[17-19] The theory of "phase transitions" provides a theoretical framework within which all the previously obtained results are interpreted correctly and are clearly understood (see

---

[a]This work was supported in part by NASA grants NAGW-1510, NAGW-2447, NAGW-2376, and NAGW-3839, by National Science Foundation Grant AST-9008166, and by grants from the San Diego Supercomputer Center and the National Center for Supercomputing Applications. One of the authors (I. S.) acknowledges support from the Gauss Foundation.
[e]Gauss Foundation Fellow.

below). However, owing to its simplicity and novel approach, the same theory bears an additional wealth of unanticipated results whose impact reaches far beyond astrophysical instabilities and star formation and into the fields of classic fluid and stellar dynamics (e.g., structural stability of fluids and evolution of stellar systems) and low-temperature superfluids (e.g., the $\lambda$-transition[20,21] of liquid $^4$He).[1] Furthermore, applied to the instabilities of astrophysical systems, the theory of "phase transitions" provides additional insight to the concepts of symmetry-breaking and topology-breaking during structural transformations/instabilities in evolving physical systems and indicates how the various types of hydrodynamical transitions/transformations/changes may be interpreted within the frameworks of Thom's "catastrophe theory"[22,23] and the Ginzburg–Landau "phase-transition theory."[24]

In what follows, we discuss first in some detail the structural stability and the $\lambda$-transitions of astrophysical fluids. We conclude our review with a summary of all the different types of $\lambda$-transitions found in the course of our investigation and with a discussion of the implications of our results for superfluids,[21] catastrophes,[23] thermodynamical phase transitions, and the breaking of symmetry and topology.[1]

## STRUCTURAL STABILITY

The term "structural stability" has been invented in mechanics to describe drastic changes in the stability of a system caused by the introduction of an additional force of very small (or infinitesimal) magnitude.[23,25] A commonly quoted example is the stability properties of viscous versus perfect (i.e., nonviscous) fluids.[25] In many cases, it is found by linear stability analyses that consideration of a small viscous force leads to fundamentally different behavior and stability characteristics in a fluid compared to the results obtained in the complete absence of viscosity. Obviously, the introduction of a small viscous force leads to a dramatic *qualitative* change in the eigenvalues and oscillatory modes of a perfect fluid instead of causing only a small *quantitative* change. Furthermore, in many cases, the results for a perfect fluid are not recovered in the limit of zero viscosity. In fact, Drazin and Reid[25] go as far as to suggest that this behavior indicates a deficiency in the classic nonviscous Euler equations of motion, as opposed to the Navier–Stokes equations, in describing the hydrodynamics of evolving fluids.

What has not been realized in previous investigations of structural stability in fluids, and is explained in detail in our work,[1] is that there exists a much deeper reason for such dramatic qualitative differences, other than the mere introduction of an additional, small viscous term into the equations of motion: an infinitesimal viscous force may be negligible compared to pressure, inertial, or gravitational forces but, at the same time, it is all it takes to "destroy" the conservation law of circulation that is valid in a perfect fluid. This simple fact has been known for a long time[26,27] and has been mentioned in various astrophysical investigations.[5–8,14,15] But the full implications of the implicit elimination of a conservation law from a stability analysis and the resulting, qualitatively different outcomes have not been previously clarified or discussed in various related problems. Herein we provide an example of structural instability caused by the nonconservation of either angular momentum or circulation. This example, the stability properties of incompressible Maclaurin spheroids subject to $m = 2$ toroidal modes of disturbance,[5] has been discussed to some extent in relation to conservation laws[5–8,14,15] and, in considerably more detail, in relation to structural stability and *catastrophic morphogenesis*[1] (i.e., the creation of new forms and structures as a result of a catastrophe or, equivalently, a phase transition).

The secular and dynamical instability of Maclaurin spheroids[5] occur above a meridional eccentricity of $e = 0.81267$ and $e = 0.95289$, respectively. The point of dynamical instability is singled out only for a perfect-fluid Maclaurin sequence of spheroids for which both angular momentum and circulation are assumed to be conserved. On the other hand, the point of secular instability manifests itself only if, alternatively, the conservation law of angular momentum or circulation is eliminated from the analysis, assuming that the corresponding physical quantity is no longer conserved due to the presence of, for example, gravitational radiation or viscous forces, respectively.

Both types of instabilities can be interpreted physically as typical *second-order phase transitions.*[1,9,12] A Maclaurin spheroid with $e > 0.95289$ evolves on a dynamical time scale toward the so-called $x = +1$ self-adjoint sequence of Riemann S-type ellipsoids, which bifurcates from the Maclaurin sequence at $e = 0.95289$. The transition occurs only if *mass, angular momentum*, and *circulation are conserved* during evolution. The final equilibrium state is an $x = +1$ Riemann S-type ellipsoid because such an object has lower energy than a Maclaurin spheroid of the same mass, angular momentum, and circulation with $e > 0.95289$.

A viscous Maclaurin spheroid with $e > 0.81267$ evolves on a secular time scale, dictated by the magnitude of viscosity, toward the Jacobi sequence of zero-vorticity ($x = 0$) ellipsoids, which bifurcates from the Maclaurin sequence at $e = 0.81267$. The transition occurs only if *mass and angular momentum are conserved while circulation is destroyed* by viscosity during evolution. The final equilibrium state is a Jacobi ellipsoid because such an object has lower energy and no circulation/vorticity compared to a Maclaurin spheroid of the same mass and angular momentum with $e > 0.81267$.

A perfect-fluid Maclaurin spheroid with $e > 0.81267$ that suffers angular momentum losses due to, for example, gravitational radiation or external tidal torques evolves on a secular time scale, dictated by such loss mechanisms, toward the Dedekind sequence of nonrotating ($x = \pm\infty$) ellipsoids, which bifurcates from the Maclaurin sequence also at $e = 0.81267$. The transition occurs only if *mass and circulation are conserved while angular momentum is decreased* during evolution. The final equilibrium state is a Dedekind ellipsoid because such an object has lower energy and no angular momentum/pattern rotation compared to a Maclaurin spheroid of the same mass and circulation with $e > 0.81267$.

We note that the secular and dynamical $m = 2$ bar-forming instabilities of *stellar* "Maclaurin" systems can also be analyzed in the same manner within the framework of phase transitions.[1] Some caution should, however, be exercised with respect to two important differences. First, the definitions of kinetic energy due to ordered rotational motions have been traditionally different between fluid and stellar systems. As a result, although the "Jacobi" sequence bifurcates from the "Maclaurin" sequence at the same point in fluid and stellar systems, the stellar "Dedekind" sequence does not. Despite this "difference," markedly analogous instabilities toward the same sequences do occur in stellar systems as well. Second, the phase transition toward the stellar "Jacobi" sequence is effectively *dynamical*, that is, unlike in fluids, it does not depend on frictional forces and dissipation. This dynamical transition is explained by the *ab initio* elimination of the conservation law of circulation in stellar systems because of the off-diagonal terms of the stress-tensor gradient in the Jeans equations of motion.[28] The preceding differences and other issues related to the stability of stellar systems and to the significance of the Ostriker–Peebles[29] criterion for stability to $m = 2$ modes are discussed in more detail in references 1 and 30.

## $\lambda$-TRANSITIONS

The term "$\lambda$-transition" has been introduced in low-temperature physics to specify a particularly puzzling phase transition discovered in liquid $^4$He at a temperature of $T \approx 2.2$ K.[20,21] This transition appears to share some common characteristics with Bose–Einstein condensation (BEC), a phase transition that is found in a hypothetical, idealized system, the so-called "ideal Bose gas" of noninteracting particles.[21] But the $\lambda$-transition of liquid $^4$He is not quite the same as BEC. The fact that $^4$He molecules are not ideal bosons, but actually interact with one another, leads to some extraordinary properties exhibited by $^4$He at low temperatures. In particular, the specific heat diverges at the $\lambda$-point $T \approx 2.2$ K while it is continuous at the BEC point $T \approx 3.1$ K of the ideal Bose gas. Furthermore, liquid $^4$He below the $\lambda$-point exhibits "superfluid" properties, such as absolutely no molecular viscosity (allowing it to leak out through the walls of a solid container) and quantized vortices in a state of rotation (usually characterized as a macroscopic manifestation of quantum effects).

Our study has revealed what appear to be dynamical analogs of BEC and the $\lambda$-transition of liquid $^4$He in the dynamics of rotating, self-gravitating, incompressible fluids. Both types of phase transitions appear as typical axisymmetric ($m = 0$) instabilities encountered along the Maclaurin sequence of spheroids above two different values of the eccentricity ($e = 0.99892$ and $e = 0.98683$, respectively). A $\lambda$-transition also exists on the Jacobi sequence of ellipsoids. The identification of these and other types of $\lambda$-transitions in uniformly rotating, self-gravitating, incompressible fluids is exact and quantitative. It is based on a *phase-transition theory of instabilities*[1] in which dynamical quantities are identified with corresponding thermodynamical variables. Specifically, the energy $E$ of an evolving, generally out-of-equilibrium, system plays the role of the free-energy function whose minima identify stable equilibria and maxima identify unstable equilibria. The inverse of the total angular momentum $L$ and the inverse of the rotation frequency $\Omega$ correspond to thermodynamical temperature $T$ and internal energy $U$. The change in "specific entropy" $dS$ and the "specific heat" $C_v$ are obtained from the first and second derivative, respectively, of the "free-energy function" $E$, viz.,

$$dS = -I d(\ln \Omega), \tag{1}$$

and

$$C_v = I^2 \frac{d\Omega}{dL}, \quad (2)$$

where $I$ is the specific moment of inertia. Equation (1) leads to an expression for the "latent heat" $\mathscr{L}$ of a transition, viz.,

$$\mathscr{L} \equiv L^{-1}dS = d(\Omega^{-1}), \quad (3)$$

where $L^{-1}$ is to be regarded as "temperature" and $d(\Omega^{-1})$ denotes the change in the "internal energy" of a system. This correspondence can be easily realized in the case of an "ideal-gas" system by a comparison between the standard ideal-gas thermodynamical relation

$$U = C_v T, \quad (4)$$

and the equivalent "ideal-gas" dynamical relation $L = I\Omega$ written in the form

$$\Omega^{-1} = IL^{-1}. \quad (5)$$

In this "ideal" case, the specific moment of inertia $I$ is equivalent to the thermodynamical specific heat. More details about the preceding formulation can be found in references 1 and 12. Below, we summarize the most important properties of the hydrodynamical $\lambda$-transitions in Maclaurin spheroids and Jacobi ellipsoids.

### Bose–Einstein Condensation and $\lambda$-Transition of Maclaurin Spheroids

Two sequences involving mainly toroidal axisymmetric equilibria have been found to bifurcate from the Maclaurin sequence of spheroids. The *one-ring sequence* was found[10] under the assumption that the new equilibria are rotating uniformly. The *Maclaurin sequence of toroids* was found[13] under the assumption that the distribution of specific angular momentum is the same as that of Maclaurin spheroids. A transition between the Maclaurin and the one-ring sequence may only take place while rotation remains uniform. Thus, such a transition is *secular* and viscosity must be assumed sufficiently strong to maintain uniform rotation. A transition between the Maclaurin sequences of spheroids and toroids is *dynamical* since fluid elements conserve their specific angular momentum and, equivalently, their circulation. Such fluid elements are effectively "ideal" (i.e., noninteracting) elements. In contrast, fluid elements in an out-of-equilibrium, one-ring torus are interacting through viscous forces.

The shapes of the two toroidal bifurcating sequences in the $(L, \Omega)$ and the $(L, E)$ planes of the parameter space are fundamentally different. In the $(L, \Omega)$ plane, the Maclaurin sequence of toroids has a C-shape (here $\Omega$ denotes an averaged rotation frequency for the differentially rotating objects), while the one-ring sequence has an inverted S-shape. In the $(L, E)$ plane, both sequences bifurcate toward higher energies relative to the Maclaurin sequence of spheroids, but eventually turn around toward higher $L$-values and cross below the Maclaurin sequence to lower energies.

The differences in shape and in conservation laws are sufficient to allow for a *dynamical phase transition* at the spheroid–toroid bifurcation[31] and to delay a *secular phase transition* and place it somewhere beyond the bifurcation point of the one-ring sequence.[1]

From the dynamics standpoint, these two phase transitions appear as analogs of BEC and the $\lambda$-transition of liquid $^4$He. In particular, the specific heat is continuous at the BEC transition point because the Maclaurin sequence of toroids bifurcates exactly at that point. However, the transition does not proceed along the bifurcating sequence that initially unfolds toward higher energies. Instead, it proceeds at constant $L$ directly toward a toroidal equilibrium on the lower energy branch of the Maclaurin sequence of toroids. Axisymmetry is preserved during this transition but topology is broken since the resulting toroidal equilibrium is no longer simply connected. On the other hand, the specific heat diverges at the $\lambda$-point of the secular transition because the one-ring sequence does not bifurcate there. Again, the $\lambda$-transition proceeds at constant $L$ directly toward a uniformly rotating, toroidal equilibrium on the lower energy branch of the one-ring sequence, axisymmetry is preserved, and topology is broken with the opening of the central hole.

The dynamical BEC of the Maclaurin spheroids was captured by linear stability analysis,[31] but the secular $\lambda$-transition toward the one-ring sequence has not been previously detected by linear stability techniques. This is because the latter transition is "discontinuous" (i.e., it does not occur at a traditional bifurcation point). We also note that, in both cases, *first-order phase transition points* also appear at points of the Maclaurin sequence of spheroids with somewhat lower $L$-values relative to those of the previously discussed $\lambda$-points. These issues are discussed in much more detail in reference 1.

### The $\lambda$-Transition of Jacobi Ellipsoids

A discontinuous $\lambda$-transition between uniformly rotating sequences also appears on the Jacobi sequence of ellipsoids. This $\lambda$-transition leads to the lower energy branch of the equal-mass binary sequence. Just as in the case of Maclaurin spheroids, this secular transition can take place only if viscosity maintains uniform rotation; the $\lambda$-point occurs somewhere beyond the bifurcation point of the so-called *dumbbell-binary sequence*;[11] the instability is not captured by linear stability analysis; the specific heat diverges at the $\lambda$-point; symmetry is preserved; and topology is broken when a Jacobi ellipsoid becomes an equal-mass binary.

The $\lambda$-transition on the Jacobi sequence is important because it provides *the only possible mechanism* through which a viscous, incompressible, Jacobi ellipsoid can abandon its sequence and undergo fission during slow, quasi-static evolution. Specifically, a third-harmonic dynamical instability found by Cartan[4,5] on the Jacobi sequence before the $\lambda$-point is reached is irrelevant to the classic fission hypothesis[2-8] of binary star formation. This mode of instability was found for a perfect-fluid Jacobi ellipsoid, and it does not occur at all in the presence of viscosity of any nonzero magnitude (see reference 1 for details). Furthermore, the *pear-shaped sequence*[11] of uniformly rotating objects that bifurcates at this instability point is itself unstable; it stands entirely at higher energies relative to the Jacobi sequence and it is thus completely inaccessible from the Jacobi sequence through quasi-static or dynamical evolution. Therefore, the *classic fission hypothesis of binary star formation*[2] is viable but the evolution proceeds toward the binary branch of the dumbbell-binary sequence through the $\lambda$-point of the Jacobi sequence.

## TYPES OF HYDRODYNAMICAL $\lambda$-TRANSITIONS

Our study has identified altogether *four types* of $\lambda$-transitions along the Maclaurin sequence of spheroids:

*Type 1:* The $\lambda$-transition toward the Maclaurin sequence of toroids was discussed earlier. This transition corresponds to dynamical instability. Its thermodynamical analog is BEC. Spatial symmetry is preserved, topology is broken, and the "specific heat" is continuous at the $\lambda$-point.

*Type 2:* The $\lambda$-transition toward the one-ring sequence was discussed earlier. This transition corresponds to secular instability. Its thermodynamical analog is the $\lambda$-transition of liquid $^4$He. Spatial symmetry is preserved, topology is broken, and the "specific heat" diverges at the $\lambda$-point.

*Type 3:* This transition originates at the point of third-harmonic dynamical instability on the Maclaurin sequence ($e = 0.96696$) and leads to a binary system. It has been observed in the smoothed-particle hydrodynamical simulations of Miyama et al.[32] Its thermodynamical analog may possibly be the order–disorder transition in binary alloys.[21] Spatial symmetry is broken first, topology is broken second, and the "specific heat" diverges at the $\lambda$-point.

*Type 4:* This transition is related to the so-called *two-ring sequence* that bifurcates from the Maclaurin sequence at $e = 0.99375$. The $\lambda$-transition takes place at $e = 0.99802$ and corresponds to secular instability. Its thermodynamical analog may possibly be the Curie-point transition in ferromagnetism.[21] Spatial symmetry is preserved, topology is broken, and the "specific heat" diverges at the $\lambda$-point. In contrast to the above type-2 transition, no first-order phase transition exists prior to the $\lambda$-point.

## ADDITIONAL IMPLICATIONS OF THE RESULTS

The results described in the preceding sections have additional implications that are not only interesting in astrophysical applications but they may also be relevant to applications of catastrophe theory[22,23] in a variety of physical problems involving the elimination of different conservation laws by various physical processes; to investigations of the roles of symmetries and topological changes in instabilities/phase transitions/catastrophes; to the classification of thermodynamical phase transitions; and to the macroscopic description of superfluid liquid $^4$He at very low temperatures.[21]

## Relation to Catastrophe Theory

Thom's catastrophe theory[22,23] can also be used efficiently to classify phase transitions. Many of the transitions just discussed occur along different evolutionary paths in the *control planes* of various *elementary catastrophes*. For example, some paths in the control plane of the *cusp catastrophe* can be identified with a first-order or a second-order phase transition, while other paths can be identified with a type-1 or a type-4 $\lambda$-transition.[1] A description of secular and dynamical instabilities along such paths can also be incorporated easily since static catastrophe theory is usually combined with *conventions*[23] in order to describe time-dependent phenomena. The *Delay convention* is adopted to describe quasi-static evolution, while the *Maxwell convention* is adopted to describe dynamical evolution.

Interpreting phase transitions and instabilities in the framework of catastrophe theory[33] also lends physical significance to the *control parameters* of the theory. Such parameters are functions of the integrals of motion, some of which may be strictly conserved, while others may be allowed to vary during evolution (see references 1, 9, 33). Thus, the conserved and nonconserved integrals of motion determine the variations of the control parameters during evolution which, in turn, mark specific evolutionary paths in the control plane of a catastrophe and specify accurately which phase transition, if any, will actually take place under an adopted *convention*.

Type-2 $\lambda$-transitions are *not* elementary catastrophes. This is easily explained since the disappearance of the free-energy barrier at the $\lambda$-point is discontinuous and the actual transition is only allowed slightly beyond the $\lambda$-point. (The barrier is still in place precisely at the $\lambda$-point and disappears suddenly beyond that point.[1] Note that, after the sudden disappearance of the barrier, a type-2 $\lambda$-transition *appears as* a typical second-order phase transition.) This implies that a type-2 $\lambda$-transition does not take place on a *fold line* in the control plane of an elementary catastrophe, but at an infinitesimal distance away from that line. Thus, this type of discontinuous transition cannot be described by any elementary catastrophe because, in all elementary catastrophes, the transition points always appear and the barriers always disappear on fold lines.[23]

## Topology versus Symmetry Breaking

The preceding list of types of $\lambda$-transitions suggests clearly that neither the symmetry properties nor the behavior of the "specific heat" at each $\lambda$-point can be used to define the four types of $\lambda$-transitions as a class. (Symmetry may or may not break and the "specific heat" may or may not diverge at the $\lambda$-point.) The common property among all types is the *breaking of topology* (i.e., the transformation from a simply connected to a multiply connected system). In contrast, topology is not broken during the second-order phase transitions toward the Jacobi, Dedekind, and $x = +1$ Riemann sequences. This distinction between apparent second-order phase transitions (i.e., $\lambda$-transitions) and true second-order phase transitions appears to be fundamental. It appears to be related to the following difference: in a true second-order phase transition, the lower energy state *is created* beyond the transition point; in a $\lambda$-transition, the (preexisting) lower energy state *becomes accessible* beyond the

$\lambda$-point. (For example, Jacobi ellipsoidal equilibria do not exist prior to the bifurcation point of the Jacobi sequence, but one-ring, Maclaurin-toroid, and binary equilibria of lower energy do exist prior to the corresponding $\lambda$-points, although they are not naturally accessible from the "mother" sequences.)

An analogous distinction does not exist between $\lambda$-transitions and first-order phase transitions. In both cases, a preexisting lower energy state becomes accessible at the transition points. This leads us to speculate that topology is broken (not necessarily symmetry, as is usually stated) in first-order phase transitions, just as in $\lambda$-transitions.

## Classification of Thermodynamical Phase Transitions

The preceding macroscopic property (i.e., the breaking of toplogy) and classification scheme, as well as the insight gained from the study of hydrodynamical $\lambda$-transitions, may also be relevant to problems involving thermodynamical phase transitions. For example, they may help us understand why *renormalization group theory* classifies the $\lambda$-transition of liquid $^4$He alone in one *universality class* ($d = 3$, $n = 2$), but the ubiquitous gas-to-liquid first-order phase transition together with the $\lambda$-transitions in binary alloys and in ferromagnetism in another *universality class* ($d = 3$, $n = 1$).[34] (Here $d$ is the *dimensionality of space* and $n$ is the *phase-set number*.) If the superfluid transition is of type 2 and the other two $\lambda$-transitions are of type 3 and 4, respectively, then we can interpret these two universality classes of renormalization group theory as follows:

(a) The superfluid transition belongs to the ($d = 3$, $n = 2$) class all by itself because it is the only $\lambda$-transition in which the free-energy barrier disappears discontinuously at the transition point. In other words, this is the only transition that is not an elementary catastrophe.

(b) The other two $\lambda$-transitions belong to the ($d = 3$, $n = 1$) class along with the gas-to-liquid first-order phase transition because the free-energy barrier disappears smoothly at the transition point. All the transitions in this class are elementary catastophes.

If the earlier speculation is correct and the gas-to-liquid first-order phase transition does, in some suitable sense, exhibit a breaking of topology, then topology breaking is a common characteristic of all the transitions in the ($d = 3$, $n = 1$) class. We believe, however, that topology breaking is a common characteristic also of all the transitions in the ($d = 3$, $n = 2$) class. (More details about the transitions in this universality class can be found in reference 34.) Thus, topology breaking does not appear to be the property that defines the different universality classes.

We conclude therefore that renormalization group theory does not classify phase transitions primarily according to symmetry breaking, or topology breaking, or the behavior of the specific heat at the transition point. Instead, its universality classes are grouped primarily according to the way (continuous or discontinuous) the free-energy barrier disappears at the transition point. And this continuity/discontinuity property of the free-energy barrier is equivalent to the existence/absence of an evolutionary path describing the phase transition in the control plane of an elementary catastrophe.

## Superfluid Liquid $^4He$

If the preceding macroscopic descriptions are applicable to the $\lambda$-transition of liquid $^4$He, then we can deduce the following properties.[1] The superfluid transition of $^4$He is:

(a) Bose–Einstein condensation modified by an unusual intermolecular interaction that occurs only between the BEC point $T \approx 3.1$ K and the new $\lambda$-point $T \approx 2.2$ K and gives rise to "roton" excitations;[21,35]
(b) a type-2 $\lambda$-transition with infinite specific heat at the $\lambda$-point;
(c) discontinuous in the sense that the free-energy barrier disappears suddenly slightly beyond the $\lambda$-point;
(d) not an elementary catastrophe;
(e) characterized by a breaking of the (phase-space) topology and not of the symmetry.

## REFERENCES

1. CHRISTODOULOU, D. M., D. KAZANAS, I. SHLOSMAN & J. E. TOHLINE. 1995. Astrophys. J. **446**: 472–520. (Papers I–IV.)
2. POINCARÉ, H. 1885. Acta Math. **7**: 259–380.
3. JEANS, J. H. 1961. Astronomy and Cosmogony. Dover. New York.
4. CARTAN, H. 1928. Proc. 1924 Int. Math. Congress, Vol. 2. Toronto, pp. 2–17.
5. CHANDRASEKHAR, S. 1969. Ellipsoidal Figures of Equilibrium. Yale Univ. Press. New Haven, Conn.
6. LEBOVITZ, N. R. 1972. Astrophys. J. **175**: 171–183.
7. LEBOVITZ, N. R. 1977. In Applications of Bifurcation Theory, Vol. 1, P. H. Rabinowitz, Ed.: 259–284. Academic Press. New York.
8. LEBOVITZ, N. R. 1987. Geophys. Astrophys. Fluid Dyn. **38**: 15–24.
9. BERTIN, G. & L. A. RADICATI. 1976. Astrophys. J. **206**: 815–821.
10. ERIGUCHI, Y. & D. SUGIMOTO. 1981. Prog. Theor. Phys. **65**: 1870–1875.
11. ERIGUCHI, Y., I. HACHISU & D. SUGIMOTO. 1982. Prog. Theor. Phys. **67**: 1068–1075.
12. HACHISU, I. & Y. ERIGUCHI. 1983. Mon. Not. R. Astron. Soc. **204**: 583–589.
13. ERIGUCHI, Y. & I. HACHISU. 1985. Astron. Astrophys. **148**: 289–292.
14. LAI, D., F. A. RASIO & S. L. SHAPIRO. 1993. Astrophys. J. Suppl. **88**: 205–252.
15. LAI, D., F. A. RASIO & S. L. SHAPIRO. 1994. Astrophys. J. **423**: 344–370.
16. NEW, K. C. B., D. M. CHRISTODOULOU & J. E. TOHLINE. 1995. Astrophys. J. In preparation.
17. TASSOUL, J.-L. 1978. Theory of Rotating Stars. Princeton Univ. Press. Princeton, N.J.
18. DURISEN, R. H. & J. E. TOHLINE. 1985. In Protostars and Planets, Vol. II., D. C. Black and M. S. Matthews, Eds.: 534–575. Univ. of Arizona Press. Tucson.
19. BODENHEIMER, P. 1992. In Star Formation in Stellar Systems, Vol. 1, G. Tenorio-Tagle, M. Prieto, and F. Sánchez, Eds.: 1–65. Cambridge Univ. Press. Cambridge, England.
20. ATKINS, K. R. 1959. Liquid Helium. Cambridge Univ. Press. Cambridge, England.
21. HUANG, K. 1963. Statistical Mechanics. Wiley. New York.
22. THOM, R. 1975. Structural Stability and Morphogenesis. Benjamin. Reading, Mass.
23. GILMORE, R. 1981. Catastrophe Theory for Scientists and Engineers. Dover. New York.
24. LANDAU, L. D. & E. M. LIFSHITZ. 1986. Statistical Physics, Part 1. Pergamon Press. New York.
25. DRAZIN, P. G. & W. H. REID. 1981. Hydrodynamic Stability. Cambridge Univ. Press. Cambridge, England.
26. LAMB, H. 1932. Hydrodynamics. Dover. New York.
27. CHANDRASEKHAR, S. 1981. Hydrodynamic and Hydromagnetic Stability. Dover. New York.

28. BINNEY, J. & S. TREMAINE. 1987. Galactic Dynamics. Princeton Univ. Press. Princeton, N.J.
29. OSTRIKER, J. P. & P. J. E. PEEBLES. 1973. Astrophys. J. **186**: 467–480.
30. CHRISTODOULOU, D. M., I. SCHLOSMAN & J. E. TOHLINE. 1995. Astrophys. J. **443**: 551–569. (Papers I–II.)
31. BARDEEN, J. M. 1971. Astrophys. J. **167**: 425–446.
32. MIYAMA, S. M., C. HAYASHI & S. NARITA. 1984. Astrophys. J. **279**: 621–632.
33. POSTON, T. & I. N. STEWART. 1978. Catastrophe Theory and its Applications. Pitman. London.
34. BUCKINGHAM, M. J. 1988. *In* Near Zero, Vol. 1, J. D. Fairbank, B. S. Deaver, Jr., C. W. F. Everitt, and P. F. Michelson, Eds.: 152–189. Freeman. New York.
35. LANDAU, L. D. & E. M. LIFSHITZ. 1986. Statistical Physics, Part 2. Pergamon Press. New York.

# Some New Understandings on Nonlinear Stellar Pulsation

### YANQIN WU

*130–33, Astronomy Department*
*California Institute of Technology*
*Pasadena, California 91125*

Why are variable stars pulsating with finite amplitudes? What determines the magnitudes of these amplitudes, and why are they different from variable to variable? Why do some stars pulsate in only one or two modes (like RR Lyrae stars, some Cepheids), while others pulsate in many modes simultaneously (like the sun, pulsating white dwarfs)? Can this be fully explained by the different numbers of overstable modes in these different types of stars? Why do some of these stars pulsate smoothly with constant amplitudes, while some others vary more erratically?

These questions are among those that one hopes to answer when studying nonlinearity in stellar pulsations. Pulsating white dwarfs attracted our special attention, both because of the simplicity in their structure and the vast amount of observational data accumulated to date (especially during the campaigns of the Whole Earth Telescope, see, e.g., Winget et al.[1]). Therefore, we have started our study on nonlinear stellar pulsation by investigating the multimode behavior of pulsating white dwarfs. The following is a brief report on the current status.

A star pulsating in many modes can be classified as a weakly nonlinear, weakly dissipative dynamical system. Under the assumption of weak nonlinearity, the system can still be studied in the eigenspace spanned by its linear eigenmodes, which include both driven and damped ones.

In our investigation, we assume that nonlinearity originates mostly from low-order resonant and near-resonant mode–mode couplings; for multimode systems ($N \gg 3$), this takes the form of 3-mode coupling. The reason we favor the nonlinearity from mode coupling over the nonlinearity resulting from the nonlinear response of luminosity to temperature and density variations, is that the amplitude above which nonlinearity from mode coupling becomes significant decreases strongly with increasing number of modes (as will be demonstrated later in this report). This is in contrast to the nonlinear response of the stellar material, which is only a function of the total amplitude. In a multimode system, 3-mode coupling is favored over 4-mode coupling on the basis of its low order, even if there are more chances of exact resonance when four modes are involved. Furthermore, we argue that the nonadiabatic properties of the modes (both driven and damped ones) are important in understanding the nonlinear dynamics of the system. A nonzero driving (or damping) rate provides the mode with a finite linewidth in the system's normal mode spectrum, thus enhancing the probability of it interacting with other modes. Lastly, as will become clear later, even though damped modes usually have negligible energy, they are nevertheless affecting the dynamics prominently both through their number density in the frequency spectrum and through their linewidths induced by damping.

The formalism we adopt here for studying mode–mode coupling is the amplitude equation (AE). When only lowest order nonlinearity is considered (3-mode coupling), and when the coupling coefficient is taken to be constant for any interactions, the AE can be written as,[2]

$$\frac{dA_i}{dt} = (\alpha_i + i\omega_i)A_i + \frac{i3\kappa\omega_i}{\sqrt{8}} A^2, \quad (1)$$

where, $A_i$ is the complex amplitude of mode $i$'s relative density variation, including the fast oscillating part introduced by its frequency $\omega_i$; $\alpha_i$ is the mode's growth rate (positive for overstable modes); $\kappa$ the constant 3-mode coupling coefficient mentioned earlier; and $A$ is the summation over real parts of all the normal modes,

$$A \equiv \sum_j (A_j + A_j^*). \quad (2)$$

In real stars, $\kappa$ is not a constant, but rather a function of all quantum numbers of the modes involved in a particular triplet, since it reflects the overlap of their wave functions. We have derived it analytically for $g$-modes in a polytropic model using the variational principle (Wu, unpublished). Its magnitude depends sensitively on the similarity of the wave functions involved, which also implies that interaction is local. Therefore if we only use modes that are close to each other in the 3-dimensional space of quantum numbers $(n, l, m)$, a constant coupling coefficient will not only describe the dynamics accurately but also reduce the computation drastically. An additional advantage of using constant coupling coefficients is that, because they are independent on spatial eigenvectors (i.e., the horizontal and vertical wave vectors), the 3-dimensional star is automatically reduced to a one-dimensional system, and the usual 3-dimensional resonance condition for mode–mode coupling is reduced to the frequency resonance condition.

A transformation of $\tilde{A} = \kappa A$ will produce the following equivalent equation:

$$\frac{d\tilde{A}_i}{dt} = (\alpha_i + i\omega_i)\tilde{A}_i + \frac{i3\omega_i}{\sqrt{8}} \tilde{A}^2, \quad (3)$$

where $\tilde{A} \equiv \sum_j (\tilde{A}_j + \tilde{A}_j^*)$. Constant $\kappa$ disappears from this new equation. This implies that, in our case of constant $\kappa$, we can set $\kappa$ to an arbitrary number for the convenience of the calculation.

As a small detour, we consider the problem of assigning appropriate driving/damping rates to the modes. This is directly related to the question of the excitation mechanism. For white dwarfs, the conventional $\kappa$-mechanism (a driving mechanism that depends on the opacity increase upon compression in an ionizing region) probably does not work very effectively, since in these stars the ionization zones (of hydrogen in a ZZ Ceti star, helium in a DB star, and probably of carbon/oxygen in a DO star) are located inside the convection zone, where only a few percent of the total flux is carried in the form of radiation. To solve this problem, Pesnell[3] proposed the "convective blocking" mechanism, and Brickhill[4–6] further developed the theory. To improve our physical insight, we derived an analytical version of Brickhill's model, combining the effects of $\kappa$-mechanism and "convective blocking," and applied it successfully to the model atmosphere of ZZ Ceti stars (Goldreich, unpublished). The maximum period for overstable modes in this model is $\sim 1000$ s, which is consistent with the observations. The driving rates in the model can be written down in the following order-of-magnitude manner,

$$\alpha_i \sim \frac{1}{n_i \tau_i}, \tag{4}$$

where $n_i$ is the number of radial nodes for mode $i$, and $\tau_i$ the thermal time scale at the first node point of the horizontal wave function. The numerical calculation we carried out yields agreeing results.

Once the normal-mode spectrum of a system and the nonadiabatic properties of its modes are known (precise knowledge of the frequency spectrum is especially important), the pulsation amplitudes and nonlinear behavior could in principle be calculated from the AEs. Also, a rough estimate of the total energy in the system can be given as a function of these parameters. If we assume that there is no phase correlation among modes (i.e., complete phase randomization), which is a good approximation when the number of modes is large, we can average over phase in the AEs and write down a simplified equation for energy. It is logical to assign a negligible energy to damped modes, and similar energy ($E$) to every driven mode. This leads to the following energy equation for any driven mode:

$$\frac{dE}{dt} \simeq 2\alpha_e E - \sum \beta E^2, \tag{5}$$

where $\alpha_e$ is the driving rate. The summation is carried out over all triplets that consist of two driven modes and one damped mode, independent of their proximity to resonance. These triplets are the ones that work most efficiently to transport energy from the source to the sink, thus constricting the overstable modes to finite amplitudes. The coupling strength $\beta$ for such a triplet is given by

$$\beta \simeq 9\kappa^2 \omega^2 \left[ \frac{\alpha}{\alpha^2 + (\Delta\omega)^2} \right], \tag{6}$$

where $\alpha = |\alpha_1| + |\alpha_2| + |\alpha_3|$, with modes 1 and 2 driven, and mode 3 damped. In general, damping is much stronger in magnitude than driving, that is, $\alpha \simeq |\alpha_3| \simeq \alpha_d$. $\Delta\omega$ is the amount of detuning from frequency resonance for the triplet. The part in parenthesis in (6) is a Lorentzian profile, which accounts for this detuning. When $|\Delta\omega| = 0$, the coupling is called *exact resonance;* when $|\Delta\omega| \leq \alpha$, it is *near-resoannce;* otherwise, it is *nonresonance*.

In an equilibrium state, the energy of every driven mode is approximately given by

$$E \simeq \frac{2\alpha_e}{\sum \beta}, \tag{7}$$

and the total energy in the system by

$$E_{\text{tot}} \simeq N_e E \simeq N_e \frac{2\alpha_e}{\sum \beta}. \tag{8}$$

Here, $N_e$ is the number of excited modes in the system.

Two different situations exist. First, when none of the combinations is close to resonance, that is, when all $|\Delta\omega| \gg \alpha$, the dynamics is dominated by the one that comes closest to resonance. For this case,

$$E_{\text{tot}} \simeq \frac{2}{9} \frac{N_e (\Delta\omega)^2_{\min} \alpha_e}{\kappa^2 \omega^2 \alpha_d}. \tag{9}$$

When chances for resonance and near-resonance are very good, that is, many $|\Delta\omega|_{\min} \ll \alpha$, the summation is necessary and can be substituted by multiplying $\beta$ with an estimate of the number of such (near-)resonances.

A simple one-dimensional system is designed to test these conclusions, as well as to familiarize us with the kinds of nonlinear behavior that are present in a real star. The system consists of $N_e$ driven modes, and $N_d$ damped ones, whose frequencies are chosen randomly over a range of $\Delta\omega_e$ and $\Delta\omega_d$, respectively, where $\Delta\omega_e$ falls within $\Delta\omega_d$. The driving rates for $N_e$ driven modes distribute randomly between zero and some positive number, with central value $\alpha_e$; similarly for the $N_d$ damped modes, except with a negative $\alpha_d$. Under these conditions,

$$|\Delta\omega|_{\min} \sim \frac{\Delta\omega_e}{N_e N_d}. \tag{10}$$

We define

$$C \equiv \frac{|\Delta\omega|}{\alpha}, \quad C_{\min} \equiv \frac{|\Delta\omega|_{\min}}{\alpha}, \tag{11}$$

$C_{\min} \gg 1$ represents the first situation discussed earlier, we have

$$E_{\text{tot}} \sim \frac{2}{9} \frac{N_e \alpha_e \alpha_d}{\kappa^2 \omega^2} C^2_{\min}, \tag{12}$$

while when $C_{\min} \ll 1$, the number of near-resonances (with $C \leq 1$) is large and can be approximated by $1/C_{\min}$, which leads to

$$E_{\text{tot}} \sim \frac{2}{9} \frac{N_e \alpha_e \alpha_d}{\kappa^2 \omega^2} C_{\min}. \tag{13}$$

It is interesting to notice the different dependences for these two results on $N_e$ and $N_d$. With the help of (10), the energy (12) scales as

$$E_{\text{tot}} \propto \frac{1}{N_e N_d^2}, \tag{14}$$

while in (13),

$$E_{\text{tot}} \propto \frac{1}{N_d}. \tag{15}$$

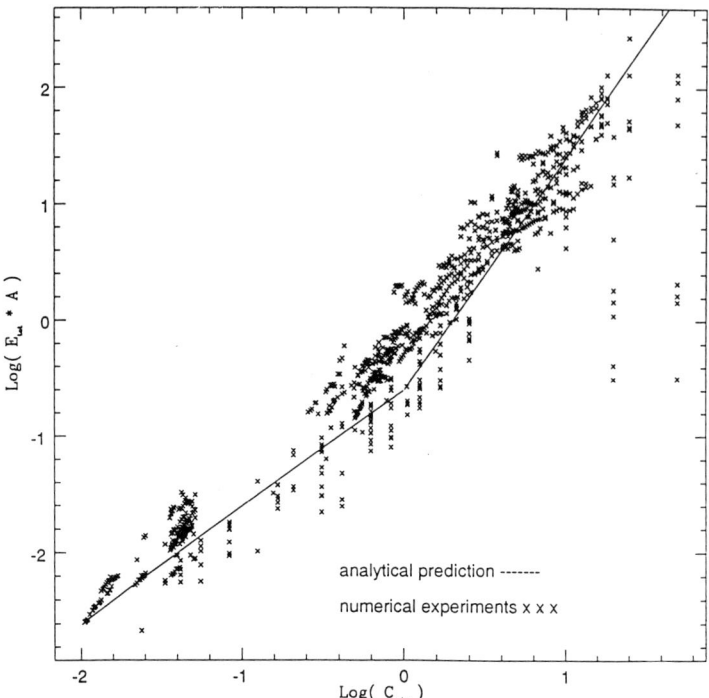

**FIGURE 1.** Comparison between theoretical prediction and results from many numerical simulations. The *solid line* reflects (**12**) and (**13**), with a somewhat arbitrary break at the point of $C_{\min} = 1$. Here, $(1/A) = \frac{2}{9}(N_e \alpha_e \alpha_d)/(\kappa^2 \omega^2)$. The large scatter at $C_{\min} \gg 1$ is associated with the small values of $N_e$ there (mostly with $N_e = 1$); since the analytical results only describe the simulations statistically, a bigger $N_e$ (or $N_d$) provides a better fit, as is shown in the rest of the plot.

The independence of $E_{\text{tot}}$ on $N_e$ after $C_{\min}$ passes unity from above is due to saturation of the number of possible combinations between any two driven modes; before this happens, the bigger the number of driven modes, the smaller the total energy.

Our numerical experiments yield results that agree satisfactorily with the estimates just given, as is shown in FIGURE 1.

In summary, we have investigated the properties of a weakly nonlinear, weakly dissipative dynamical system that resembles a multimode pulsating star. Three-mode coupling is shown to be very effective in transporting energy from driven modes to damped modes, thus limiting the pulsation amplitudes. Analytical predictions on mode amplitudes is confirmed by numerical experiments.

## ACKNOWLEDGMENTS

This work has been carried out under the advisorship of Prof. P. Goldreich at Caltech.

## REFERENCES

1. WINGET, D. E., *et al.* 1991. Astrophys. J. **378**: 326.
2. KUMAR, P. & P. GOLDREICH. 1989. Astrophys. J. **342**: 558.
3. PESNELL, W. D. 1987. Astrophys. J. **314**: 598.
4. BRICKHILL, A. J. 1991. Mon. Not. R. Astron. Soc. **251**: 673.
5. BRICKHILL, A. J. 1991. Mon. Not. R. Astron. Soc. **252**: 334.
6. BRICKHILL, A. J. 1992. Mon. Not. R. Astron. Soc. **259**: 529.

# Tidal Perturbations, Gravitational Amplification, and Galaxy Spiral Arms[a]

GENE BYRD

Department of Physics and Astronomy
University of Alabama at Tuscaloosa
Tuscaloosa, Alabama 35487-0324

## 1. INTRODUCTION

Via two-dimensional (2D) simulations of a galaxy disk including self-gravity and stellar velocity dispersion, Toomre[1] showed that a symmetric tidal perturbation can be gravitationally amplified into strong spiral arms. Via simulations and examination of observational surveys, Byrd and Howard[2] found that tidal perturbations sufficient to create grand design spiral patterns in galaxies are extremely frequent. This paper extends Toomre's work by conducting three-dimensional (3D) simulations that include nonsymmetric perturbations by a variety of disturber masses, encounter orbits, and amounts of halo relative to the disk.

The simulations are then discussed analytically by extending Toomre's use of the Goldreich/Lynden-Bell (GLB) equation for tidal arms evolving toward the density wave stage. The velocity dispersion normal to the disk results in a softening of gravitation via the disk thickness. In the present paper, use of softening to specify velocity dispersion permits analytic discussion of gravitational amplification in the GLB equation. A nonaxisymmetric stability parameter, $P$, is thus obtained that is a function of local disk properties but is independent of arm multiplicity. The pitch of the maximum amplification arms is found to be a function of the relative importance of the halo to disk and $P$. The importance of bin softening's role in reducing the validity of simulations of galaxies with tightly wound, gravitationally amplified arms is evaluated. The just-mentioned analytic formulas are used to better understand the ages and morphologies of arm patterns in M51 as deduced via a detailed computer simulation.

## 2. SPIRAL ARMS IN SIMULATIONS AND GALAXIES

### 2.1. Disk Model and N-body Code

The disk surface density in our analytic discussion is

$$\mu = \left( \frac{V^2 \cos^{-1}(r/A)}{\pi^2 G(f+1)r} \right)_M \approx \frac{V^2}{2\pi G(f+1)r} \tag{1}$$

[a]This work was supported by National Science Foundation Grants NSF EPSCoR RII8996152 and NSF AST 9014137.

for a finite Mestel disk (Mestel;[3] Byrd et al.[4]). Here and henceforth, the $M$ subscript is used to indicate that the Mestel disk expression applies. The disk radius, $A$, is about three times the observed exponential disk scale length for the disk. Beyond this radius, the surface density is zero. The circular orbital velocity, $V$, is constant with radius. The right-hand expression is exact for an infinite Mestel disk out to the truncation radius, $A$. Over most of the disk, these also approximate a truncated exponential disk used in the simulations. Here $f$ is the halo-to-disk ratio (specifically the contribution of the halo to the gravitational acceleration relative to the undisturbed disk contribution). This represents the inert halo Ostriker and Peebles[5] found necessary to stabilize disk galaxies. The axisymmetric stablity parmeter, $Q$, is assumed to be one (Binney and Tremaine;[6] Toomre[7,8]).

$$\sigma_r = \sigma_{r,\,min} Q = \frac{3.36 G\mu}{\kappa} Q = \left(\frac{3.36 V}{\sqrt{2}(2\pi)(f+1)} Q\right)_M. \quad (2)$$

$N$-body simulations using 2D logarithmic polar grids were introduced by Miller[9] to study stability of small central regions of isolated disk galaxies and subsequently elaborated by Byrd et al.[4] to include simulation of a passing inert companion galaxy's tidal effects on the entire self-gravitating disk of an $S$-type galaxy. The program uses the mass of individual particles and their number in each bin of the grid to calculate the smoothed gravitational field of the entire disk on any point in

TABLE 1. Parameters of Computer Simulations for M51 System

| Units | | | | | |
|---|---|---|---|---|---|
| Length | Sky Plane Angle | Time | Mass | | |
| 20 kpc | 400" | $80 \cdot 10^6$ yr | $2 \cdot 10^{11} M_\odot$ | | |
| System Member Parameters | | | | | |
| | Total Mass | Halo/Disk | Radius | Disk Particles | Gas/Stars |
| M51 | 1.0 | 2/1 | 1.0 | 60,000 | 1/5 |
| Companion | 0.4 | 1/0 | 0.2 | 5,000 | 0/1 |
| Observational Orbit Constraints of Companion Relative to M51 | | | | | |
| Right Ascension | | Declination | | Radial Velocity | |
| +70.3" | | +255" | | 100 km/s | |
| M51 Disk Parameters | | | | | |
| Tilt | | Node Position Angle | | Rotation Speed | |
| 20° | | 170° | | 220 km/s | |
| Simulation Orbit of Companion Relative to M51. | | | | | |
| Disk crossing: Time: | | Recent $56 \cdot 10^6$ yr ago | | Previous $440 \cdot 10^6$ yr ago | |
| Distance: | | 20.8 kpc | | 22.3 kpc | |
| Position angle: from node in disk | | +15° | | -15° | |

Tilt = 73° w.r.t. disk. Apofocus = 28 kpc. Perifocus = 17 kpc. Currently 12.5 kpc beyond disk.

the grid. There is high spatial resolution in the inner disk regions that is crucial for simulation of centrally condensed disks, yet the total number of mesh points remains reasonable. The bin size in such a "spider web" grid is proportional to the distance from the bin from the grid center with the angular size of the bin and the radial spacing adjusted so that the bins are approximately square. Salo[10] introduced a new version with an interacting galaxy pair represented by two moving, mutually overlapping 2D polar grids, attached to the centers of inert "dark matter" halos. Salo also extended the grid size.

The code has a 3D polar logarithmic grid. Salo provided this code to Klaric,[11] who uses it in his dissertation and gives a complete description of it. Salo and Laurikainen[12] also use it for the Arp 86 system and give a description of it along with the arm-fitting procedure below. The bins near the disk plane are approximately cubical. The disturbance within the companion can also be studied by including its own tilted self-gravitating disk and 3D set of bins. However, for this study, the companion is assumed to be an inert, truncated isothermal sphere. The bins are 10° in angular size. Within practical limits, the gross structure is not sensitive to angular size of bins or number of particles. The disk particles are designated either "gas clouds" or "stars" with appropriate velocity dispersions (larger for the stars matching Toomre's stability criterion). A simulation typically includes 50,000 stars and 10,000 gas clouds. Details of tidal spiral structure are not sensitive to the amount of disk gas present. TABLE 1 gives parameters for a typical simulation of M51 in this case.

### 2.2. Simulations and Encounter Orbits

FIGURE 1(a), 1(b), and 1(c) give an example of one of our simulations. Here, the gas cloud distribution is shown that outlines the arms more sharply. These images can be taken to correspond to HI images or blue sensitive photographs. The encounter shown in FIGURE 1(a), 1(b), and 1(c) corresponds to an $f$ of 9, a parabolic approach distance of 1.5 disk radii from a truncated exponential disk, a 0.15 galaxy mass perturber, and an inclination of 0° relative to the disk plane. The long-term arm shape in the central portions of the disk appears to be insensitive to details of the encounter orbit. Several otherwise unchanged encounters were tried with only the close approach distance changed to 1.7 disk radii, or perturber mass changed to 0.075 or inclination to 45°. We also studied orbits that grazed the disk edge (e.g., our simulation of M51 discussed at the end of this paper). As long as the encounter is not extremely violent, all these single variations with other parameters unchanged give the same long-term arm shape. Only tidal features near the disk edge are sensitive to orbital details of the encounter.

### 2.3. Spiral Arm Fitting Process

A logarithmic spiral is fitted to the arms' pitch at the radius at which the arm amplitude is greatest. The arms are of the form

$$\phi = \phi_0 - \frac{p}{m} \ln(r), \qquad (3)$$

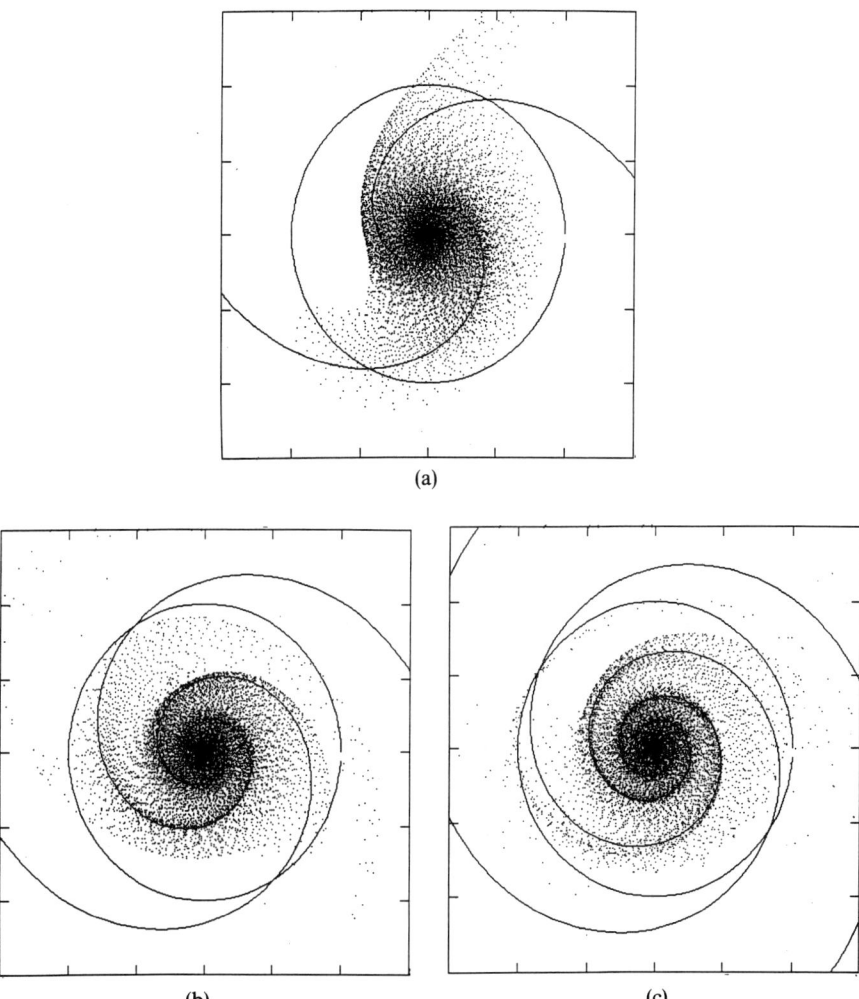

**FIGURE 1.** (a) Disturbed galaxy with $f$ of 9. Parabolic 0.15 galaxy mass disturber approached to 1.5 disk radii in the disk plane. The time is two crossing times after close approach. A crossing time is the disk radius divided by circular orbital speed. The initial disk radius is shown as a *circle*. The disk is composed of "Population II" making up 0.8 of the disk surface mass density and a "Population I" component making up 0.2 of the surface density. Both populations are assigned velocity dispersions, so $Q$ equals 1.0 for each Population I points are plotted. (b) Part (a) continued. Simulation at five crossing times is shown. (c) Parts (a) and (b) continued. Simulation at seven crossing times is shown.

where the orbital motion is in the direction of increasing disk plane position angle $\phi$ and $m$ is the arm multiplicity. To be consistent with the common observational notation, the pitch angle $i$ is used which is the acute angle of the spiral arm with a circle at the radius and angle location, $r$, $\phi$. The arms trail if the arm pitch, $p/m = \cot i = -r\, d\phi/dr$ is a positive constant with respect to radius.

At different times during the encounters, the twofold azimuthal density variation amplitude was calculated between 0.1 and 1.0 times the disk radius. The radius at which the maximum amplitude occurs was located. The simulation also plotted the position angle of the density crests versus radius to estimate the arm pitch. One crossing time before close approach ($t = 0$), these initial inner arm amplitudes are less than a few percent but have a well-defined maximum. As time passes, the maximum strength peak gets stronger and moves outward. Equation (3) arms were fitted to the simulation with $\phi_0$ and $p$ set at the values for which the arms were strongest. FIGURE 1(a), 1(b), and 1(c) show plots of this formula (curved lines) fitted to the arm patterns in our simulation at five, seven, and and ten crossing times. FIGURE 2 shows a log $r$ versus $\phi$ plot at the seven crossing times in the same simulation. In this display, (3) is a sloping straight line. These arms have reached the log form in the inner portions. As is indicated in Toomre's figures 1 and 13 simulations,[1] the particles along the line toward and opposite the perturber initially follow the arms in a "material arm" behavior. However, the inner disk arms soon begin to behave like a density wave, with particles leaving the arms and catching up to the opposite arms, and so on. FIGURE 2 shows how most of the disk has (3) density wave arms after seven crossing times.

### 2.4. Observed Spiral Arms

Patterns correspondng to (3) were fitted to fit the arms in SA galaxies as long ago as 1942 by Danver.[13] More sophisticated versions of this fitting process have since been carried out with the arm pattern being considered as the superposition of individual logarithmic spiral components in simulations by Sellwood and

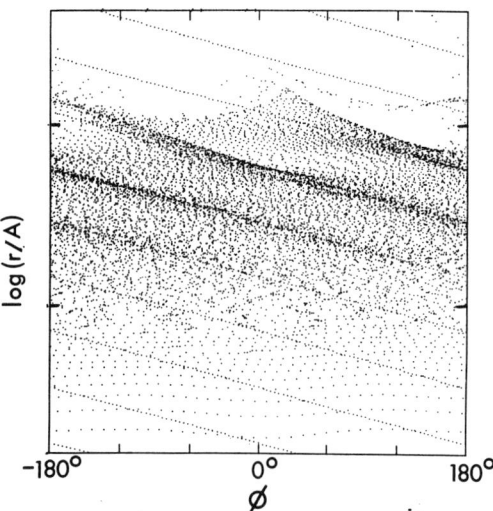

**FIGURE 2.** Seven crossing time frame of FIGURE 1(c), log $r$ versus $\phi$ display. *Sloping straight dotted lines* are fits of a logarithmic spiral. The two *tic marks* on the $y$ axes correspond to 0.1 and 1.0 of the original disk radius. *Horizontal axis* is position angle in disk plane.

Athanssoula[14] and for 16 actual galaxies by Considere and Athanassoula,[15] revealing that the two-armed component is everywhere dominant. Elmegreen *et al.*[16] fit spirals like (3) to photographs of arm patterns and obtain pitch angles $i$ of 16°, 18°, and 15° for the central disks of M81, M100, and M51, respectively.

## 3. OSCILLATION AND THE SPRING PARAMETER

To describe these simulation results analytically, consider a perturber outside the disk or perhaps a tidal disturbance in the outer disk itself that give small velocity perturbations to particles near the central region. Recall that Toomre's[1] simulation figures 1 and 13 show that particles along a line from the perturber through the disk center are initially in the arm as it winds up in "material arm" fashion. Relative to the disk center, particles on opposite sides of this line receive small velocity perturbations toward it. We will show how this purturbation will be gravitationally amplified as it is wound up by differential orbital motion in the disk.

Using notation and a discussion similar to Toomre,[1] we define local Cartesian coordinates with the usual orientation ($x$ positive outward from the disk center and $y$ positive in the direction of disk orbital motion). Define

$$\xi = x \sin \gamma + y \cos \gamma, \tag{4}$$

which is the particle's displacement from its initial position at the local coordinate origin due to the velocity impulse. This displacement is initially zero and is positive towards the arm, normal to the arm.

Here $\gamma$ is the angle of the velocity perturbation relative to the positive $y$. Note that $\gamma$ increases positively in the clockwise sense for our example particle. The angle $\gamma$ is also a measure of the arm's turning as it is sheared by the differential orbital motion of the disk. To match what is seen in the simulations and the previous discussion of the perturbation of the inner portions of the disk, take the arm to be initially radial with $\gamma = 0°$. For a particle perturbed in the direction of disk spin toward the radial line, the particle is initially at $\gamma = 0°$ and is at the initial $y_i$ and $x_i$ of its epicyclic oscillation. Note that $x_i = 0$ and $y_i = 0$ so that $\xi_i = 0$. As the particle moves toward the arm, its counterparts on the opposite side do the same and an excess surface density is created as $\xi$ and $\gamma$ both increase from zero.

In Toomre's terminology, the Goldreich/Lynden-Bell[17] (GLB) oscillation equation of a perturbed particle in the disk is

$$d^2(\xi)/dt^2 = -S\xi, \tag{5}$$

where the "spring parameter" and epicyclic frequency are, respectively,

$$S = \kappa^2 - 2\pi G \mu k \mathscr{F} + 4\Omega r \frac{d\Omega}{dr} \cos^2\gamma + 3r^2 \left(\frac{d\Omega}{dr}\right)^2 \cos^4\gamma \tag{6}$$

$$\kappa^2 = 4\Omega^2 + 2\Omega r \frac{d\Omega}{dr} = \left(\frac{2V^2}{r^2}\right)_M. \tag{7}$$

Here $r$ is the radial distance from the disk center and $\Omega$ is the local orbital angular rate. The second term in (6) is due to self-gravity and velocity dispersion and is discussed in more detail later.

## 4. NONGRAVITATING SPRING PARAMETER AND AMPLIFICATION

The enemy of an arm is the shearing of the disk that will tend to stretch the arm to nothingness. As Toomre[1] discusses, for oscillating particles, the well-known epicyclic oscillation of disturbed particles can conspire to moderate this destructive action by keeping them in the arm longer than would be expected. Self-gravitation of the disk can also moderate the shearing effects. This section concentrates on the effects of the epicyclic oscillation, saving self-gravitation for later.

For particles in the disk edge toward a perturber passing close by in an eccentric or in a tilted orbit, the perturbation is primarily one-sided. For these, the tidal force is of fairly short duration to the disk edge orbital period. Near the disk center or a perturber passing outside the disk, the tidal velocity perturbation is primarily twofold relative to the disk center. Perturbations that are quickly amplified are those toward a perturber–disk line extending outward in opposite directions from the center. Gerber and Lamb[18] have demonstrated the usefulness of direct calculation of epicyclic oscillation to simulate arms in outer disk particles without self-gravitation.

Rather than using brute force numerical calculation of epicyclic oscillation, analytic study of the size of the GLB spring parameter reflects the moderating influence of epicyclic oscillation and self-gravity on the destructive effects of shear. The smaller the spring parameter the more shear is moderated. We will first examine the spring constant without self-gravity as the arm shears or turns from an initially radial state. The GLB spring parameter without self-gravity and velocity dispersion is

$$S(\gamma) = \left(\frac{2V^2}{r^2} - \frac{4V^2 \cos^2 \gamma}{r^2} + \frac{3V^2 \cos^4 \gamma}{r^2}\right)_M.$$

The initial spring parameter of our $\gamma = 0°$ radial arm is $S = (V^2/r^2)_M$. As the arm turns, the spring parameter changes as geometry conspires to amplify density enhancements against the shearing effects of differential disk rotation. We evaluate the value of $\gamma$ for which $S$ is a minimum by taking the partial derivative of the preceding equation with respect to $\gamma$ and setting the result equal to zero. We find a minimum of $S(\gamma) = (0.67V^2/r^2)_M$ for $\gamma = -35°, +35°$. In our case initially $\gamma \approx 0°$ for a roughly radial arm that winds up through $\gamma = 35°$. As the arm winds past this minimum, $S$ rises toward $S(\gamma) = (2V^2/r^2)_M$ and the arm would greatly weaken.

The $\gamma = 0°$ arms have an initially small spring parameter compared to a more tightly wound initial disturbance. The time interval for these arms to reach a minimum of the spring parameter will be about one-tenth (35°/360°) of an orbital period. As we will see later, these reductions of the parameter are all purely geometric. As will be shown, the self-gravity term in the parameter can serve to keep the parameter small far past 35°. A small perturbation of the inner disk can be greatly amplified as the arm turns. After the initial perturbation, we will be considering cases where the oscillatory motion of the particle relative to the arm is dominated by amplification as reflected in the value of $S$.

## 5. LONG-TERM ARMS, SELF-GRAVITY, AND VELOCITY DISPERSION

### 5.1. The GLB Spring Parameter for Tightly Wound Arms

We now interpret the simulation results using self-gravity and velocity dispersion in (6) for the spring parameter. Defining the variables in the second term, the arm wave number

$$k = \frac{2\pi}{\lambda_n} = \frac{2\pi}{(\lambda_y \cos \gamma)} = \frac{2\pi}{(2\pi r/m) \cos \gamma} = \frac{m}{r \cos \gamma}, \tag{8a}$$

where $\lambda_n$ is the wavelength normal to the arm, $\lambda_y$ is the transverse wavelength, and $m$ is the multiplicity of the arm pattern of interest (usually 2). In our analytical discussion up to now, we have used $\gamma$ as a measure of the particle's oscillation and the arm's turning. To conform better with quantities used by observers, we now change to the variable $i = 90° - \gamma$ in the GLB equation. We can also use the approximation $\cot i = 1/\cos \gamma$. This is reasonable for the median observed pitch angle of spiral galaxies, 13° according to Binney and Tremaine[6] and Kennicut.[19] Equation (8a) becomes

$$k = 2\pi/r = m \cot i/r. \tag{8b}$$

The dynamics of the disk are dominated by a stellar component with a velocity dispersion $\sigma$ that can reduce the effect of self-gravity. In (6), the "reduction factor," $\mathscr{F}$, specifies how much the self-gravitational response to a perturbation is reduced below the $\mathscr{F} = 1$ of a disk with no velocity dispersion. Toomre[1] uses the complicated "hot stars" reduction factor derived by Lin and Shu[20] and by Kalnajs[21] as the most realistic to numerically study changes in $S$ as a function of the arm's turning, disk self-gravitation, and velocity dispersion. Compared to the complicated hot stars reduction factor, Binney and Tremaine[6] state that the softened gravity formulation developed by Miller[22,23] "provides close analogs to stellar disks . . . that . . . are much easier to investigate numerically." Athanassoula[24] has a very complete discussion of this question. As we shall see, softened gravity is also very convenient analytically. We will therefore deviate from Toomre's[1] theory discussion by using the softened gravity reduction factor

$$\mathscr{F} = e^{-ka}. \tag{9}$$

The softening parameter, $a$, can be taken to be equal to the scale height normal to the disk plane, $a_z$. Gilmore et al.[25] review how the observed vertical stellar number density distribution away from the middle of our galaxy's plane can be fitted by an exponential of scale height, $a_z$. The disk scale height is related to the disk surface density, $\mu$, and vertical velocity dispersion, $\sigma_z$, as follows

$$\sigma_z = \sqrt{(1)\pi G \mu a_z} \tag{10}$$

for the vertical velocity dispersion in the disk plane. The value of the parenthesis, 1, in this equation, is one extreme, with 2 being the other extreme, assuming the disk to be isothermal in the $z$ direction. The exact parenthesis value does not strongly affect $\sigma_z$ because of the square root.

When the arms are tightly wound, $i$ is small, $\gamma \approx 90°$, and the last two terms in the spring parameter equation, (6), become small relative to the first two terms and can be omitted. Using (8b) and (9), (6) becomes

$$S = \kappa^2 - \frac{2\pi G\mu \cot(i) m}{r} e^{-m \cot(i) a/r}. \quad (11)$$

### 5.2. Gravitational Amplification, Stability, and Arm Pitch

We will now see if there is a reason the spirals have the arm pattern given by (3), and the fitting procedure works so well in our simulations. After a tidal perturbation, we have an arm that winds up with time, that is, $\cot i$ increases. As we have discussed, this winding is faster at smaller $r$. As $\cot i$ increases, $S$ decreases as we have more gravitational amplification via the second term in (11). However, the negative $\cot i$ velocity dispersion exponent will ultimately limit the increase in amplification. If the spring parameter dips below zero as the arm winds up, then there is no restoring acceleration to prevent the particles from rapidly being gravitationally accelerated toward the arm center, that is, the disk is locally unstable. The velocity dispersion, $\sigma_z$ can rescue the disk from instability. The minimum value to ensure stability can be obtained from (11) by requiring $S$ to be 0, its first derivative with respect to $\cot i$ to also be zero, and its second derivative with respect to $\cot i$ to be greater than zero. Solving for the softening, $a$, or equivalently the velocity dispersion, $\sigma_z$, we obtain

$$a = a_{\min} P^2 = \frac{2\pi G\mu}{\kappa^2 e} P^2 = \left(\frac{P^2 r}{2e(f+1)}\right)_M \quad (12a)$$

$$\sigma_z = \sigma_{z,\min} P = \sqrt{\frac{2}{e}} \frac{\pi G\mu}{\kappa} P = \left(\frac{V}{2(f+1)\sqrt{e}} P\right)_M \quad (12b)$$

using (1) and (10).

In (12a) and (12b), we define a *nonaxisymmetric* stability parameter, $P$, in analogy with the axisymmetric stability parameter, $Q$, in (2). If the nonaxisymmetric stability parameter, $P \geq 1$ in (12a) and (12b), the disk is locally stable. Equation (12a) is the same as that obtained using exponential softening to study *axisymmetric* stability (e.g., as done in Binney and Tremaine[6]). Thus, $a = a_z = a_r$ and our nonaxisymmetric stability parameter, $P = Q$, the axisymmetric parameter. Using (2) and (12b), for stability, $\sigma_{z,\min} = (0.80)\sigma_{r,\min}$. The constant in the parenthesis is 1.2 if the disk is isothermal in the $z$ direction.

We can also obtain the arm pitch, $\cot i$ where amplification is greatest, that is, where $S$ is a minimum,

$$\cot i_{\max} = \frac{r}{ma} = \frac{r}{ma_{\min} P^2} = \frac{\kappa^2 er}{2\pi G\mu m P^2} = \left(\frac{2e(f+1)}{mP^2}\right)_M = \frac{p}{m}. \quad (13)$$

Note that the maximum amplification pitch is the same for all radii in a Mestel disk. A constant pitch corresponds to a log arm pattern as in (3). If a disk does not have a sufficient velocity dispersion, that is, $P$ in (12) is less than 1, then violent relaxation should occur when excited arms wind to an arm pitch where $S < 0$. Once the dispersion reaches the (12b) value with $P = 1$, the velocity dispersion heating should stop. If, via other heating mechanisms, $P$ in (12) is greater than 1, then tidally excited arms should wind to the (13) pitch for $P > 1$, at which point gravitational amplification will decrease.

## 6. LONG-TERM ARMS IN SIMULATIONS AND THEORY

### 6.1. Material Arms and Density Wave Domains

The inclination decreases with time in our simulations, as can be seen in FIGURE 1(a), 1(b), and 1(c). The arms that appear on the disk edge shear like material arms and appear to be discontinuous relative to the inner arms. As can be seen in FIGURE ((a), 1(b), and 1(c), log arms (straight lines on the FIGURE 2 plot) appear first in the inner disk, then extend outward to encompass more of the disk. In our simplified theoretical discussion, the initial behavior at any radius is like a material arm. We took the initial arm pitch $i$ to be around $90°$ in our theoretical discussions. This is an initially invisible radial arm toward which the particles are perturbed in velocity. Each of these arms shears like a material arm as the particles move toward it and the arm becomes a density enhancement. The shearing and gravity will amplify the inner arm since the orbital periods are shorter for that portion. We initially expect a material arm pitch variation with radius with an approach to a strong log arm as observed in our simulations. Starting at the time of close approach, 0 crossing times, we have material arm pitch as a function of time

$$\cot i = r \, d\phi/dr = \frac{r(d\Omega/dr) \, dr}{dr} t = \left(\frac{Vt}{r}\right)_M. \quad (14)$$

In our simulations, the outer arm pitch of (14) approaches the (13) pitch of maximum amplification. We combine these two to obtain a relation between time, $t$, and the radius, $r$, between log arms and the material arm domain,

$$t \approx \frac{\cot i_{max}}{r(d\Omega/dr)} = \left(\frac{r \cot i_{max}}{V}\right)_M = \left(\frac{e(f+1)r}{VP^2}\right)_M. \quad (15)$$

This radius will be larger the longer the time since perturbation. Exterior to this radius, the arms will behave more like material arms. Interior to this radius, the arms are more like classic density waves discussed by Lin and Shu,[20] Lin et al.,[26] and Lin and Bertin's paper in the present proceedings.[27]

### 6.2. Lifetime of the Tidal Spiral Arms

As we shall see, our simulations contain unavoidable defects. However, they indicate a rather long lifetime for a tidal grand design pattern. A simulation is carried out like FIGURE 1's, except that $f = 3$ and time is carried all the way through

20 crossing times. The last frames of the simulation are shown in FIGURE 3(a), 3(b), 3(c) and FIGURE 4(a), 4(b), 4(c) at 10, 15, and 20 crossing times, respectively. FIGURE 3(a), 3(b), and 3(c) show the disk as it would be seen face-on. In FIGURE 4(a), 4(b), and 4(c) we again use a log $r$ versus $\phi$ plot. By step 10, the strongest parts of the arms are of shallow pitch and have moved toward the outer parts of the disk. At time 15 in our long-term FIGURES 3 and 4 simulation, an interesting thing has happened. The strongest parts of the arms have reached the outer parts of the disk. The gas cloud arms are decidedly lumpy and adjacent portions of the different arms are

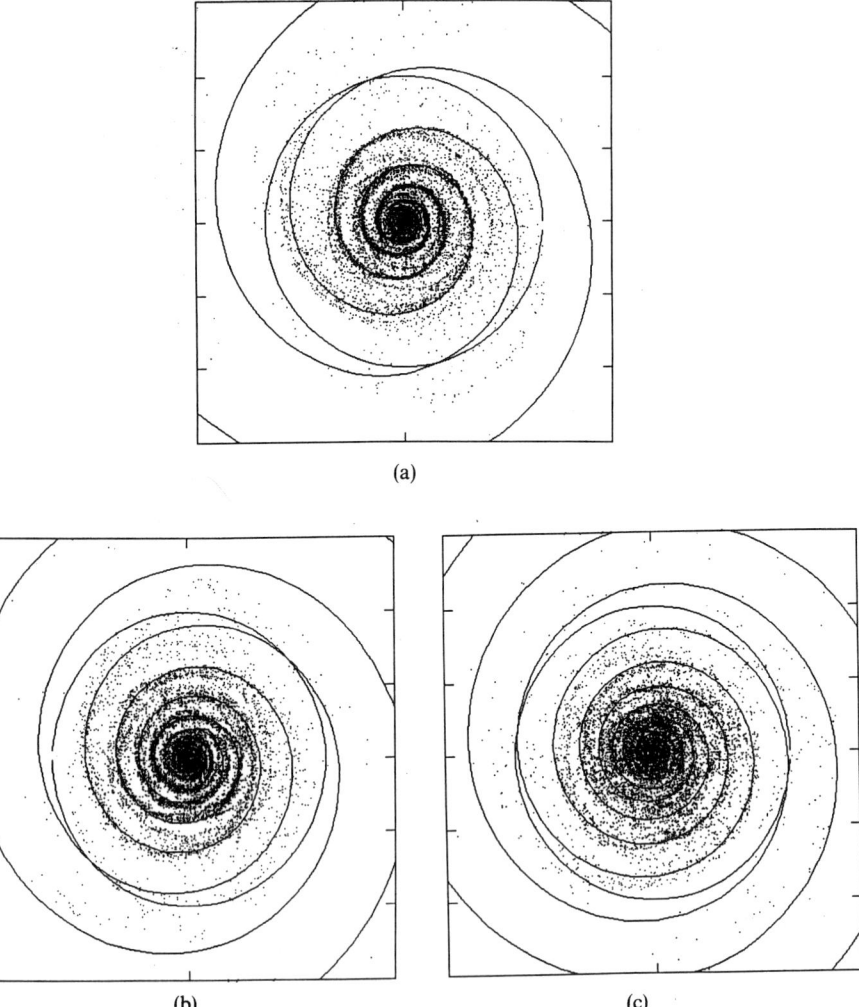

**FIGURE 3.** (a) Simulation like FIGURES 1(a), 1(b), 1(c), but $f = 3$. The frame is at ten crossing times. (b) Part (a) continued. Frame at fifteen crossing times. (c) Parts (a) and (b) continued. Frame at twenty crossing times.

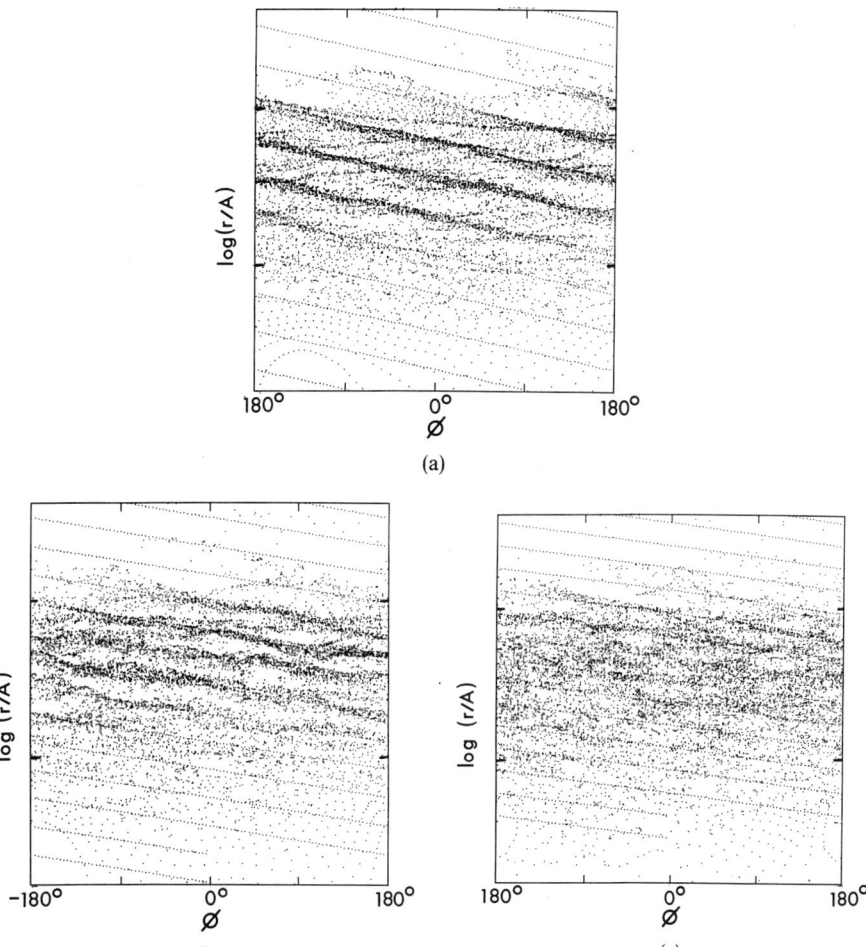

**FIGURE 4.** (a) Simulation like FIGURE 3(a) ($f = 3$), but log $r$ versus $\phi$ display. Frame at ten crossing times. (b) Simulation like FIGURE 3(b) ($f = 3$), but log $r$ versus $\phi$ display. Frame at fifteen crossing times. (c) Simulation like FIGURE 3(c) ($f = 3$), but log $r$ versus $\phi$ display. Frame at twenty crossing times.

moving toward one another. By time 20, the lumps in the inner disk have disappeared and the arms have dissolved over the entire disk. The FIGURE 4 plots show the breakup very well in the log $r$ versus $\phi$ plot. One disk edge rotation is $2\pi$ crossing time units. So our simulation's arms die out over the entire disk after about three disk edge rotations, almost 2 billion years for a 20 kpc $= A$ and $V = 200$ km/s galaxy. This long lifetime and the shorter interval between perturbations sufficient to create tidal arm patterns[2] indicate that interactions between galaxies are extremely important in generating global spiral structure.

## 7. LONG-TERM PITCH, HALO/DISK RATIO, AND CODE SOFTENING

It is clear that approximations were made in the derivation using the GLB equation. After examining the impressive length of time covered by the FIGURES 3 and 4 simulation, one is tempted to consider the simulations to be much more realistic than the simplified analytic theory. However, long-term simulations in particular have real problems that can be pointed out with the GLB equation spring parameter. In a polar grid or tree $n$-body code, the use of bins or softening is necessary because of the small number of particles in the simulation compared to an actual galaxy. Byrd et al.[4] and, most recently, a thorough discussion by Romeo[28] have cautioned that this softening is equivalent to a given axisymmetric stability parameter, $Q$. This $Q$ can be significantly larger than that due to the velocity dispersion of the particles. It could lead to misinterpretation of simulation results.

In FIGURES 1 and 2 we showed results for an $f = 9$ Mestel disk simulation. In FIGURES 3 and 4 we display long-term results for a lesser halo $f = 3$ Mestel disk simulation. According to our analysis of the GLB equation spring parameter, these two simulations should differ in the arm pitch for maximum amplifications, that is, the first simulation's long-term log arms should be wound more tightly than the second. Equation (13) gives a maximum amplification pitch of $2°$ for the $f = 9$ FIGURE 1, 2 simulation. In contrast, the $f = 3$ simulation in FIGURES 3 and 4 should have a maximum amplification pitch angle of $5°$ according to (13). However, we find that both of these simulations have pitch angels of $9°$ at ten crossing times, larger than the expected result for the FIGURES 3, 4 simulation and much too large for the FIGURES 1, 2 simulation! On the other hand, an $f = 1$ simulation quickly reaches a steeper log arm pitch of about $16°$ at six crossing times, just as would be expected if $f$ is decreased in (13).

It turns out that bin softening gives a ready explanation of why the simulations with the two larger $f$'s behave so similarly while the smaller $f$ simulation behaves more like we would expect. In our simulations, we use a 3D polar grid that has a gravitational softening of half the bin size, or $a_{bin} = 0.087r$. We can use the left-hand side of (13) with $a$ equal to the code softening to evaluate its effect on the value of the maximum amplification pitch. In (13), $\cot i_{max} = r/(m\, a_{bin}) = 5.7$ due to bin softening or $i_{max} = 10°$ for arms with the theoretical equation, (13), $i_{max}$ tighter that this, any gravitational amplification effects are artificially weakened.

The softening acts like a velocity dispersion generated $P$ or $Q$. Equation (12a) can be used to ascertain the corresponding nonaxisymmetric $P$ parameter due to bin softening. In our case, the bin $P = 2$ for the FIGURE 1(a), 1(b), 1(c) simulation. The bin $P = 1.3$ for the FIGURE 3(a), 3(b), 3(c) simulation. Besides the bin softening there is also an "absolute softening" of 1/20 disk radius in our simulation that is greater than the bin softening near the disk center. The effect of the absolute softening is to simulate a real galaxy where $P$ is high in the central regions of the disk and declines going outward, the log arms would be initially steep and wind up with time in such a galaxy as seen in our simulations.

At this point the reader might think that galaxy simulators are in a dilemma. Their inability to use "billions and billions" of stars means that bin softening must be used. Conversely, the use of softening can lead to unrealistic simulation results. However, as with analytic theory, simulations are useful as long as one keeps in mind the assumptions that have been made. The bad effects of bin softening apply to arms in which gravitational amplification plays a key role as the spacing between the arms approaches the bin softening. For a polar grid, a simulation of an observed galaxy will be valid as long as $\cot i < r/(ma)$, where $a$ is half the bin size. For the grid we have used in the FIGURES 1(a), 1(b), 1(c) and 3(a), 3(b), 3(c) simulations, $\cot i_{bin} \approx 6$. The grid we use is good for simulating gravitationally amplified arms that have an inclination, $i > 9°$. For more tightly wound arms where the disk has a significant amount of self-gravity, the simulation cannot be trusted.

Kinematic arms resulting from direct forcing in a disk with little or no self-gravity will be simulated correctly in a highly softened (or even nongravitating) simulation. An example would be the Toomre and Toomre[29] simulations of the outer arms of M51. In the following section, we describe self-gravitating simulations of M51 where the observed log arm $i = 15°$, well within the capabilities of our grid. Simulations of galaxies with 9° arms or tighter will require smaller bins. The small bins will require the use of many more particles to avoid statistical number variation in the bins.

## 8. SPRING PARAMETER THEORY AND SIMULATIONS OF M51

The encounter history of M51 can be deduced from simulation duplication of outer disk tidal features. These features appear via direct action of the passage of the disturber rather than any excitation or amplification process. The simulations also produce the inner log arm pattern seen in M51. The arm history according to the GLB spring parameter formulas can be compared to the simulation results.

The M51 system (NGC 5194/5) is an open, almost face-on, two-armed grand-design spiral galaxy with a companion north of the disk's center. The left frame in FIGURE 5 is from an optical photograph of M51. The pitch of each arm is steep in the outer portions of the disk and shallow toward the center. There are two "kinks" in M51's arms north and south of the nucleus at a radius of 140″. HI ratio observations by Rots et al.[30] and Appleton et al.[31] of M51 show a long, curving "far tail" that extends to the south and east well away from the optical regions of the system. They also show clumpy features north of the companion. Both these features show anomalous Doppler shifts relative to the rotation pattern of the visible disk.

Non-self-gravitating simulations by Toomre and Toomre[29] support a highly inclined passage of NGC 5195 that tidally perturbed its primary NGC 5194 (M51) during this passage. The Toomres indicated that the orbit of NGC 5195 has an inclination of $\approx 70°$, with an eccentricity, $e \geq 0.8$. The "bridge" on the near side to NGC 5195 seen in photographs is merely an optical illusion, a disk spiral arm optically superimposed on NGC 5195, which is now well beyond the disk plane. Using non-self-gravitating simulations, Appleton et al.[31] proposed that the 21-cm far tail may also have been created in the passage of the companion.

Even for self-gravitating simulations, Hernquist[32] and Howard and Byrd[33] find that the time interval since perturbation for the type of companion orbit found by the Toomres was too short for interior spiral arms to be well developed and for the far tail to appear. Hernquist[32] proposed a revised single passage, high eccentricity orbit that does create a spiral pattern over the disk and a far tail. However, in the

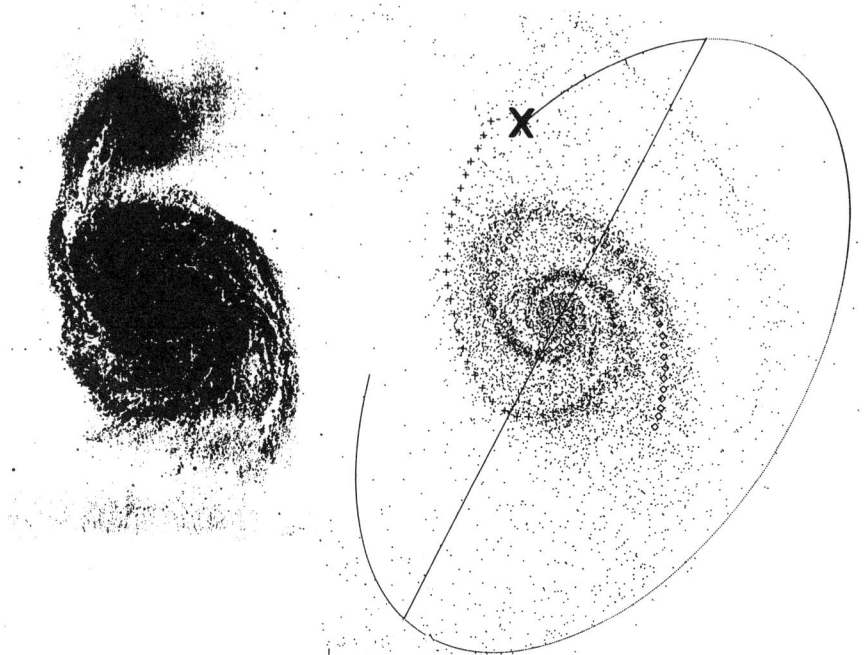

**FIGURE 5. Left side:** A blue sensitive photo of M51 with the *dark areas* indicating regions of young stars and *light areas* in the arms due to dust lanes. **Right side:** A plot of M51 simulated disk gas clouds (*dots*). The observed dust lane arms are marked (*squares* and x's) to show scale and orientation relative to the photo.

words of Hernquist, "the inner spiral structure in the model . . . is only a rough approximation to that of M51, in phase or position angle as viewed on the sky." Hernquist's model assumes the companion to half the mass of M51 rather than the one-third of Toomre and Toomre.[29] Toomre[34] has proposed a similar single-passage model.

In contrast, Howard and Byrd[33] proposed and demonstrated that the inner density wave and the 21-cm far arm could have the same dynamical origin. Byrd and Salo[35] have recently extended this idea and obtained a careful simulation morphology and Doppler shift match to the far tail, north clump, and the inner visible disk density wave pattern. The arm kinks are even duplicated. The right side of FIGURE 5 gives an image of the morphology of our M51 simulation. To provide scale and orientation, small ×'s and squares are plotted to indicate the observed dust lanes on the right side of FIGURE 5. The highly tilted orbit and line of nodes are shown as solid lines. The parameters of the simulation are given in TABLE 1. For more plots and a complete description, see Byrd and Salo.[35] The companion has cut through the disk twice, with the prior passage creating the far arm by direct perturbation and the inner density wave via tidal excitation. Finally the clump north of the companion is created by the second passage. FIGURE 6 shows a log $r$ versus $\phi$

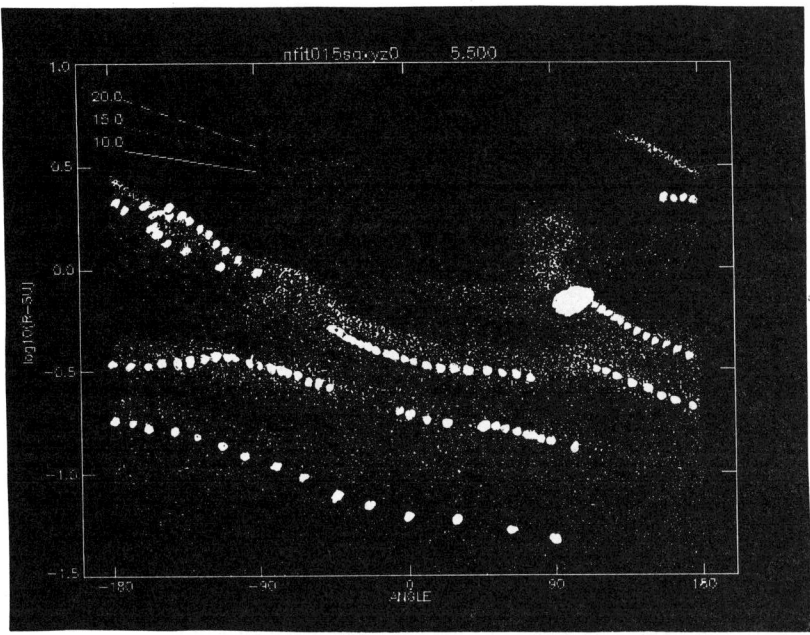

**FIGURE 6.** Display in log $r$ versus $\phi$ form of FIGURE 5 M51 simulation. The two large *tic marks* on the $y$ axes correspond to 0.1 and 1.0 of the original disk radius. *Horizontal axis* is position angle in disk plane. The gas clouds are *small white dots*. The observed dust lanes are larger symbols. The companion is the irregular ellipse. Long arm slopes are shown for 10°, 15°, 20°.

plot of the simulation gas clouds (small white dots) versus the dust lane arms (larger symbols). The inner log arms and the break with the outer "material" or tidal arms are particularly clear in FIGURE 6. In both FIGURES 5 and 6, the plotted points are the gas clouds with which the young stars are associated. As can be seen in the photo of M51 in FIGURE 5, these stars are displaced in angle from the dust lane arms. These stars and gas clouds are similarly located in our simulation.

The GLB spring parameter results are consistent with the simulations of M51. Shu *et al.*[37] and Tulley[38] applied the QSSS density-wave theory to M51. Shu *et al.*[37] propose that a density wave extends over the entire visible disk of M51. They attribute defects in their fit to possible disturbing effects of the companion on the prexisting density wave pattern. Equations **(13)**, **(14)**, and **(15)** with a circular rotation speed of 220 km/s can be used to estimate ages of material arms and density waves in M51. Using a log arm pitch of cot 15° with **(13)** and **(15)**, to have a density wave extend over the entire visible disk of M51 ($\sim 20$ kpc) would require that the log arm pattern be older than 340 million years. The simulation match for M51 indicates that M51 did indeed have a tidal encounter with its companion about 440 million years (a bit less than five crossing times) ago, which now results in a log arm pattern.

However, Elmgreen et al.[16] and also Howard and Byrd[33] find that an inner 15° log arm pattern of M51 matches the arms well only out to 6.5 kpc (140″). There is a discontinuity at this radius. Rather than a log arm, Howard and Byrd[33] find that a material arm pitch versus radius fits best beyond this radius. Equation (14) gives an age of about 70 million years. The simulations do indicate a second passage of the companion about 60 million years ago that has distorted the outer disk arm pattern to create the steepness. This agrees with our crude (14) estimate.

## 9. CONCLUSIONS

The general process of tidal spiral arm formation has been studied via 3D simulations that include nonsymmetric perturbations by a variety of perturber masses, encounter orbits, and amounts of halo relative to the disk. The edge of the disk near the perturber is strongly disturbed, forming a separate tidal bridge arm apart from arms nearer the center. The central arms appear very quickly as trailing arms. Similar to Toomre's[1] 2D symmetrically perturbed simulations, the radius at which these central arms are most amplified moves outward from the center as the time increases. Logarithmic arm density waves are seen interior to this maximum strength radius.

The present paper analytically extends Toomre's use of the GLB equation for arms prior to the density wave stage. The velocity dispersion normal to the disk results in a softening of gravitation in the GLB equation via the disk thickness. Use of this softening is easier than Toomre's numerical use of the more complicated Lin/Shu/Kalnajs reduction factor. In (12), a nonaxisymmetric stability parameter, $P$, is obtained that is a function of local disk properties but is independent of arm multiplicity. For stability, $P$ must be greater than or equal to 1. Equation (13) is derived for the pitch of the maximum amplification arms as a function of the relative importance of the halo to disk and $P$.

Due to bin softening, the GLB spring constant indicates a strong cautionary note in using simulations to match actual galaxies whose log arms are tightly wound. Our simulations of M51 plus the observed spiral pattern of this galaxy show that (13) for the maximum amplification arm pitch, (14) for the material arm age, and (15) for the extent of the log arm pattern can be useful in making rough estimates of a particular galaxy's arm age and history. Both the formulas and the simulation match to M51 indicate a multiple-encounter history for M51.

## ACKNOWLEDGMENTS

The author acknowledges interesting and spirited discussions with C. C. Lin and with Alar Toomre. A careful reading and comments by Tarsh Freeman are appreciated. Finally, Heikki Salo was very helpful in making the computer illustrations for this article while the author was on a sabbatical visit to Finland and in supplying plots for our collaborative work on M51.

# REFERENCES

1. TOOMRE, A. 1981. *In* Structure and Evolution of Normal Galaxies, S. M. Fall and D. Lynden-Bell, Eds.: 67. Cambridge Univ. Press. Cambridge, England.
2. BYRD, G. G. & S. HOWARD. 1992. Astrophys. J. **103**: 1089.
3. MESTEL, L. 1963. Astrophys. J. **126**: 553.
4. BYRD, G. G., M. J. VALTONEN, B. SUNDELIUS & L. VALTAOJA. 1986. Astron. Astrophys. **166**: 75.
5. OSTRIKER, J. P. & P. J. E. PEEBLES. 1973. Astrophys. J. **186**: 467.
6. BINNEY, J. & S. TREMAINE. 1987. Galactic Dynamics: **364**: 415. Princeton Univ. Press. Princeton, N.J.
7. TOOMRE, A. 1964. Astrophys. J. **13**: 1217.
8. TOOMRE, A. 1974. *In* Highlights of Astronomy, G. Contopoulos, Ed.: 457. Reidel. Dordrecht, The Netherlands.
9. MILLER, R. H. 1976. J. Comput. Phys. **21**: 400.
10. SALO, H. 1991. Astron. Astrophys. **243**: 118.
11. KLARIC, M. 1992. The Role of Cloud Collisions in N-body Simulations of Disc Galaxies, Ph.D. thesis. Univ. of Alabama, Tuscaloosa.
12. SALO, H. & E. LAURIKAINEN. 1993. Astrophys. J. **410**: 586.
13. DANVER, C. G. 1942. Lund Ann., No. 10.
14. SELLWOOD, J. A. & E. ATHANASSOULA. 1986. Mon. Not. R. Astron. Soc. **221**: 195.
15. CONSIDERE, S. & E. ATHANASSOULA. 1988. Astron. Astrophys. Suppl. **76**: 365.
16. ELMEGREEN, B., D. M. ELMEGREEN & P. E. SEIDEN. 1989. Astrophys. J. **343**: 602.
17. GOLDREICH, P. & D. LYNDEN-BELL. 1965. Mon. Not. R. Astron. Soc. **130**: 125.
18. GERBER, R. A. & S. A. LAMB. 1994. Astrophys. J. **431**: 604.
19. KENNICUT, R. C. 1981. Astron. J. **86**: 1847.
20. LIN, C. C. & F. H. SHU. 1966. Proc. Nat. Acad. Sci. U.S.A. **55**: 229.
21. KALNAJS, A. J. 1965. The Stability of Highly Flattened Galaxies, Ph.D. thesis: 129 pp. Harvard Univ., Cambridge, Mass.
22. MILLER, R. H. 1971. Astrophys. Space Sci. **14**: 73.
23. MILLER, R. H. 1974. Astrophys. J. **190**: 539.
24. ATHANASSOULA, E. 1984. Phys. Rep. **114**: 319.
25. GILMORE, G., I. KING & P. VAN DER KRUIT. 1990. *In* The Milky Way as a Galaxy: 154, 217. Univ. Science Books. Mill Valley, Calif.
26. LIN, C. C., C. YUAN & F. H. SHU. 1969. Astrophys. J. **155**: 721.
27. LIN, C. C. & E. BERTIN 1995. On GlobaL Wave Patterns in Galaxies: Their Generation and Maintenance. This issue.
28. ROMEO, A. 1994. Astron. Astrophys. **286**: 799.
29. TOOMRE, A. & J. TOOMRE. 1972. Astrophys. J. **178**: 623.
30. ROTS, A. H., A. BOSMA, J. M. VAN DER HULST, E. ATHANASSOULA & P. C. CRANE. 1990. Astron. J. **100**: 387.
31. APPLETON, P. N., P. A. FOSTER & R. D. DAVIES. 1986. Mon. Not. R. Astron. Soc. **221**: 393.
32. HERNQUIST, L. 1990. *In* Heidelberg Conference on Dynamics and Interactions of Galaxies, R. Wielen and A. Toomre, Eds.: 108. Springer-Verlag. Berlin, Germany.
33. HOWARD, S. & G. BYRD. 1990. Astron. J. **99**: 1978.
34. TOOMRE, A. 1994. Brouwer Award Lecture, American Astron. Soc. Meeting, Minneapolis, Minn.
35. BYRD, G. G. & H. SALO. 1995. Astrophys. Lett. Commun. **31**: 193.
36. SALO, H. & G. BYRD. 1994. *In* Mass Transfer in Galaxies, I. Shlosman, Ed.: 412. Cambridge Univ. Press. Cambridge, England.
37. SHU, F. H., R. V. STACHNIK & J. C. YOST. 1971. Astrophys. J. **166**: 465.
38. TULLY, R. B. 1974. Astrophys. J. Suppl. Ser. **27**: 415, 437, 449.

# Gaseous Vortices in Barred Spiral Galaxies[a]

MARTIN N. ENGLAND[b] AND JAMES H. HUNTER, JR.[c]

[b]*IUE Observatory*
*NASA/GSFC*
*Greenbelt, Maryland 20771*

[c]*Department of Astronomy*
*University of Florida*
*Gainesville, Florida 32611*

## INTRODUCTION

During the course of examining many two-dimensional, as well as a smaller sample of three-dimensional, models of gas flows in barred spiral galaxies, we have been impressed by the ubiquitous presence of vortex pairs, oriented roughly perpendicular to their bars, with one vortex on each side. The vortices are obvious only when viewed in the bar frame, and the centers of their velocity fields usually are near Lagrangian points $L_{4,5}$. In all models that we have studied, the vortices form on essentially the same time scale as that for the development of gaseous spiral arms, typically two bar rotations. Usually the corotation radius, $r_c$, lies slightly beyond the end of the bar. A brief discussion of vortex pairs may be found in a paper by Hunter *et al.*[1] on NGC 3992. More recently, Athanassoula[2] has described vortices that appeared in her models. Depending upon the mass distributions of the various components, gas spirals either into, or out of, the vortices. In the former case, the vortices become regions of high density, whereas the opposite is true if the gas spirals out of a vortex. The models described in this paper have low-density vortices, as do most of the models we've studied. Moreover, usually the vortex centers lie approximately at $r_c$, within $\pm \simeq 15°$ of $L_{4,5}$.

In the stellar dynamic limit, when pressure and viscous forces are absent, short-period orbits exist, centered on $L_{4,5}$. These orbits need not cross and therefore their morphology is that of gas streamlines, that is, vortices. We believe that the gas vortices in our models are hydrodynamic analogues of closed, short-period, libration orbits centered on $L_{4,5}$.

## REPRESENTATIVE MODELS

As stated previously, when viewed in the bar frame, gaseous vortices are present in all models we have studied that have gas arms of significant amplitude and angular extent. This is true even for contrived models that are not self-consistent. In

---

[a]This work was supported in part by National Science Foundation Grant AST9022827.

the present section, we consider four different types of models that we have studied in the past for various reasons.

Disks having self-consistent oval distortions and massless, non-self-gravitating gas components are models that were used in early studies of barred spiral galaxies.[3,4] The oval distortions of the density and potential are exact solutions of Poisson's equation. Then the pattern speed of the "bar" is adjusted until the most satisfactory gas response is achieved. Using Fourier–Bessel transformations, exact solutions can be developed for disks having simple analytical forms for their potentials, $\phi_0$, rotation laws, $V_0$, and surface densities, $\Sigma_0$. As a practical and familiar example, we consider Toomre's $n = 1$ disk.[5] For this disk, the solutions are:

$$V_1(a, r) = \frac{C_1 a^{1/2} r}{(a^2 + r^2)^{3/4}} \tag{1}$$

$$\Sigma_1(a, r) = \frac{C_1^2 a^2}{2\pi G(a^2 + r^2)^{3/2}}, \tag{2}$$

and

$$\phi_1(a, r) = \frac{C_1^2 a}{(a^2 + r^2)^{1/2}}, \tag{3}$$

where $a$ is a characteristic length and $C_1$ has the dimensions of velocity. The rotation curve has a maximum velocity, $V_m$, when $r = \sqrt{2}\,a$. Hence, $V_m = \sqrt{2}(3)^{-3/4} C = 0.620 C$. The total disk mass, $M = 2\pi \int_0^\infty r\Sigma_0(a, r)\, dr = C^2 a/G = 3^{3/2} V_m a/(2G) = 2.598 V_m^2 a/G$. We write the surface density as

$$\Sigma(r, a, \beta, \theta) = \Sigma_1(a, r) + \Sigma_2(r, a, \beta, \theta) = \Sigma_1(a, r)[1 + A_2 \cos(2\theta)], \tag{4a}$$

where

$$A_2 = \frac{\varepsilon_{20} r^2 \beta^{2(l-1)}(a^2 + r^2)^{3/2}}{(\beta^2 + r^2)(l + \tfrac{3}{2})}. \tag{4b}$$

with $l = 1$ or $2$, and $\beta$ is a length scale characterizing the distortion. The corresponding gravitational potential,

$$\phi(r, a, \beta, \theta) = \phi_1(a, r) + \phi_2(r, a, \beta, \theta). \tag{5a}$$

If $l = 1$,

$$\phi_2 = \frac{\varepsilon_{20}\, C_1^2 a^2 r^2 [\beta + 2\sqrt{\beta^2 + r^2}]}{3(\beta^2 + r^2)^{3/2}[\beta + \sqrt{\beta^2 + r^2}]^2} \tag{5b}$$

and, if $l = 2$,

$$\phi_2 = \frac{\varepsilon_{20}\, C_1^2 a^2 \beta r^2}{5(\beta^2 + r^2)^{5/2}} \cos(2\theta). \tag{5c}$$

In these expressions, $\varepsilon_{20}$ is a dimensionless amplitude, gauging the strength of the distortion. The amplitude, $\varepsilon_{20}$ is related to the dimensionless amplitude, $\varepsilon_{02}$, appearing in Hunter et al.[1] by the expression, $\varepsilon_{20} a^2 = \varepsilon_{02} \beta^2$. Exactly the same logic may be applied to calculate the self-consistent distortions of generalized Mestel disks (GMDs) as discussed in this reference. Interior to disk radius $R$, the rotation law of a GMD reads,

$$V_0(a, r) = C_0 \left[ 1 - \frac{a}{(a^2 + r^2)^{1/2}} \right]^{1/2}, \qquad (6)$$

where $a$ is a shape parameter and $C_0$ is the asymptotic velocity as $(r/a) \to \infty$.

FIGURES 1 and 2 show the density response and vortex velocity vectors for a distorted $n = 0$ disk with $l = 1$. The disk parameters are $C = 290$ km s$^{-1}$, $a = 1.75$ kpc, and $\beta = 4.2$ kpc. The disk mass interior to $r = 15$ kpc is $2.61 \times 10^{11}$ $M_\odot$. The maximum relative density perturbation $\Sigma_2/\Sigma_0 = 0.085$. The bar pattern speed $\Omega_p = 52.4$ km s$^{-1}$ kpc$^{-1}$ and the corotation radius $r_c = 4.3$ kpc. The radius to the vortex center, $r_v$, is 4.4 kpc and its position angle, $\psi$, measured from the bar to the vortex center in the direction opposite that of the bar rotation $= 89°$.

FIGURES 3 and 4 show the same quantities for a strongly distorted $n = 1$ disk with $l = 1$. The model characteristics are $C = 352$ km s$^{-1}$, $a = 8.38$ kpc, $\beta = 4.19$ kpc, $\Omega_p = 24.7$ km s$^{-1}$ kpc$^{-1}$, and $r_c = 8.5$ kpc. The total disk mass is $2.42 \times 10^{11}$ $M_\odot$ and, for the vortex centers, $r_v = 8.5$ kpc and their position angles relative to the bar, $\psi = 91°$. The maximum relative density "perturbation" $= 0.367$ at $r = 4.42$ kpc. The vortices develop at the same locations in models having bars of smaller amplitudes but that otherwise are identical.

Hunter et al.,[1] studied an extensive network of models of the barred spiral galaxy NGC 3992. Although the models were not self-consistent, the structure of all components (disk, bar, etc.) were derived from surface photometry at several wavelengths, combined with detailed VLA HI observations and some CO data. Velocity

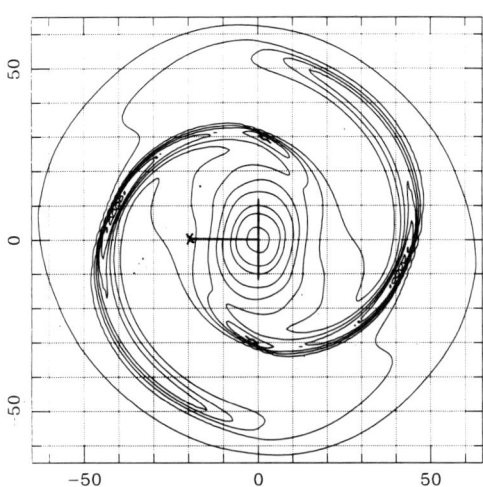

**FIGURE 1.** The surface density of an $n = 0$ Toomre disk having an $l = 1$ oval distortion and the characteristics described in the text. The major axis of the distortion and vortex center are shown also.

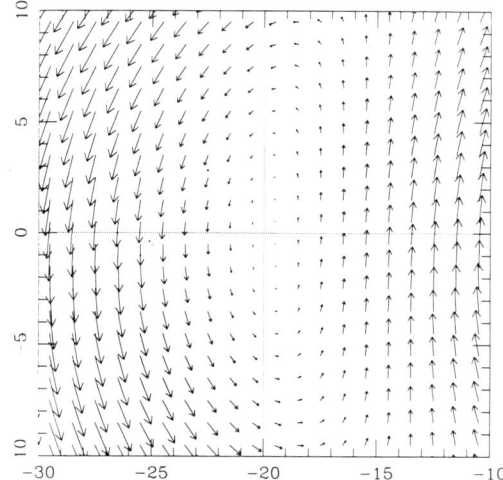

**FIGURE 2.** Velocity vectors in the bar frame for one of the vortices for the model shown in FIGURE 1.

vectors, viewed in the bar frame, of the gas flow in the bar region of their best model (#3) are illustrated in FIGURE 5. The vortices are obvious, but their centers are displaced inward 0.5 kpc from the corotation radius, $r_c = 4.3$ kpc and $r_v = 3.8$ kpc, and $\psi = 80°$.

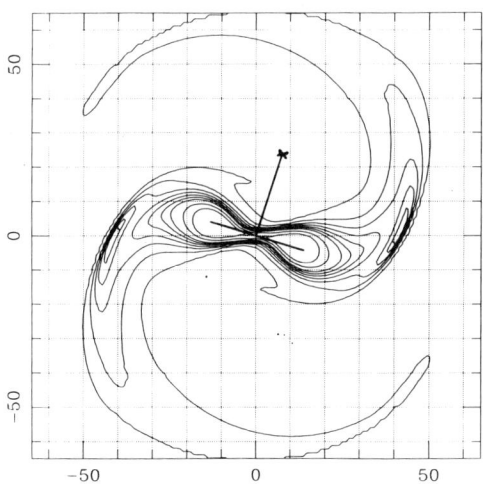

**FIGURE 3.** The surface density of the gas in an $n = 1$ Toomre disk having an $l = 1$ oval distortion and the characteristics described in the text. The major axis of the distortion and the vortex center are shown also.

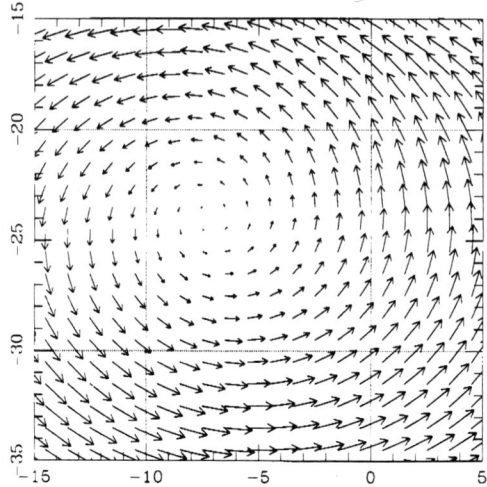

**FIGURE 4.** Velocity vectors in the bar frame of one of the vortices of the model in FIGURE 3.

The gas in the models described thus far settled into nearly stationary configurations with respect to their bar frames within a few bar rotation periods. However, this happy condition is not guaranteed. Some seemingly reasonable, but non-self-consistent, models result in irregular oscillatory gas behavior. We have calculated

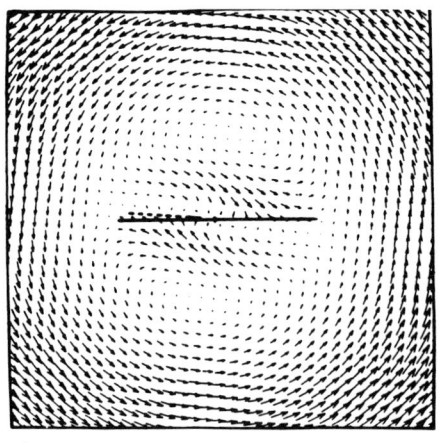

300 km/s

**FIGURE 5.** Velocity vectors in the bar frame for one of the vortices in the best model of NGC 3992 discussed in reference 1.

the gas response to a disk and rotating bar of the type considered by Contopoulos et al.[6] The model was intended to be a crude, but minimal component representation of NGC 3992. The background potential and rotation law are

$$\phi_0 = C^2\left[\ln r + E_1\left(\frac{r}{a}\right)\right], \tag{7}$$

and

$$V_0 = C[1 - e^{-(r/a)}]^{1/2}, \tag{8}$$

where the asymptotic rotation speed $C = 280$ km s$^{-1}$, disk scale length $a = 2.5$ kpc, and $E_1$ is the first exponential integral. The bar potential is

$$\phi_2 = Aq_1\left[q_2 - \cos\left(\frac{\pi r}{r_*}\right)\right]\cos(2\theta), \quad (r < r_*) \tag{9a}$$

and

$$\phi_2 = \frac{Ae^{-0.4r}}{r^2}(q_3 r - 1)\cos(2\theta) \quad (r_* < r). \tag{9b}$$

The transition radius, $r_*$, occurs where the radial force along the bar is zero. For these models, $r_* = 1.9523$ kpc and, consequently, $q_3 = 0.79983$ kpc$^{-1}$. $A$ and $q$ are amplitude factors for the bar; for the models shown in this paper, $A = 97{,}422$ km$^2$ s$^{-2}$ kpc$^2$ and $Aq = 2409.3$ km$^2$ s$^{-2}$. By equating $\phi_1$ and $\phi_2$ at $r = r_*$, we find $q_2 = 1.72822$. In our calculations, the numerical viscosity was reduced by roughly a factor of 2 relative to that of our oval distortion models, and the models were run for longer times. After spiral arms had formed, vortex pairs were evident at all later

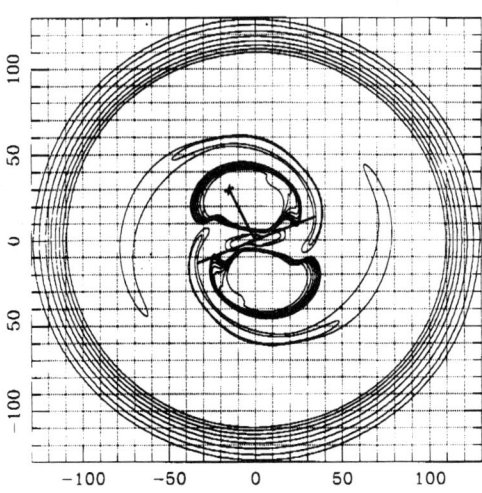

**FIGURE 6.** The surface density of the gas in the very approximate model of NGC 3992 described in the text, after 3.94 bar rotations.

times. However, $r_v$ and $\psi$ oscillated in an irregular fashion in the vicinity of $L_{4,5}$; $78° \lesssim \psi \lesssim 104°$ and $2.8 \text{ kpc} \lesssim r_v \lesssim 3.8 \text{ kpc}$.

An apparent characteristic of the oscillations is that the vortices expand as the gas arms become more prominent, and conversely. This behavior is shown in FIGURES 6 and 7. Moreover, as shown in FIGURE 8, at seemingly irregular times two vortices appeared in each low-density vortex region. We attribute this irregular

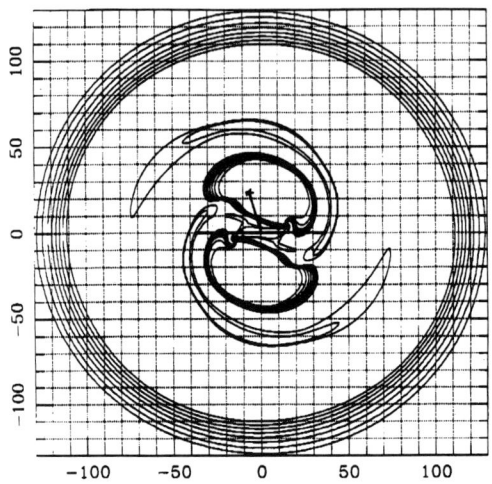

FIGURE 7. The surface density of the gas and vortex center for the model described in FIGURE 6 after 4.00 bar rotations.

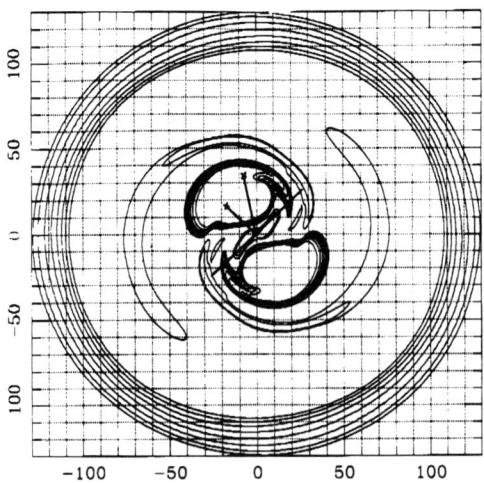

FIGURE 8. The surface density of the gas and vortex centers for the model described in FIGURE 6 after 3.88 bar rotations. Note that the two vortices are situated within each of the low-density regions.

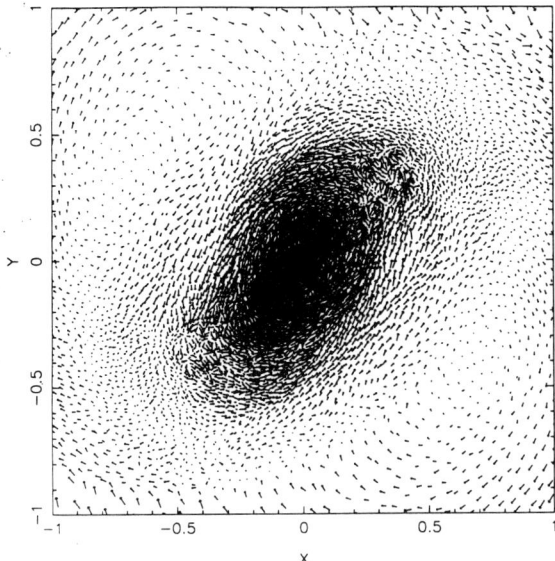

**FIGURE 9.** The velocity vectors of the gas in a self-consistent model (including both stars and gas), viewed in the bar frame. Vortices are present in the vicinity of $L_{4,5}$.

behavior to the lack of self-consistency between the imposed bar and background disk. (The oscillations can be quenched, but only if the viscosity is at least doubled.)

Thus far, the models we have discussed have had varying degrees of self-consistency, or lack thereof, and the gas was assumed to be a massless component. To further augment our understanding, we have constructed fully self-consistent models, including both stars and gas, using modified versions of two SPH tree codes, kindly given to us by Hernquist, and by Heller and Shlosman. The velocity field of one such model is shown in FIGURE 9 at $t = 4.2 \times 10^8$ years. Vortices obviously are present, located in the vicinity of $L_{4,5}$. The model characteristics are $M = 1.37 \times 10^{11}\ M_\odot$, $R_D = 25$ kpc, and $\Omega_p = 26.02$ km s$_{-1}$ kpc$_{-1}$. The initial density distribution is that of a Toomre $n = 1$ disk.

## CONCLUSION AND A SUGGESTION

We conclude that, in addition to forming spiral arms, the gas in a wide variety of galaxy models form vortices, nearly centered upon $L_{4,5}$. Moreover, the vortex pairs appear in calculations having degrees of sophistication, ranging from heuristic, non-self-consistent to fully self-consistent models. However, these ubiquitous vortices are apparent *only* when viewed in the bar frame. This fact leads to a practical suggestion: If a gas-rich, barred spiral galaxy could be observed in HI at $\lambda\ 21$ cm with sufficient resolution and signal–noise ratio, and if the gas velocities in that region are transformed so that they are "observed" in a rigidly rotating frame, having its origin at the galaxy's center, a pair of vortices will appear when the angular velocity of the frame equals the pattern speed of the bar. Based upon synthetic observations of the gas in our models, we estimate that the pattern speed and corotation radius can be deduced to within roughly 10 percent of their true values. At present, we are trying to do this experiment with some of the barred galaxies we have observed in HI at the VLA.

## ACKNOWLEDGMENTS

We thank G. Contopoulos for many helpful discussions on this subject; especially for suggesting the analogy between closed, short period libration orbits and hydrodynamic streamlines.

## REFERENCES

1. HUNTER, J. H., JR., R. BALL, J. M. HUNTLEY, M. N. ENGLAND & S. T. GOTTESMAN. 1988. Astrophys. J. **324**: 721.
2. ATHANASSOULA, E. 1992. Mon. Not. R. Astron. Soc. **259**: 345.
3. SANDERS, R. H. & J. M. HUNTLEY. 1976. Astrophys. J. **209**: 53.
4. HUNTLEY, J. M., R. H. SANDERS & W. W. ROBERTS. 1978. Astrophys. J. **221**: 521.
5. TOOMRE, A. 1963. Astrophys. J. **138**: 385.
6. CONTOPOULOS, G., S. T. GOTTESMAN, J. H. HUNTER, JR. & M. N. ENGLAND. 1989. Astrophys. J. **343**: 608.

# The Pattern Speed of the Barred Spiral Galaxy NGC 1398[a]

E. M. MOORE[b,d] AND S. T. GOTTESMAN[c]

[b]Department of Astronomy
Boston University
Boston, Massachusetts 02215

[c]Department of Astronomy
University of Florida
Gainesville, Florida 32611

## INTRODUCTION

We attempt to determine the pattern speed of NGC 1398, a large, isolated, barred spiral galaxy, using 21-cm observations and the location of prominent morphological features. For galaxies like NGC 1398, the HI velocity field coupled with the location of resonance rings should make the determination of the corotation radius and pattern speed possible. In addition, we believe it useful to attempt to apply current models to a specific, well-behaved system. HI data have been used to provide constraints on gas and stellar dynamical models of barred galaxies;[1-3] however, many areas concerning the properties of such systems are still poorly understood. Our understanding tends to be qualitative; generic models have been created but it is proving difficult to reproduce self-consistently the observations of specific systems. More observational data are needed at the fine scale offered by contemporary models. The observations discussed here are presented with this goal in mind.

## THE GALAXY

NGC 1398 is a southern barred spiral galaxy classified by Sandage and Tammann[4] as an SBab(r)I and by de Vaucouleurs et al.[5] in the *Third Reference Catalogue of Bright Galaxies* (hereafter RC3) as a (R')SB(r)ab. Optical parameters of the galaxy are given in TABLE 1.[6-11] Optical photographs (FIG. 1, reproduced from the Hubble Atlas[12]), show a strikingly symmetric system with an inner ring surrounding a bright central bulge and bar. Further out the galaxy has been classified as having an outer pseudoring. However, Buta[11] recently reclassified it as an $(R_1R_2')SB(rs)ab$ system, where $(R_1R_2')$ galaxies are a subclass of ringed galaxies that have double outer pseudorings.

If an inclination of 43°[7,13] is assumed, the inner ring is elongated with its major axis aligned parallel to the bar, as is typical for the inner rings of barred spiral

---

[a]This material is based upon work supported by the National Science Foundation under grants AST-9022827 and AST-9116525.
[d]Current address: Department of Physics and Astronomy, Rutgers, The State University of New Jersey, P.O. Box 849, Piscataway, New Jersey 08855-0849.

**FIGURE 1.** An optical photograph of NGC 1398 reproduced from the Hubble Atlas. North is up and east is to the left.

**TABLE 1.** Optical Parameters of NGC 1398

| Source | NGC 1398 |
|---|---|
| Type | SBab(r)I[a] |
| RA (1950) | 03 36 45.0 |
| Dec (1950) | −26 29 55 |
| Photometric $D_{25}$ (') | 7.1 × 5.4 |
| Bar length (") | 78[b,c] |
| Ring diameter (") | 98[d] |
|  | 101[b] |
|  | 115[e] |
| Corrected blue magnitude (B) | 10.35 |
| Distance (Mpc) | 16.1[f] |
| Position angle (°) | 96.4[g] |
| Inclination (°) | 43[c] |

[a]Sandage and Tammann.[4]
[b]Kormandy.[6]
[c]Ohta et al.[7]
[d]de Vaucouleurs and Buta.[8]
[e]Pedreros and Madore.[9]
[f]Tully.[10]
[g]Buta.[11]
*Note:* Values are from the RC3 unless otherwise noted.

TABLE 2. Characteristics of the Beam and Cleaned Channel Maps for the Various Array Configurations

|  | DnC Array | CnB Array | Combined Arrays |
|---|---|---|---|
| FWHP of synthesized beam (") | 57.3 × 59.2 | 14.6 × 16.4 | 20.3 × 22.3 |
| FWHP of synthesized beam (kpc) | 4.47 × 4.62 | 1.14 × 1.28 | 1.56 × 1.72 |
| Beam area (str), $\Omega$ | $9.04 \times 10^{-8}$ | $6.39 \times 10^{-9}$ | $1.20 \times 10^{-8}$ |
| rms noise (mJy per beam) | 0.704 | 0.454 | 0.383 |
| rms noise (K) | 0.12 | 1.13 | 0.51 |
| Peak brightness temperature (K) | 7.3 | 16.1 | 12.3 |
| Peak S/N ratio | 59 | 14 | 24 |

galaxies.[14] The ring is bright and well defined, but does not appear to be a continuous structure (FIG. 1). Instead, it seems to be composed of several spiral segments and therefore Buta[11] reclassified it as an inner pseudoring. The brightest segment starts in the northeast just beyond the bar minor axis and extends through three-quarters of a revolution. The corresponding symmetric component, which begins in the southwest just before the bar minor axis, is neither as continuous nor as prominent. In the outer disk, the galaxy has long, narrow spiral arms, each of which completes almost a full revolution, possibly forming pseudoring structure. The arms are accented by bright, discrete areas but are faint compared to the nuclear region. In the outermost part of the optical galaxy, they appear as dim, flocculent structures. Buta[15] points out that when the inner ring of a galaxy is well-defined and conspicuous, the outer ring is usually quite faint, and vice versa.

## OBSERVATIONS

NGC 1398 was observed at 21-cm at the Very Large Array (VLA)[e] in the DnC hybrid configuration in February 1991 and at the higher resolution CnB hybrid configuration in February 1992. Because of the low declination of the galaxy, hybrid configurations, with an extended north arm, were used to obtain greater uniformity in the $(u, v)$ coverage and a more circular beam. The two arrays provided spatial resolutions of 15" and 60". A bandwidth of 3.125 MHz (663 km s$^{-1}$) centered on a velocity of 1409 km s$^{-1}$ (optical definition) was used. The band was divided into 32 channels with a velocity separation of 20.7 km s$^{-1}$ and resolution of 25.3 km s$^{-1}$. A total of 10.5 hours of data on source was collected for NGC 1398, approximately 4 hours in the DnC configuration and 6.5 in the CnB.

The individual data sets were edited and calibrated using the NRAO Astronomical Image Processing System (AIPS). The calibrated amplitudes and phases (the $u, v$ data sets) were then combined and Fourier transformed and CLEANed. Characteristics of the CLEANed beam and channel maps for all three data sets are given in TABLE 2. Details of the observations, calibration, and mapping techniques are given in Moore and Gottesman.[13]

[e]The Very Large Array is an instrument of the National Radio Astronomy Observatory, which is operated by Associated Universities, Inc., under cooperative agreement with the National Science Foundation.

## THE NEUTRAL HYDROGEN DISTRIBUTION

The channel maps, after correction for the primary beam attenuation, were used in a moment analysis to obtain the density and temperature weighted mean radial velocity of the neutral hydrogen. FIGURE 2 shows the global distribution of the neutral hydrogen of NGC 1398 for the DnC array (top) and the CnB data (bottom). FIGURE 3 shows the combined data set superimposed on an optical photograph. The gas has a nonuniform distribution with a severe depletion toward the center of the galaxy. Beyond this, high-density segmented structure, coincident with the tightly wound optical spiral arms, is apparent in the high-resolution images. Regions that are prominent in the optical image, mainly the bright bar, ring, and bulge, are weak in the radio, while the weak optical spiral zone is quite prominent in the HI. The HI gas extends well beyond the outer optical disk and the densest gas features are coincident with the prominent optical arms or lie slightly beyond them, overlying the flocculent structure. The high gas density region does not completely encircle the galaxy, but weakens in the southwest. In addition, there is a third zone of enhanced HI beyond the optical spiral structure. Given the constraints of 20" resolution, the optical spiral structure is well aligned with the more intense regions of HI. However, a substantial amount of high-intensity HI lies beyond the optical spirals.

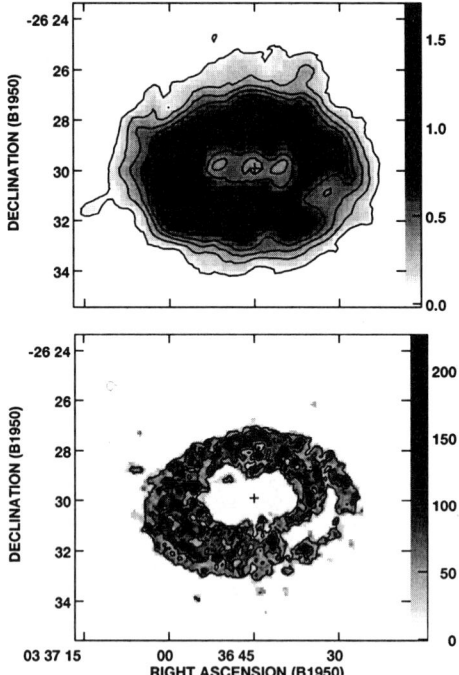

**FIGURE 2. Top:** Greyscale with contours of the HI surface density of the low-resolution DnC data. Contours are percents (5, 15, 25, 35, 50, 65, 80, and 90) of the peak flux, which corresponds to a column density of $5.4 \times 10^{20}$ cm$^{-2}$. The $2\sigma$ level is at 5 percent of the peak. North is up and east is to the left for this and all subsequent figures. **Bottom:** The higher resolution CnB data. Contours are at 20, 40, 60, 80, and 90 percent of the peak. The $2\sigma$ level is at the 20 percent level. The peak column density is $10.3 \times 10^{20}$ cm$^{-2}$.

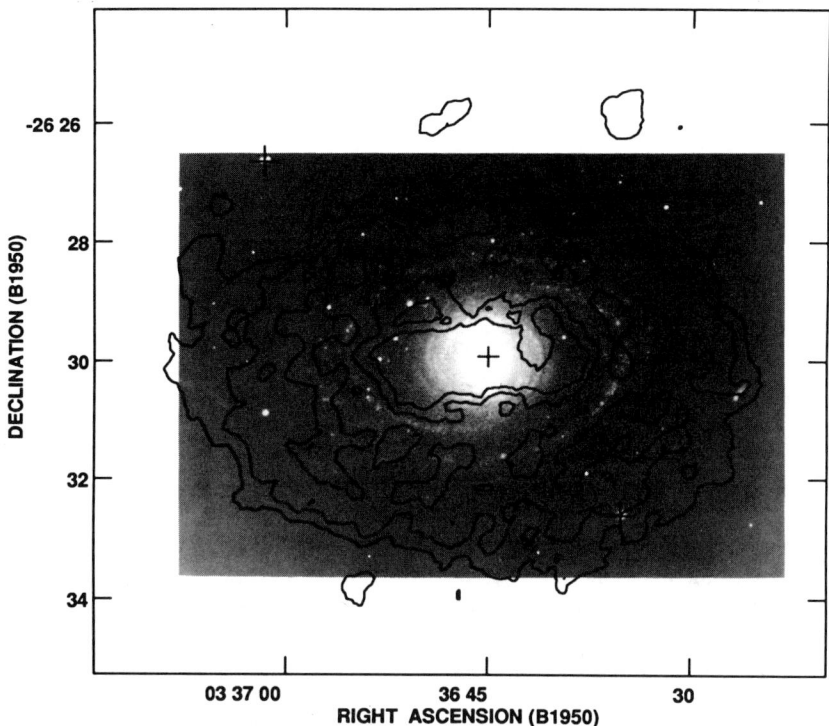

**FIGURE 3.** HI density contours superimposed on an optical photograph. The HI data are of the combined data sets. Contours are at 10, 30, 60, and 90 percent of the peak HI flux. The lowest contour marks the $2\sigma$ level.

The positions of the HI peaks are seen clearly in a radial density profile of the high-resolution data (FIG. 4), determined by integrating the HI density in concentric annuli in the plane of the galaxy. The inclination and position angle used in the calculation are best fit values derived from the velocity field (as discussed in the next section). The low central density is apparent, especially in the higher resolution data. The prominent optical arms to the west lie at radii of approximately 140" and 165" (along the line of nodes). These have been marked with arrows on FIGURE 4 and agree well with the peaks found in the CnB data set. Beyond this, the third gas peak (the HI outer arm) is seen at approximately 220". Integrated fluxes and other characteristics of the first moment maps from all data sets are listed in TABLE 3.

## *The Central Region*

FIGURE 5 (top) shows an enlarged view of the inner region of the galaxy from the DnC data. *B*-band photometry (kindly provided by R. Buta) of the inner ring and bar area is superimposed on the same HI image in the bottom of FIGURE 5. The three central depressions in the HI are at a level of 20 percent of the peak brightness

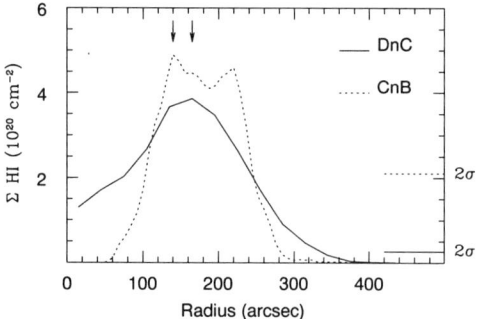

**FIGURE 4.** The radial density profile of the HI made using the DnC (*solid line*) and CnB data (*dotted line*). The $2\sigma$ noise levels are marked on the right-hand axis. Arrows mark the positions (along the line of nodes) of the prominent optical arms. The outer HI arm, which lies at the edge of the optical disk, is seen at 220".

(corresponding to the $5\sigma$ level). Like the gas disk itself, the central low-density region is not symmetric about the center of the galaxy; it extends farther to the east and is elongated nearly perpendicular to the optical bar. Furthermore, it is not aligned with the line of nodes. The HI disk has an axial ratio of approximately 0.83 as measured on the higher sensitivity DnC image. The axial ratio of the optical galaxy is 0.76 (RC3). The central low-density region has an axial ratio of 0.34, significantly different from that of the galaxy as a whole. The difference is most likely caused by gas associated with the optical ring (FIG. 3). Not only is a tongue of HI seen overlying the ring but there is HI coincident with the northern and southern edges of the ring; the net effect is a pinching of the evacuated region. FIGURE 5 also shows an association of HI gas on the eastern and western sides of the ring.

## THE NEUTRAL HYDROGEN KINEMATICS

The CnB velocity field of NGC 1398 is superimposed on a density image in FIGURE 6 (bottom). Only in the DnC data is the velocity field in the center of the

TABLE 3. Characteristics of the Moment Maps for the Various Array Configurations

|  | DnC Array | CnB Array | Combined Arrays |
|---|---|---|---|
| $2\sigma$ column density ($10^{20}$ cm$^{-2}$) | 0.26 | 2.1 | 0.94 |
| Max HI column density ($10^{20}$ cm$^{-2}$) | 5.4 | 10.3 | 8.74 |
| Integrated flux (Jy km s$^{-1}$) | 43.4 ± 1.1 | 40.4 ± 4.0 | 45.9 ± 2.5 |
| Total hydrogen mass ($10^9 M_\odot$) | 2.7 ± 0.07 | 2.5 ± 0.03 | 2.8 ± 0.02 |
| Reduced HI mass ($10^6 M_\odot$ Mpc$^{-2}$) | 10.2 ± 0.5 | 9.5 ± 1.9 | 10.8 ± 1.2 |
| $\Delta V$ (20%) (km s$^{-1}$) | 470 | — | — |
| $V_{sys}$ (km s$^{-1}$) | 1398 | 1397 | 1397 |
| Total galaxy mass ($R \leq 310''$) ($10^{11} M_\odot$) | 5.1 | 4.1 | 4.3 |
| $2\sigma$ HI diameter (') | 10.7 × 8.7 | 8.6 × 6.0 | 9.6 × 6.9 |

*Note:* $\Delta V$ is the velocity width of the galaxy at a width of 20 percent of the peak flux. The total galaxy mass was calculated assuming a Keplarian galaxy.

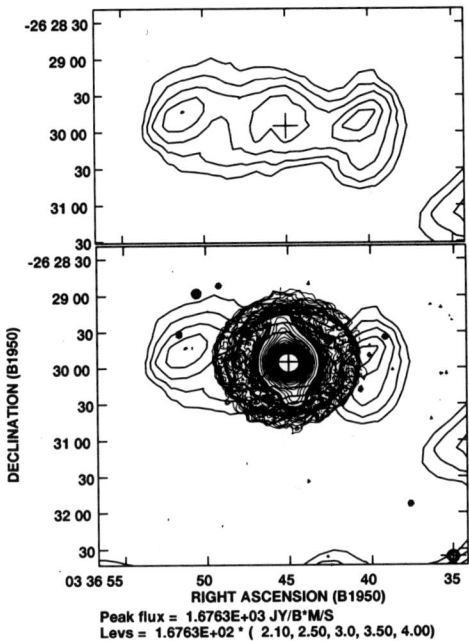

**FIGURE 5. Top:** The inner low-density region of the galaxy from the low-density DnC data. The contours are in percent of the peak flux and the outermost contour is at 40 percent, the innermost at 21 percent. **Bottom:** The same gas data with *B*-band photometry of the inner bulge and ring structure superimposed. Photometry contours are not equally spaced, but where chosen to highlight the ring structure.

galaxy detected (FIG. 6, top). The velocity field appears symmetric and ordered. There is no evidence of large-scale noncircular motion. Where the line of sight crosses the spiral arms in the high-resolution data, irregularities are expected and observed. In general, they are less than 20 km s$^{-1}$. The contours at either end of the major axis close, indicating that the rotation curve is falling in the outer part of the galaxy. As there is no indication that the HI disk is warped, these effects are assumed to be caused by a turnover in the rotation curve.

The second moment map, the velocity dispersion, is seen in FIGURE 7 for the DnC data. The dispersion is low in the outer part of the galaxy and high in the center. This is expected if the rotation curve is relatively flat in the outer galaxy and changing rapidly in the inner region. This map is also symmetric, arguing that the velocity field of the galaxy is well behaved. The line of nodes is quite apparent.

The velocity field was used to determine orientation parameters and a galactic rotation curve. A least squares fit was performed to determine the position of the kinematic center, the inclination, the position angle of the line of nodes, the systemic velocity, and a rotation curve. All parameters are calculated to symmetrize the velocity field. Solutions were found to be stable over a range of radii and opening angles. In addition, the rotation curves for the receding and approached halves of

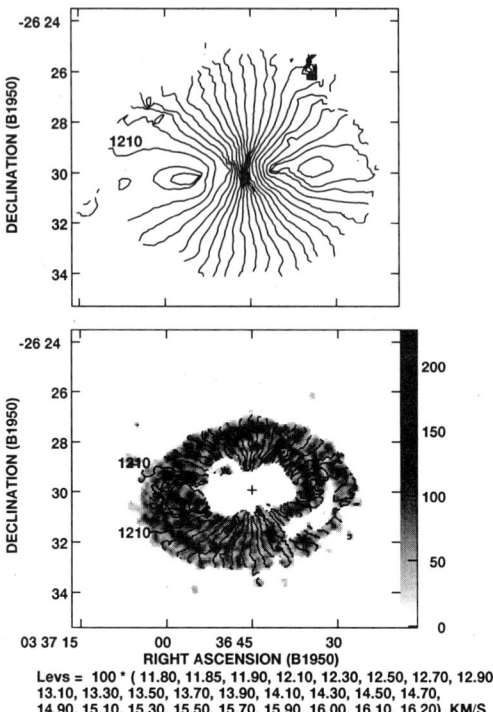

Levs = 100 * ( 11.80, 11.85, 11.90, 12.10, 12.30, 12.50, 12.70, 12.90, 13.10, 13.30, 13.50, 13.70, 13.90, 14.10, 14.30, 14.50, 14.70, 14.90, 15.10, 15.30, 15.50, 15.70, 15.90, 16.00, 16.10, 16.20) KM/S

**FIGURE 6. Top:** Radial velocity contours of the DnC data. Contours are 20 km s$^{-1}$ apart except at the velocity extremes. The contours 11.8 and 16.2 are *not* seen in this image. The contour 12.10 (1210 km s$^{-1}$) is labeled. **Bottom:** Radial velocity contours made using the higher resolution CnB data superimposed on the HI surface density.

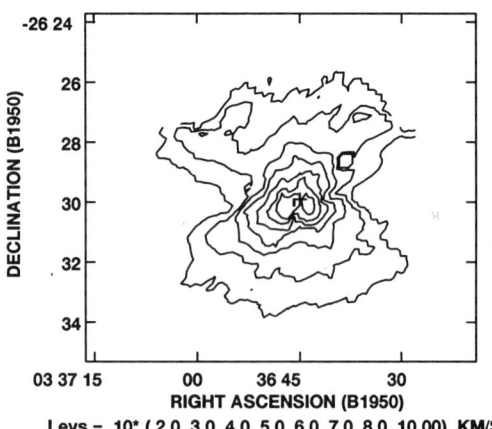

Levs = 10* ( 2.0, 3.0, 4.0, 5.0, 6.0, 7.0, 8.0, 10.0) KM/S

**FIGURE 7.** The second moment map, made from the DnC data. The outer contour is at 20 km s$^{-1}$ and the contour interval is 10 km s$^{-1}$.

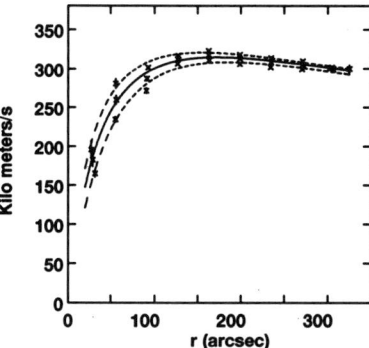

**FIGURE 8.** Rotation curves made from the DnC data set. The *solid line* includes the entire galaxy in the solution, the *bottom dashed line* is the approaching half of the galaxy, the *top dashed line* is the receding half.

the galaxy are similar. The extreme regularity of the kinematically determined parameters is impressive. There is no hint of a warp nor of multiple kinematic systems.

The rotation curve from the DnC array is shown in FIGURE 8 for the entire galaxy and for the receding and approaching halves. All peak at a radius roughly at the midpoint of the gas disk. In the section that follows, we employ these data to study some of the dynamical consequences of the bar as reflected in the kinematics and structure of the HI.

## RESONANCES

Of fundamental interest in the study of galaxy dynamics is the value of the pattern speed and the positions of the major resonances. In the case of NGC 1398, the inner ring is expected to be the signature of either the inner 4/1 resonance (Schwarz;[16] Byrd *et al.*[17]) or of corotation. Gas inside the corotation radius will lose angular momentum as it interacts with the density pattern. If the ring marks the 4/1 resonance, gas moving radially inward will enter this region and enhance the resonance zone; a gas ring forms. The increased density encourages star formation, and the ring should be blue. The ring in NGC 1398 is blue in color; the narrow high-intensity feature is outstanding in *B*-band, but almost lost in the background in *I*-band. However, the ring appears to be only weakly associated with HI.

If instead the ring is located at corotation, it is more likely to be a stellar feature rather than a strong gaseous one. If the bar is weak, the ring is probably associated with stable orbits surrounding the Lagrange $L_4$ and $L_5$ points (Contopoulos and Grosbøl[18]). If the bar is strong, the ring is supported primarily by chaotic orbits between corotation and the outer 4/1 resonance (Contopoulos *et al.*;[19] Kaufmann[20]).

In order to determine the pattern speed and resonances, we have used the DnC array rotation curve to calculate the angular velocity of rotation, $\Omega(R)$. This is

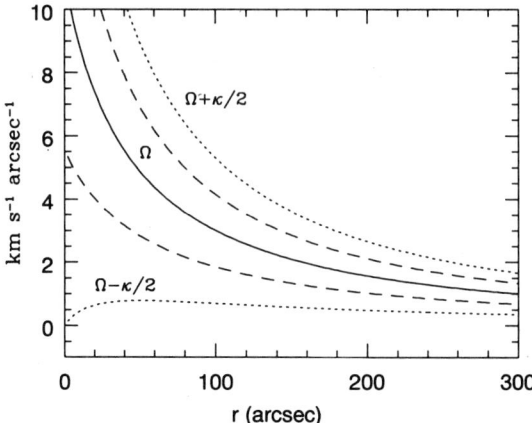

**FIGURE 9.** The angular velocity, inner, and outer Lindblad resonances (*dotted lines*) and the inner 4/1 and outer 4/1 resonance (*dashed lines*). Values were determined from the best fit to a rotation curve of the DnC data set.

shown in FIGURE 9 along with the inner 4/1 resonance and the inner (ILR) and outer Lindblad resonances (OLR). We use this angular velocity to determine a pattern speed by associating prominent optical features with resonances. Using the inner optical ring and the outer pseudoring(s) as markers of possible resonances, several kinematic models are explored, none of which is completely satisfactory.

## *Model 1*

In the first model we assume that the inner ring marks the position of the inner 4/1 resonance. This has been found in numerical simulations of the gaseous component.[16,17] Published values of the radius of the ring range from 50" to 57". Using 57" as the position of the inner 4/1 resonance gives a pattern speed of 2.56 km s$^{-1}$ arcsec$^{-1}$. This places corotation at 119" and the OLR at 204", close to the edge of the optical disk. For positions of resonances in all models, see TABLE 4. Most of the optical spiral structure lies between the corotation resonance and the OLR (FIG. 10, top left), in accord with stellar theory.[18] The outer gas arm lies at approximately 220", slightly beyond the OLR.

However, the bar lies within the 4/1 resonance at a radius of 45 percent of corotation. Stellar dynamical analysis maintains that the bar is supported primarily by the orbital families $x_1$, and the 4/1 periodic orbits (the $x_1(2)$ family, which starts at the 4/1 resonance and extends beyond it[18]). Furthermore, strong bars are expected to show boxy shapes, possibly reflecting the existence of the 4/1 family. This is observed in the photometric images, and therefore we expect the bar to end just within the corotation radius. This is clearly not the case for this pattern speed. However, it has been found that the 4/1 family does not contribute significantly to

TABLE 4. Values of the Pattern Speed and Major Resonances for Each Model Discussed in the Text

| Model | $\Omega_p$ (km/s/″) | 4/1 (″) | CR (″) | OLR (″) |
|---|---|---|---|---|
| 1 | 2.56 | 57 | 119 | 204 |
| 1 | 2.73 | 50 | 110 | 193 |
| 2 | 4.40 |  | 57 | 122 |
| 2 | 4.75 |  | 50 | 112 |
| 3 | 3.93 | 17 | 68 | 137 |
| 4 | 4.40 |  | 57 | 122 |
| 4 | 2.50 |  | 122 | 209 |

*Note:* Models 1 and 2 list two values of the resonances calculated using a ring radius of 57″, and 50″. For Model 4, the first values are for the inner pattern, the second for the outer.

the bar in some self-consistent numerical models.[20,21] These authors believe the boxy shape of the bar can be explained by libration of the $x_1$ family.

A rectified V image of the outer portion of the galaxy is shown in FIGURE 11. The image has been smoothed to enhance the spiral structure. At the top left of

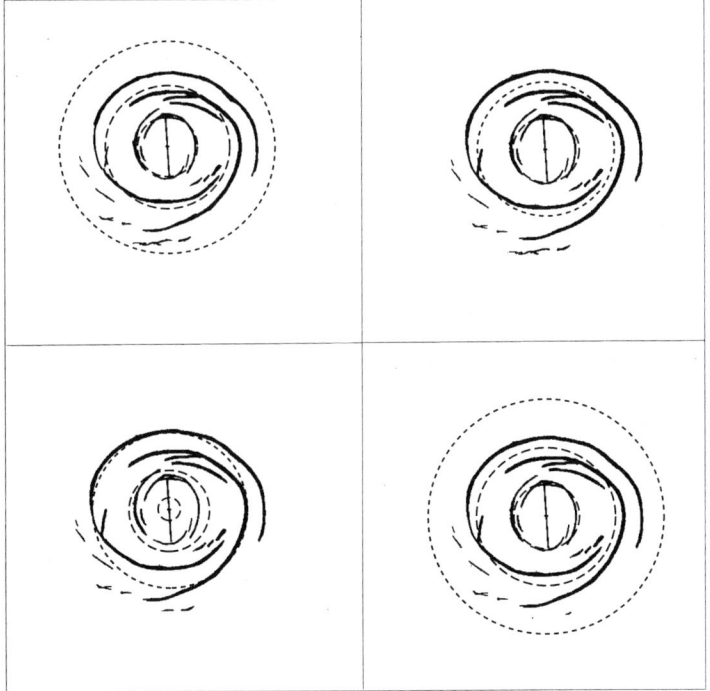

**FIGURE 10.** Comparison of the observations with the four models discussed in the text. The **upper left** refers to Model 1, the **upper right** to Model 2, the **lower left** to Model 3, the **lower right** to Model 4. In each panel the *dotted lines* refer to the resonances (see text), the *solid lines* to the spiral structure seen in FIGURE 11.

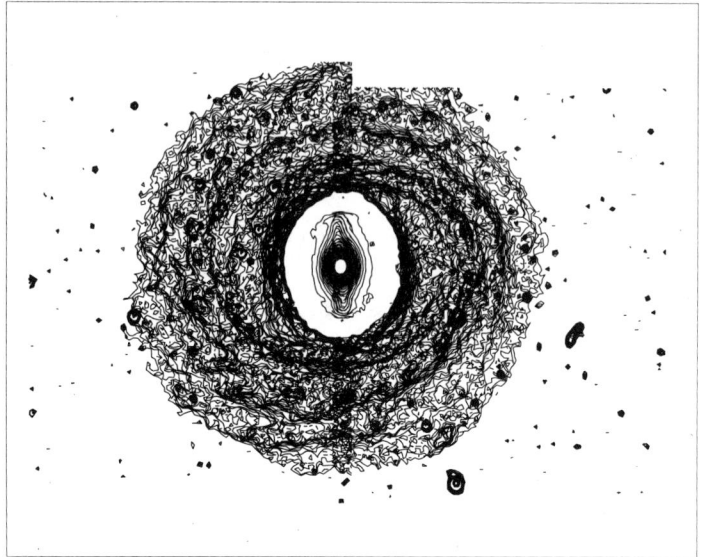

**FIGURE 11.** A rectified V photometric image. The data have been smoothed to enhance the outer spiral structure. The inner ring region has been removed for clarity and is shown in FIGURE 12. The *bar* is drawn in (at different contour intervals) for reference.

FIGURE 10 we show the relationship between the spiral structure and the resonances according to Model 1. The principal spiral arms and the inner ring have been drawn and the location of the resonances indicated.

### *Model 2*

In light of the problem associated with the first model, we assume in Model 2 that the ring marks the corotation radius ($\Omega_p = 4.40$ km s$^{-1}$ arcsec$^{-1}$ if a ring radius of 57″ is used). The bar now ends just within corotation. The OLR lies at a radius between 112″ and 122″ for ring radii of 50″ to 57″. The disadvantage of this model (FIG. 10, top right) is that almost all of the optical spiral structure lies beyond the OLR. If, as is often argued,[18,19] the OLR marks the termination of the stellar spiral structure, this model is only valid if the spirals are primarily gaseous. It is not clear if this interpretation is correct; the spirals appear as prominent in the *I*-band image as in the *B*.

### *Model 3*

In a catalogue of ringed galaxies, Buta[11] classifies NGC 1398 as an $(R_1 R'_2)$ galaxy. The OLR is the site of two families of periodic orbits, the first of which is elongated perpendicular to the bar and has an average radius just within the OLR ($R_1$ or $R'_1$ galaxies). The second has an average radius just outside the OLR and orbits elongated parallel to the bar ($R_2$ or $R'_2$ galaxies). $R_1 R'_2$ systems have double outer pseudorings corresponding to both orbit families. Buta claims that these systems provide some of the clearest evidence for the OLR. The average of the major and minor axes of the two outer pseudorings for NGC 1398 places the OLR at 137". Corotation then lies at 68" ($\Omega_p = 3.93$ km s$^{-1}$ arcsec$^{-1}$) and the inner 4/1 resonance at 17", close to the radius of the nuclear bulge (FIG. 10, bottom left).

This third model has several advantages. As mentioned, the inner 4/1 resonance appears to mark the inner bulge. In addition, it is possible that corotation marks the position of the inner ring. As discussed earlier, the inner ring structure is complex. FIGURE 5 shows a high-resolution image of the ring in the $B$-band that looks very similar to the optical photograph (FIG. 1). However, the V image in FIGURE 11 has been smoothed to bring out the outer spiral structure. FIGURE 12 shows a rectified V image of the inner region of the galaxy at that lower resolution. The ring now appears to extend beyond the dust lane on the west and lies at a radius in agreement with the corotation radius of this model. However, this is probably not a realistic alignment; stellar models that predict a corotation ring place the resonance at the ring center, not the edge. In addition, it is not clear that the ($R_1 R'_2$) classification is correct. Optical images seem to show continuous spiral, not pseudoring structure.

**FIGURE 12.** A rectified V image of inner ring. On the west (**right-hand side**), the middle spiral corresponds to the strong blue feature seen at the edge of the ring in FIGURE 5 (**bottom**).

### Model 4

In order to overcome the shortcomings of Models 1–3, we consider here a two-pattern velocity model (Sellwood;[22] Sellwood and Sparke[23]). As with Model 2, we place corotation of the bar pattern at a radius of 57", within the ring complex. The OLR of this pattern (122") becomes the corotation of the outer pattern. In Sellwood's notation this serves as a "groove" that generates both an inward and outward traveling spiral density wave (see also Lovelace and Hohlfeld[24]). This clearly fits the observations (FIG. 10, bottom right); the spiral pattern coincides with this resonance, extending both inward and outward. The OLR of the outer pattern lies at a radius of 209", beyond all prominent optical spiral structure, and close to the outermost gas spiral. Another point in favor of this model is that the inward propagating spirals do not end at any preferred angle with respect to the bar.

## DISCUSSION

NGC 1398 was chosen for study at the VLA because we believed it would be an uncomplicated system. It appeared extremely symmetric. In addition it has several observational advantages; it is large, has a high HI surface brightness, and the bar is prominent. There are no obvious close companions to interact with or to complicate the local kinematics. These initial expectations have been confirmed by analysis of the reduced and calibrated data; the plane of the system appears ideally undisturbed.

In trying to understand NGC 1398, the presence of an inner ring provides added interest. This distinctive feature should mark the presence of a resonance in the disk, either the inner 4/1 or corotation with the pattern. Furthermore, RC3 and, in more detail, Buta[11] suggest that the outer spiral structure reflects the presence of an OLR. Since we possess a velocity field defined by high-quality data, we hoped to couple this information with the locations of these structural features and determine the pattern velocity, one of the important but unknown quantities in spiral galaxy dynamics.

However, as the discussion in last section revealed, despite the fact that NGC 1398 is in many ways an unremarkable barred spiral system, the pattern speed and resonances were not easy to determine. If the ring signifies the inner 4/1 resonance, then there are implications for the stellar orbits that support the bar. In addition, such a model contradicts the findings of many computational investigations of barred galaxies (see Sellwood and Wilkinson[25] for a review). Furthermore, the inner 4/1 resonance is expected to be gas rich. This is not seen in the HI observations.

However, any model that places the ring at the corotation radius must cope with the implications of having the OLR in the middle of the disk. Only if the spiral structure is essentially gaseous would we expect the arms to propagate through the resonance. Furthermore, this ring is expected to be stellar and as such it is difficult to explain the multiple structures observed (FIG. 12). It would be useful to have a $K$-band image to see if the individual "spiral-like" features disappear leaving only a broad continuous stellar ring. Such a broad band could be a signature of chaotic orbits supporting the resonance feature.[20] Any gas clouds present would be expected to follow the stellar orbital streamlines that cross. Thus the gas clouds will collide, forming shocks and encouraging star formation. Such a mechanism could explain the presence of the prominent blue component of the ring, but not necessarily the narrow, sharp structure of the blue ring.

A model with two pattern velocities avoids many of these difficulties. The bar ends just within the corotation radius, the OLR lies beyond the spiral pattern, and the resonances are associated with obvious structures in the galaxy. In many ways this appears to be the most attractive model, but it is not clear that any of these models reveal the unique relationship we had hoped to find between structures, resonances, and the kinematics of NGC 1398.

## ACKNOWLEDGMENTS

We would like to thank G. Contopoulos for many helpful discussions and R. Buta for providing B, V, and I images of NGC 1398.

## REFERENCES

1. BALL, R. 1992. Astrophys. J. **395**: 418.
2. ENGLAND, M. N. 1989. Astrophys. J. **337**: 191.
3. KAUFMANN, D. E. 1993. Ph.D. dissertation, Univ. of Florida, Gainesville.
4. SANDAGE, A. & G. A. TAMMANN. 1986. A Revised Shapley-Ames Catalog of Bright Galaxies, 2nd ed. Carnegie Institution of Washington. Washington, D.C.
5. DE VAUCOULEURS, G., A. DE VAUCOULEURS, H. G. CORWIN, JR., R. J. BUTA, G. PATUREL & P. FURGUE. 1991. Third Reference Catalogue of Bright Galaxies. Springer-Verlag. New York.
6. KORMANDY, J. 1979. Astrophys. J. **227**: 714.
7. OHTA, K., M. HAMABE & K. WAKAMATSU. 1990 Astrophys. J. **357**: 71.
8. DE VAUCOULEURS, G. & R. BUTA. 1980. Astron. J. **85**: 637.
9. PEDREROS, M. & B. F. MADORE. 1981. Astrophys. J. **45**: 541.
10. TULLY, R. B. 1988. Nearby Galaxies Catalog. Cambridge Univ. Press. Cambridge, England.
11. BUTA, R. 1995. Astrophys. J., Suppl. Ser. **96**: 39.
12. SANDAGE, A. 1961. The Hubble Atlas of Galaxies. Carnegie Institution of Washington. Washington, D.C.
13. MOORE, E. M. & S. T. GOTTESMAN. 1995. Astrophys. J. **447**: 159.
14. BUTA, R. 1986. Astrophys. J. Suppl. **61**: 609.
15. BUTA, R. 1990. In Galactic Models, J. R. Buchler, S. T. Gottesman, and J. H. Hunter, Eds.: 58. N.Y. Acad. Sci. New York. **596**.
16. SCHWARZ, M. P. 1984. Mon. Not. R. Astron. Soc. **209**: 93.
17. BYRD, G., P. RAUTIAINEN, H. SALO, R. BUTA & D. A. CROCKER. 1994. Astrophys. J. **108**: 476.
18. CONTOPOULOS, G. & P. GROSBØL. 1989. Astron Astrophys Rev. **1**: 261.
19. CONTOPOULOS, G., S. T. GOTTESMAN, J. H. HUNTER & M. N. ENGLAND. 1989. Astrophys. J. **343**: 608.
20. KAUFMANN, D. E. 1992. In Astrophysical Disks, S. F. Dermott, J. H. Hunter, Jr., and R. E. Wilson, Eds.: 140. N.Y. Acad. Sci. New York. **675**.

21. SPARKE, L. S. & J. A. SELLWOOD. 1987. Mon. Not. R. Astron. Soc. **225:** 653.
22. SELLWOOD, J. A. 1990. *In* Galactic Models, J. R. Buchler, S. T. Gottesman, and J. H. Hunter, Jr., Eds.: 101. N.Y. Acad. Sci. New York. **596**.
23. SELLWOOD, J. A. & L. S. SPARKE. 1988. Mon. Not. R. Astron. Soc. **231:** 25p.
24. LOVELACE, R. V. E. & R. G. HOHLFELD. 1978. Astrophys. J. **221:** 51.
25. SELLWOOD, J. A. & A. WILKINSON. 1993. Rep. Prog. Phys. **56:** 173.

# Index of Contributors

**A**bernathy, R. A., 168–188

**B**almforth, N. J., 55–69, 80–94
Bertin, G., 125–144
Bradley, B. O., 168–188
Buchler, J. R., 1–13
Byrd, G., 302–319

**C**hristodoulou, D. M., 285–295
Contopoulos, G., 145–167, 189–204, 221–230

**D**avies, C. L., 231–241
Dvorak, R., 221–230

**E**fthymiopoulos, C., 145–167
England, M. N., 320–328

**G**ottesman, S. T., 329–344
Grousouzakou, E., 145–167

**H**ardee, P. E., 14–31
Hunter, C., 111–124
Hunter, J. H., Jr., vii, 32–43, 231–241, 320–328

**I**pser, J. R., 256–260

**K**andrup, H. E., 168–188, 221–230
Kazanas, D., 285–295
Kolláth, Z., 1–13
Korycansky, D. G., 261–276

**L**ichtenberg, A. J., 205–220
Lin, C. C., 125–144
Lovelace, R. V. E., 32–43, 277–284

**M**ahon, M. E., 168–188
Moore, E. M., 329–344
Morrison, P. J., 80–94

**P**apaloizou, J. C. B., 261–276

**R**omanova, M. M., 277–284

**S**erre, T., 1–13
Shlosman, I., 285–295
Siopis, C. V., 32–43, 221–230
Smith, H., Jr., 189–204
Spiegel, E. A., 55–69
Sridhar, S., 44–54

**T**erquem, C., 261–276
Tohline, J. E., 285–295

**V**ishniac, E. T., 70–79
Voglis, N., 145–167

**W**hitaker, R. W., 32–43
Willmes, D. E., 242–255
Wilson, R. E., vii
Wu, Y., 296–301

**Y**ecko, P., 95–110